The Reptiles
of Tennessee

The Reptiles of Tennessee

Edited by Matthew L. Niemiller, R. Graham Reynolds,
and Brian T. Miller

THE UNIVERSITY OF TENNESSEE PRESS / KNOXVILLE

Support for this book was provided in part by the Tennessee Wildlife Resources Agency, Division of Environmental Services.

Library of Congress Cataloging-in-Publication Data

The reptiles of Tennessee / edited by Matthew L. Niemiller, R. Graham Reynolds,
and Brian T. Miller. — First edition.
      pages cm
Includes bibliographical references and index.
ISBN 978-1-57233-949-1 (pbk.) — ISBN 1-57233-949-7 (pbk.)
 1. Reptiles—Tennessee.
 2. Reptiles—Tennessee—Classification.
  I. Niemiller, Matthew L.
 II. Reynolds, R. Graham.
III. Miller, Brian T.

QL653.T3R47 2013
597.909768—dc23
2012046688

Contributors

TODD S. CAMPBELL
Department of Biology
The University of Tampa

ARTHUR C. ECHTERNACHT
Department of Ecology and Evolutionary Biology
The University of Tennessee–Knoxville

BRAD M. GLORIOSO
U.S. Geological Survey
National Wetlands Research Center, Lafayette, Louisiana

LISA POWERS
Froghaven Farm
Bon Aqua, Tennessee

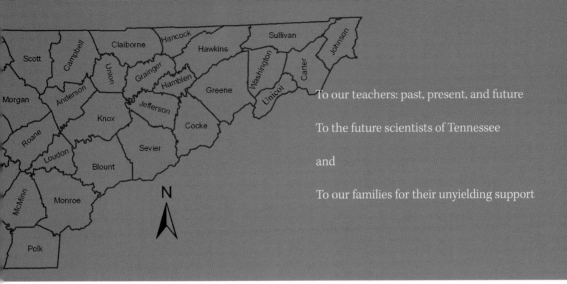

To our teachers: past, present, and future

To the future scientists of Tennessee

and

To our families for their unyielding support

All the phenomena of Nature need to be seen
from the point of view of wonder and awe.

Henry David Thoreau

Contents

Field Notes

Foreword

The Earth is an amazingly diverse planet and can be hard to understand in all of its complexity. Over the 4.5 billion years that it has been in existence, continents have appeared, changed shape, merged, split, and moved. River systems and lakes have formed and disappeared. Mountains have risen and been worn down. Global climate has changed and, with that change, local weather patterns have fluctuated.

On this ever-changing abiotic template, living organisms arose and populated the land, lakes, rivers, streams, and seas, evolving into myriad species that have themselves evolved and gone extinct, some leaving descendants that inhabit the Earth today. With the possible exception of some bacteria and viruses, none of the species that have ever populated the Earth are cosmopolitan in distribution. That is, no species is distributed over the entire planet. The distribution of each species is limited by history, physical barriers (land, for example, might be a barrier for many fish, water for fully terrestrial species), their own tolerances to factors such as temperature, and their ecological interactions with other species. The species making up ecological communities are directly or indirectly linked through food webs. Understanding food webs has been a preoccupation of ecologists in part because, increasingly, humans are part of these webs. Before there can be an understanding of the interactions of species in a food web, the identities of the species must be known, and that is complicated by the fact that we have identified only a small fraction of the species that inhabit the Earth. Of the species we have identified, we know the full extent of the geographic ranges of an even smaller fraction.

Whereas the study of the distribution and abundance of species is primarily the occupation of professional biologists, the study of species' distributions can and has received valuable contributions from nonbiologists. These are individuals who, for any number of reasons, have developed an affinity for one or another group of organisms, whether they be flowering plants, trees, beetles, fishes, amphibians, reptiles, mammals, or birds . . . especially, it seems, birds. They have reached the point that they can call attention to the occurrence of a species in an area where it has never been known to occur. For any given area, the starting point for acquiring this knowledge is often a local field guide. This is true for Tennessee.

Tennessee isn't a large state. At 42,146 square miles, it ranks only 36th among the 50 states in terms of size. It is relatively narrow, less than 120 miles at its widest from north to south, but stretches about 440 miles from east to west. The highest elevations are in the east, topping out at Clingman's Dome at 6,643 feet above sea level and decreasing irregularly westward across the Cumberland Plateau and Middle Tennessee to a low of about 178 feet at the Mississippi River. This general pattern is interrupted by major rivers that flow north-south (e.g., the Tennessee and Mississippi rivers) and east-west (e.g., Cumberland, Clinch, and Duck rivers). Four climatic divisions are recognized: East, Cumberland Plateau, Middle, and West. These patterns of landscape and climate have led to the recognition of 18 ecoregions and a diverse flora and fauna.

Tennessee's biotic diversity has been documented in field guides to the wildflowers (Horn and Cathcart 2005), trees, shrubs, and woody vines (Wofford and Chester 2002), mussels (Parmalee and Bogan 1998), fishes (Etnier and Starnes 1993), and birds (Nicholson 1997). Glaringly absent have been equivalent statewide guides to the herpetological fauna of Tennessee. Those of us interested in the amphibians and reptiles of Tennessee have had to make do with guides with much broader geographic coverage but less detail as it applies to Tennessee, such as that of R. Conant and J. T. Collins (1998) or the excellent *Amphibians of Great Smoky Mountains National Park* (Dodd 2004), which as the name implies covers but a tiny part of the state. For almost 30 years, I have been encouraging some of my well-qualified colleagues to fill this void. Although there have been a few starts, no one has reached the finish line.

Along the way, a solid base was provided by W. H. Redmond et al. (1990), W. H. Redmond and A. F. Scott (1996), Scott and Redmond (2002), and Redmond and Scott (2011). Recently, Matthew Niemiller and Graham Reynolds filled half of the void with their guide to the amphibians of Tennessee (Niemiller and Reynolds 2011), but still lacking has been a guide to Tennessee's reptile fauna. Now Matt and Graham, together with Brian Miller, have come through again with the volume you are now reading.

But who are "you"? My guess would be that you are a mother or father, or grandmother or grandfather, trying to identify the latest turtle, lizard, or snake that your son or daughter has brought home or that you found wandering around your yard. You could also be an ardent hiker interested in being able to identify the animals you see on your treks. Or you could be a K–12 teacher, or even a college or university professor, trying to answer the persistent questions of (hopefully) eager young students. Maybe you are one of those students. Then again, you might be employed by a privately owned environmental consulting firm contracted by a developer and charged with performing an environmental impact statement required by government regulations. You might be a wildlife biologist or conservation biologist employed by a city, county, state, or federal agency or by a nongovernmental conservation organization. It's even possible that you are a college or university professor thinking of expanding your research to include amphibians or reptiles and needing a brief introduction to the biology of local species. My point is that this book, and the companion volume on amphibians, should find a wide audience.

Matt, Graham, and Brian are well qualified to have undertaken the production of Tennessee's first guides to the amphibian and reptile faunas. Matt conducted his master's thesis at Middle Tennessee State University on the systematics of the Tennessee Cave Salamander complex and simultaneously conducted research on the ecology of other species of salamanders and other organisms. He developed a deep (no pun intended) and enduring fondness for caves which carried over into his doctoral research on "The Evolution, Speciation, and Conservation of Amblyopsid Cavefishes" at UTK. Graham's first published paper was on assortative mating in Panamanian poison-dart frogs and was based on research he had conducted while an undergraduate at Duke University. Bypassing a master's degree, he enrolled as a doctoral student at the University of Tennessee–Knoxville (UTK), where his dissertation focused on reptiles of the Turks and Caicos Islands: "Islands, Metapopulations, and Archipelagos: Genetic Equilibrium and Nonequilibrium Dynamics of Structured Populations in the Context of Conservation." Brian earned his bachelor's and master's degrees at the University of Missouri and then proceeded to earn his Ph.D. from Washington State University. He has been a professor of biology for about 25 years at Middle Tennessee State University, where he teaches vertebrate biology and herpetology courses. His research has focused on the natural history, morphology, and conservation of amphibians and reptiles, especially in Tennessee. Brian was Matt's master's degree advisor and has been a scientific mentor to both Matt and Graham.

While at UTK, Graham and Matt shared an office and a research adviser, Dr. Ben Fitzpatrick, in the Department of Ecology and Evolutionary Biology. Before long, Graham was assisting Matt in his cave studies, Matt was assisting Graham in his research in the West Indies, and both were beginning to talk about the absence of field guides to the amphibians and reptiles of Tennessee. They decided to take on the task of producing these volumes themselves. No doubt they could have written both books entirely by themselves, but they chose to enlist the support of colleagues to contribute accounts of species for which they were expert. Even so, Matt and Graham wrote most of the accounts in the books, as well as the supporting material, and most of the photographs are theirs. What they have produced will stand as a great introduction for the general public to the identities and fascinating lives of the amphibians and reptiles of Tennessee, and as a dateable baseline documenting the state's herpetofauna, the volumes will be useful for teachers, professional biologists, and anyone having a stake in the conservation of biodiversity and the natural heritage of our state and nation. I'd like to think that the guides Matt and Graham have produced will be the standard reference on the herpetofauna of Tennessee for years to come, but given the rapid increase in the rate of introduction of exotic species to the country's fauna, I'll look forward to a second edition in about 10 years.

Arthur C. Echternacht, Ph.D.
Department of Ecology and
Evolutionary Biology
University of Tennessee–Knoxville
July 14, 2012

Acknowledgments

This book was greatly improved by the expertise of the contributors, Todd Campbell, Sandy Echternacht, Brad Glorioso, and Lisa Powers, who assisted in the writing of this book, and we wish to acknowledge them for their outstanding contributions. The manuscript was greatly improved by two reviewers—anonymous and John Jensen—and we are especially appreciative of the great deal of time they took to read and comment on each page.

We would also like to thank the many volunteers, friends, and family who have joined us in our field adventures documenting and photographing amphibians in Tennessee or assisted with various aspects of the present endeavor (in alphabetical order): Kyle Barrett, Bob Biddix, John Bills, Crystal Bishop, Josh Campbell, Vince Cobb, Phil Colcough, Noa Davidai, Chris Davis, Terry Deal, Joe Douglas, Dylan Dittrich-Reed, Sarah Duncan, Bob English, Pandy English, Dante Fenolio, Ben Fitzpatrick, Heather Garland, Brad and Melita Glorioso, Troy Glorioso, Matt Gray, Erin Gray, Jeff Green, Nathan Haislip, Kevin Hamed, Kerry Hansknecht, Heather Hensley, Cory Holliday, Jason Hoverman, Darrin Hulsey, J. R. Jones, Denise Kendall, Chris Kerr, Richard Kirk, Katie Koss, John Lamb, Michael Lawton, Thany Mann, Laura Marsh, Roger McCoy, Jacob Miller, Joshua Miller, Joyce Miller, Avis Moni, Gerald Moni, Brad Moxley, Stephen Nelson, Trent Niemiller, Michael Ogle, Michael Osbourn, Stesha Pasachnik, Lisa Powers, Rebecca Quasney, Elizabeth Reed, Jennifer Reynolds, Cheyenne Reynolds, Bob Reynolds, Kim Reynolds, Steve Samoray, Floyd Scott, Premal Shah, Jeremy Spiess, Kyle Sykes, Jason Todd, Megan Todd-Thompson, Mark Thurman, Steve Tilley, Nathan and Jess Turnbough, Murlin Varner, Bill Walter, Jessica Welch, David Withers, Richie Wyckoff, Elizabeth Young, scores of vertebrate zoology and herpetology students from Middle Tennessee State University and the University of Tennessee–Knoxville, and any others we have forgotten to mention.

We also are grateful to the staff and personnel we have interacted with during the course of this project in the Tennessee Wildlife Resources Agency, Tennessee Department of Environment and Conservation–Division of Natural Heritage, TDEC–State Parks, Tennessee Department of Agriculture–State Forests, Tennessee Valley Authority, Tennessee Cave Survey, Tennessee Herpetological Society, National Park Service, U.S. Forest Service, and U.S. Fish and Wildlife Service. We also thank all those in the Department of Biology at Middle Tennessee State University and the Department of Ecology and Evolutionary Biology at the University of Tennessee–Knoxville. Data gathered for this book by the editors have been associated with various projects funded by the Tennessee Wildlife Resources Agency, National Speleological Society, American Society of Ichthyologists and Herpetologists, Nature Conservancy, Department of Biology at Middle Tennessee State University, and Department of Ecology and Evolutionary Biology at the University of Tennessee–Knoxville.

We also thank Dick Bartlett, Phil Colclough, Andrew Durso, Arthur "Sandy" Echternacht, Dante Fenolio, David Frymire, Gary Gerald, Brad Glorioso, Pierson Hill, Collin Jaeger, Ritchie King, Stephen Nelson, Todd Pierson, Lisa Powers, Floyd Scott, Greg Sievert, Ken Wray, and Richie

Wyckoff for generously providing excellent photographs used in this book.

Brian thanks his wife, Joyce, for her decades of tolerance of his somewhat unorthodox lifestyle, and his sons, Joshua and Jacob, for allowing him to rediscover the wonders of nature through the eyes of a child when each was young, and for countless hours of companionship in the field as they aged. He thanks his parents, Roy and Marie, and siblings, Dave, Debi, and Greg, for establishing and nurturing his love of nature, and his childhood and lifelong friend, Matt Donovan, who knows why, or should. He also thanks the many colleagues with whom he has interacted over the past few decades who have in some way influenced his philosophy of life as it intermingled with career, including Bob Powell, Joe Beatty, Richard Daniel, Chris Peterson, Bob Wilkinson, Matt Donovan, Don Diesel, Bob Herrington, Paul Hendricks, Ken Gasser, Chuck Costa, John Larsen, and, especially, the late Dean Metter. Furthermore, he is grateful to the many undergraduate and graduate students whose contagious enthusiasm for life and nature has given him hope that conservation of wild places and of wildlife, although a challenging endeavor, has a chance. And finally, he offers a special thanks to Matt Niemiller, whose dedication and motivation for discovery, and commitment to excellence, serve as constant reminders of why he decided to enter academia.

Matt thanks all his family and friends who have tolerated his obsessive passion for biology and, in particular, herpetology. He thanks his parents, Donna and Mark, his grandparents, Lorna and Mary Jane, his younger siblings, Ryan, Trent, and Kayla, his many aunts and uncles, particularly Kirk and Carolyn, who have been a second set of parents, and his many cousins, especially his "big brothers," Tom and Eric, for their unconditional love and support. He also thanks all his past teachers, particularly Kristi Slaby (Kankakee Valley High School), Peter Scott and John Whitaker (Indiana State University), Brian Miller (Middle Tennessee State University), and Ben Fitzpatrick (University of Tennessee), whose guidance and support ultimately led him into a career in biology and academia. He would especially like to thank his good friends Brad Glorioso, Richie Wyckoff, and Graham Reynolds for their continued friendship, encouragement, and support. This book would not have been written without them. Finally, Matt wishes to thank Brian Miller, who has served as a teacher, advisor, mentor, colleague, and, most important, friend. Matt's appreciation of the natural world and ultimately his career in academia has been inspired, fostered, and greatly influenced by Brian.

Graham wishes to thank his wife Jennifer, who continues to tolerate his frequent herping excursions to all parts of the globe and his late-night returns home from road cruising with Matt (which elicit a cacophony from alarmed pets when she has to get up for work in the morning). He also dedicates this book to his daughters Charlotte and Emmeline, who will arrive as the book goes to press. Graham is especially grateful to his family for encouraging the pursuit of his passion for science, wildlife, and nature, as well as to his many teachers, particularly his advisors Joanne Bartsch (Carolina Day School), Erika Deinert (Organization for Tropical Studies), Susan Alberts (Duke University), Ben Fitzpatrick (University of Tennessee), and Liam Revell (University of Massachusetts–Boston), for their knowledge and encouragement. He is also grateful for the friendship and scientific collaboration with Matt and Brian and the opportunity to spend so much time in the study and appreciation of nature with them.

Lastly, a big thank you to the staff of the University of Tennessee Press—especially Kerry Webb, Scot Danforth for understanding our vision and for being receptive to a second volume, Thomas Wells, Gene Adair, Chad Pelton, Barbara Karwhite, Cheryl Carson, Tom Post, and freelance copy editor Karin Kaufman—for their expert help with producing these volumes.

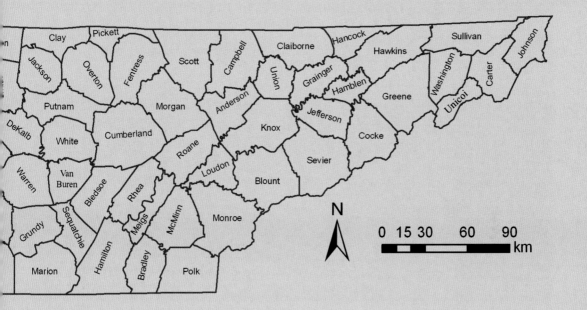

Part 1
Reptiles, Habitats, and Conservation

Mississippi Green Watersnake (*Nerodia cyclopion*), Lake County. (Photo by Collin Jaeger)

1

What Is a Reptile? Reptile Anatomy and Life History

Modern reptiles constitute members of the class Reptilia, a diverse group of vertebrate animals generally characterized by having scales and an amniotic egg. Crocodilians, snakes and lizards, turtles, tuataras, and even birds are all modern reptilian representatives; however, reptiles are an ancient group dating back to the Carboniferous period some 320 to 350 million years ago. The diversification of reptiles, as well as mammals, was initiated by the evolution of the amniotic egg—a nutrient-rich, membranous egg with a protective shell covering. This development allowed the ancestors of reptiles to lay eggs away from water, completely divorcing them from the need to reproduce in moist areas like their distant relatives, the amphibians. Reptiles have since evolved into myriad forms, from the dinosaurs and pterosaurs of the Mesozoic era to the species we find in our backyards today. The study of reptiles has been exciting in the last few decades, as new molecular methods have allowed us to peer ever deeper into the relationships of reptiles and, accordingly, our understanding of what constitutes a reptile has changed. Previously, reptiles were organized according to their morphologies, or physical characteristics, and birds were birds and lizards were lizards. However, because of the molecular innovation, we now have a much better understanding of reptilian relationships. Currently, reptiles are divided into two groups: the Archosauria and the Lepidosauria. The archosaurs include turtles, crocodilians, birds, di-nosaurs, and pterosaurs. Birds and crocodilians are each other's closest living relatives and, along with the turtles, are the surviving members of Archosauria after the extinction of dinosaurs 65 million years ago. The lepidosaurs include the snakes, lizards, amphisbaenians, and tuataras in the order Squamata. Tuataras are a unique group of lizard-like reptiles, once far more diverse but now composed of just two species restricted to islands off northern New Zealand. These unusual creatures resemble lizards, but they constitute a much more ancient family of reptiles, Sphenodontidae, which evolved prior to the lizards. The snakes and lizards, once considered different enough to be placed into separate suborders, now are known to be a paraphyletic group. In other words, some lizards are more closely related to snakes than they are to other lizards.

Reptiles worldwide consist of 9,547 species occupying terrestrial habitats on nearly every continent (except Antarctica) and in all of the world's seas. They range in size from the world's smallest amniote, the 16-mm-long (0.6-in) Jaragua Dwarf Gecko from the Dominican Republic, to the Indo-Pacific Crocodile of the South Pacific with a mass over 1,000 kg (2,204 lbs) and nearly 7 m (23 ft) long. New species of reptiles are being discovered every year, many through the use of molecular methods. Even in North America, new species are being discovered, such as the Pearl Map Turtle of Mississippi in 2010.

Unfortunately, a number of reptile species have also become extinct owing to habitat alteration by humans, and many more species are on the brink of extinction. Below we explore the three non-avian groups of reptiles found in Tennessee: the lizards and snakes, the turtles, and the crocodilians.

Lizards and Snakes

Lizards and snakes are a highly diverse vertebrate group globally, with 9,012 species belonging to the order Squamata, from the Latin squama, meaning "scale." Indeed, the presence of scales is a distinct characteristic of this group and how lizards and snakes are readily distinguished from amphibians with similar body shapes. Scales are produced from specialized cells in the epidermis of the skin and consist of a hardened keratin that serves as a small, rigid plate. The scales act as a sort of external armor when the majority of the body is covered with them. This armor protects snakes and lizards against minor abrasion and prevents water loss through the skin. However, scales serve a wide variety of purposes for lizards and snakes, from communication to enabling them to grip onto substrates.

Like other non-avian reptiles, lizards and snakes are ectotherms and must obtain their body heat from their surroundings. This limits the regions of the world in which they can live; indeed, lizards are far more common and diverse in warmer and drier habitats, such as the American Southwest, than they are in the temperate deciduous forests of the American East. The Tennessee lizard fauna is represented by a mere nine species, and several of these species approach the northern edge of their range

in our state. Some lizards, such as skinks, are able to tolerate colder temperatures and survive farther north, whereas others, such as the Green Anole, reach their northern limit in East Tennessee and North Carolina. Snakes are far more diverse in Tennessee, as we have 34 species found across the state. Some of these species, such as the Common Gartersnake, can be found in nearly any backyard; whereas others, such as the Pinesnake, are extremely localized and rare.

Lizards and snakes generally become active in Tennessee during spring, as the temperature warms enough to allow them to absorb sufficient heat. As temperatures increase, these animals become more active, with lizards peaking in activity in the summer and snakes peaking in the late spring through autumn. Indeed, more species of snakes are at the peak of their activity in late spring, which is generally the best time to look for them. Some species, such as milksnakes, are most active on the surface during spring then retire to underground or hidden areas throughout the rest of the year. Other species, such as racers, are active throughout the warmer months and are commonly seen even in the heat of summer. While lizards in Tennessee are exclusively diurnal, many snake species are nocturnal or switch from diurnal to nocturnal activity as the nighttime temperatures increase and daytime temperatures become exceedingly warm in summer.

Most lizards and snakes breed during spring and summer, and reproduction is accomplished through a variety of mechanisms. Mating usually occurs after some sort of courtship in which the male attracts the female or causes her to become reproductively available.

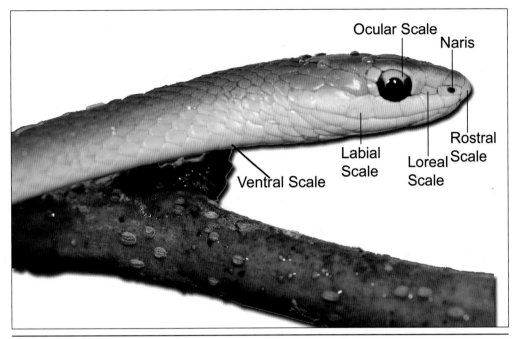

Key characteristics of a Rough Greensnake (*Opheodrys aestivus*), including major head scales used in identification. (Photo by Matthew L. Niemiller)

Sometimes this involves male territoriality and contests to establish mating rights. Male Green Anoles will establish a territory and signal to other males using extensions of their colorful dewlap and other physical displays such as pushups. If a rival fails to take notice then a fight or pursuit might ensue. Male Timber Rattlesnakes engage in a ritualistic "wrestling" match in which one male will try to pin the head of the other male to the ground, thereby establishing dominance. Male lizards and snakes have a pair of reproductive organs, called hemipenes, located in pouches posterior to the cloaca. Either hemipenis may be extended and used for copulation, depending on the position of the male's body when mating begins. Females of some species, such as Green Anoles, can store sperm

for months, requiring only one mating session to produce eggs for the rest of the season.

All lizards in Tennessee are oviparous and lay eggs, which are small with a leathery shell and generally deposited underneath cover objects or within leaf litter. Female Slender Glass Lizards and skinks actually guard the eggs by coiling around them, and female skinks of the genus *Plestiodon* protect their eggs from invertebrate predators and remove rotting eggs; lizards from other groups simply deposit the eggs and leave. Many snakes in Tennessee, such as Red Cornsnakes, are oviparous and lay clutches of eggs under cover. Other species, such as Common Gartersnakes, are in between oviparous and viviparous—a situation known as ovoviviparity, in which the female retains the eggs

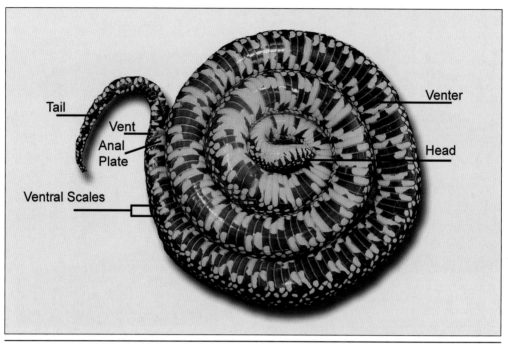

Upside-down Eastern Black Kingsnake (*Lampropeltis nigra*) showing underside external anatomy. Note that the tail begins posterior to the vent, and that ventral scales are much wider than dorsal scales. (Photo by Brad M. Glorioso)

inside of her until they hatch and are released. Newborn lizards and snakes, called hatchlings or neonates, generally look very much like their parents, though often with more vibrant colors, and they are completely independent from their parents. Venomous snakes are born with small quantities of venom already produced, so even these tiny snakes can deliver a venomous bite—a necessary adaptation, as they must begin to feed right away.

Turtles

A far less diverse group than lizards and snakes, turtles are only represented worldwide by about 323 species, though there are a few regions around the globe that have a disproportionately greater diversity of species. Southeast Asia and the southeastern United States both contain a high diversity of turtles relative to other areas of the world. In the latter region, almost 75% of the turtle species that occur in the United States are found in the southeastern states. Only Southeast Asia has a greater diversity. The Tennessee turtle fauna is represented by 16 species, nearly twice

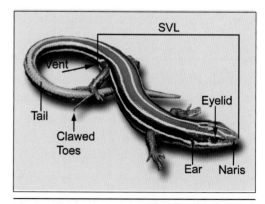

External anatomy of a Common Five-Lined Skink (*Plestiodon fasciatus*). (Photo by Brad M. Glorioso)

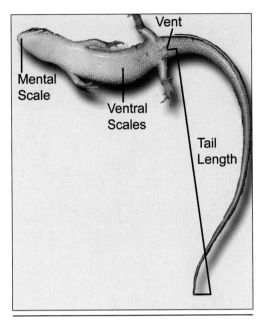

Underside anatomy of a Common Five-Lined Skink (*Plestiodon fasciatus*). Note that the tail begins posterior to the vent. (Photo by Brad M. Glorioso)

the number of lizard species. Turtles are unique among reptiles in that they have a heavily modified skeleton that includes an external shell that surrounds the shoulder and hip girdles. This shell is composed of ribs and bony plates that are covered with large scales called scutes. Shelled turtles, with their obviously excellent protective mechanism, have existed for more than 220 million years and represent an ancient lineage of unique reptiles. Turtles are frequently aquatic or semiaquatic, especially in Tennessee, where no truly terrestrial turtles (tortoises) occur. This is likely an adaptation to reduce the burden of carrying a large shell—the water provides buoyancy to reduce the weight of the shell. Many turtles have a hinged plastron and flexible necks, which allows them to tuck the soft parts of their body, such as the head, legs, and tail, into the shell

and seal them inside until danger passes. Box Turtles are exceptional in this regard—they have a hinged plastron that can be completely sealed up to protect against even the most agile predators, such as raccoons and bobcats.

Turtles are also ectotherms and are thus restricted in activity to periods when the weather is warm enough to allow them to gain enough heat from their environment to warm their muscles, enabling them to move about. However, Painted Turtles are capable of remaining active even in near-freezing temperatures and can even tolerate being frozen in a snow bank only to emerge in the spring suffering no ill effects! Most aquatic turtles bask either out of the water on logs, like the River Cooter, or in shallow water, like the Spiny Softshell. Some species, such as Box and Bog Turtles, rarely bask in the open at all. Mating takes place during spring when males often court females to induce copulation. In some species, such as Pond Sliders, males have elongated front claws, which they use to "tickle" the females to induce their sexual activity. The male will position himself directly in front of or on top of a swimming female and stroke the claws across her face and neck in a staccato motion. The copulatory organ of the male (penis) is hidden in the cloaca when not in use. Consequently, the male has a longer tail than the female, which allows him to copulate with the female while on top of her and still be able to reach the cloaca at the base of her tail. Males are generally not territorial, though in some species males will mob females to try to establish a mating position.

Females of all turtle species lay eggs in nests constructed in loose soils. Aquatic turtles must come ashore to lay eggs either on the banks of rivers and lakes or in the softer soil of

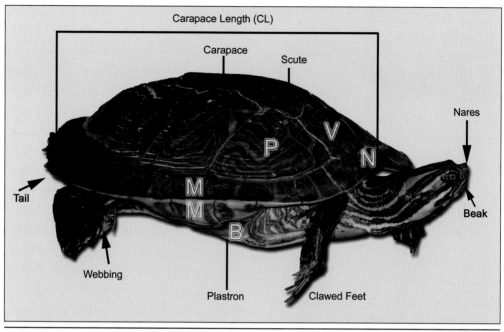

External anatomy of a River Cooter (*Pseudemys concinna*). The upper portion of the shell is the carapace, while the lower portion is the plastron. Large scales, known as scutes, cover the bony shell. Scutes are labeled with letters as follows: N = nuchal scute; V = vertebral scute; P = pleural scute; M = marginal scutes; B = bridge. (Photo by Matthew L. Niemiller)

floodplains. Females construct a nest by digging with their rear feet then depositing the eggs in the depression and covering it back up. The eggs are either round or elongate, and most are much more rigid-shelled than lizard and snake eggs. The nest sites must be carefully selected, as many species of turtles exhibit temperature-dependent sex regulation whereby the temperature at which the egg develops will determine the sex of the offspring. Generally, males are produced at lower temperatures, and females are produced at higher temperatures. Hence nests must be constructed to ensure a range of temperatures, or enough suitable nesting habitat must be available such that the females of a population produce clutches of both male and female hatchlings. Some females will cover long distances to find suitable nesting habitat, which,

unfortunately, often means having to cross a road. Hatchlings might either head straight for their parents' habitat or, as in the Snapping Turtles, disperse far and wide (often many kilometers) in search of a suitable area. Turtles, especially females, generally take longer to mature than other reptiles; hence a population can take much longer to recover from a reduction in size. Some species might take 20 years or more to reach sexual maturity.

Crocodilians

Crocodilians are one of the least diverse groups of reptiles, consisting of only 24 extant species around the world; however, they are an ancient group and have been more diverse in the past. Their closest living relatives, the birds, are far

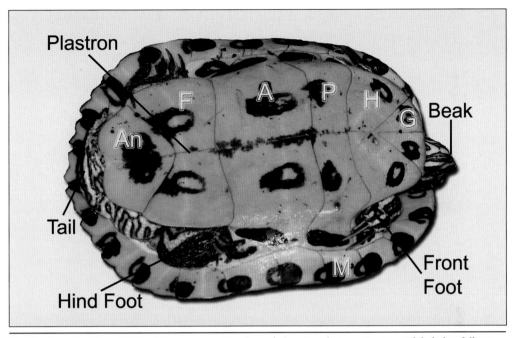

Upside-down Red-Eared Slider (*Trachemys scripta elegans*) showing plastron. Scutes are labeled as follows: G = gular scute; H = humeral scute, pectoral scute; A = abdominal scute; F = femoral scute; P = pleural scute; An = anal scute; M = marginal scute. (Photo by Matthew L. Niemiller)

more diverse, though present diversity does not necessarily translate into evolutionary success—the ancestors of modern crocodilians have been around for more than 200 million years. In the United States we have two native species of crocodilians: the American Alligator (*Alligator mississippiensis*) and the American Crocodile (*Crocodylus acutus*), with only the former occurring in Tennessee. Crocodilians are large, aquatic reptiles with a thick skin that contains hardened bony plates called osteoderms. These plates provide protection to exposed areas, such as the head, back, and tail. Crocodilians are generally cryptic creatures, ambushing their prey by surprise in a rush of motion. They are also powerful predators, capable of taking down large animals, such as deer and pigs, which are generally gripped by large teeth and powerful jaws and drowned. Large meals are often cached for later consumption or, as with the Indo-Pacific Crocodile (*Crocodylus porosus*), used as nuptial gifts to the opposite sex.

Crocodilians exhibit some complex mating behaviors, such as vocal communication and parental care, and our sole representative, the American Alligator, is an excellent example. Alligators generally become active during spring as the temperature warms, emerging from winter hibernation retreats to bask in the warming sun. Activity increases as summer approaches, and males become territorial in preparation for breeding. To establish a territory, a male American Alligator will utter a loud "bellow," sometimes described as a guttural roar—an ethereal sound to be sure. This call carries long distances and is produced by vibrating the body rapidly

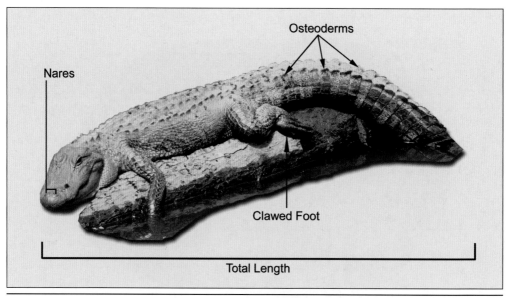

External anatomy of an American Alligator (*Alligator mississippiensis*). The nares (singular naris) are located at the front and top of the snout to facilitate breathing while the rest of the body is submerged. Note also that the eyes are high on the head to see above the water surface while the body remains below the surface. Osteoderms are thick, bony-plated scales that provide protection to the dorsal surface of the crocodilian. Note that this individual has a wound, possibly a puncture from another Alligator, on the snout between the left eye and the left naris. Pearl River, Louisiana. (Photo by Brad M. Glorioso)

as air exits the lungs. Prior to the audible call, a subaudible vibration is produced causing the body to vibrate so rapidly that water will "dance" on the male's back. Indeed, American Alligators were recently found to be one of the only natural producers of Faraday waves, or waves that move up and down above a vibrating object at half the speed of the vibrations. In fact, it is likely the distance and shape of the protruding osteoderms on the back that produce the Faraday waves, and the dancing water on the male's back plays a role in sexual signaling. Males also use "head slaps" to communicate. Males that intrude on another male's territory are subject to physical punishment, as males will violently drive them off.

Eggs are laid in early summer, when female American Alligators construct large, elaborate nests on the banks of their aquatic habitat. The eggs are carefully deposited in the nest and then covered up. Like the turtles, American Alligators have temperature-dependent sex determination; however, in crocodilians, females are produced at lower temperatures and males are produced at higher temperatures. The female will guard the nest, an important behavior, as many predators, such as raccoons, would quickly consume the eggs within. When the eggs are ready to hatch, the young inside will begin to call to the mother, and she will uncover the eggs and carry the hatchlings to the water. The young alligators are fairly precocious, being able to feed and swim on their own, but the mother will protect them for up to two years, or until they are able to defend themselves from predators, such as herons.

Reptiles are a fascinating component of the Tennessee fauna, and searching for them is a

Gray Ratsnake (*Pantherophis spiloides*) demonstrating the exceptional climbing ability of this genus, Lake County. (Photo by Brad M. Glorioso)

Common Gartersnake (*Thamnophis sirtalis*) consuming a road-killed Fowler's Toad (*Anaxyrus fowleri*), Stewart County. (Photo by Matthew L. Niemiller)

challenging and rewarding experience. Because they are generally harder to find than many other animals, many species of reptiles are only seen by the patient, determined, and skilled observer. However, the reward of finding a rarely seen snake is a true delight, as when a rock is overturned and you catch a fleeting glimpse of a crimson scarletsnake entering a hole. There are indeed many people who have discovered this joy, and we hope to convince others of the happiness and reward to be found in the appreciation of reptiles. Your adventures in search of our scaled friends will take you to some of the most beautiful corners of our state—from the ethereal cypress swamps of Reelfoot Lake to the towering forests of Big South Fork and the Smokies. We hope that you will come to appreciate reptiles and wild Tennessee as much as we do.

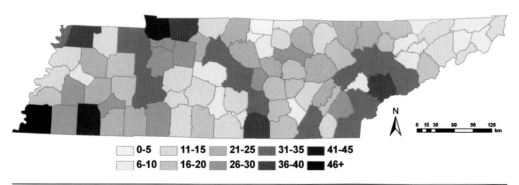

Diversity of reptile species in Tennessee based on county records. Note that low diversity in some counties (e.g., Loudon County) is likely due to limited sampling.

2
Physiography, Climate, and Habitats of Tennessee

Our state is home to an amazing diversity of turtles, snakes, lizards, and crocodilians. The rich reptile fauna in Tennessee is due in large part to the state's size and geographic orientation, and because it contains much physiographic, climatic, and habitat heterogeneity. This heterogeneity is easily witnessed as a person drives Interstate 40 from Memphis on the Mississippi River eastward across the state toward the Blue Ridge Mountains in eastern Tennessee. Although many species of reptiles are distributed statewide, several species, such as the Eastern Coachwhip, the Rough Earthsnake, and Bog Turtle, have ranges that barely extend into the borders of Tennessee or are restricted to distinct regions or rivers within specific habitats. These areas, known as ecoregions, vary with respect to geology, topography, soils, vegetation, hydrology, and climate. Maps of the state's topography, precipitation, average

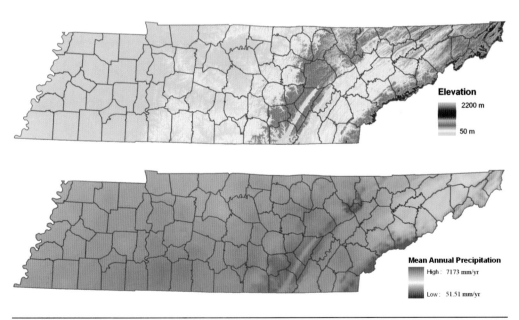

Top: Digital elevation model for Tennessee. The lowest point in Tennessee is the Mississippi River at 52 m and the highest point is Clingman's Dome in the Great Smoky Mountains National Park at 2,025 m above sea level. *Bottom:* Mean annual precipitation in Tennessee 1971–2000. Note that the range reflects high and low values for the continental United States and that mean annual rainfall varies between 1000 and 2000 mm/year in Tennessee. The region with the least precipitation is the Valley and Ridge in northeast Tennessee, while the areas receiving the most precipitation are the high Blue Ridge Mountains and portions of the Cumberland Plateau. (From: PRISM Group, Oregon State University, Corvallis; http://www.prismclimate.org/)

temperatures, and hydrology clearly illustrate the abiotic heterogeneity found within Tennessee. Accordingly, these ecoregions provide a wealth of habitats for reptiles and greatly influence the distribution and abundance of the state's reptile fauna.

In general, Tennessee has a temperate climate, with warm summers and mild winters, but a variety of climatic conditions occur because of the range of topography (elevations) across the state. Indeed, the warmest areas, which have growing seasons up to 235 days, are in the Mississippi Alluvial Plain along the Mississippi River in West Tennessee. The Nashville Basin is also quite warm, with a growing season of about 225 days. In contrast, the mountainous regions in East Tennessee (Southwestern Appalachians and Blue Ridge Mountains) have considerably cooler temperatures and shorter growing seasons (as little as 130 days). Although cooler, these regions also receive the most precipitation in the state, particularly the Blue Ridge Mountains, where precipitation exceeds 2 m (78 in) per year on average. The driest areas occur within the Ridge and Valley of northeast Tennessee, which receives less than 1.16 m (46 in) of precipitation a year on aver-

age. The warmest temperature recorded in the state was 45°C (113°F) in August at Perryville along the Tennessee River in West Tennessee, whereas the coldest temperature on record was -36°C (-32°F) in December at Mountain City in extreme northeast Tennessee.

Eight main ecoregions are recognized in Tennessee. Reptile diversity and community composition vary substantially from one ecoregion to another (Table 1). Indeed, ecoregions with greater topography, such as the Blue Ridge Mountains and Southwestern Appalachians, are cooler on average and receive more precipitation than those with less topography. These regions have lower reptile diversity than other ecoregions in the state. On the other hand, ecoregions that are warmer with less topography, such as the Mississippi Alluvial Plain and Mississippi Valley Loess Plains, support greater turtle and snake diversity.

In this section, we describe each of the major ecoregions and associated subregions found in Tennessee. In particular, we discuss the topography, geology, prominent natural vegetation, hydrology, and unique reptile fauna for each ecoregion and subregion.

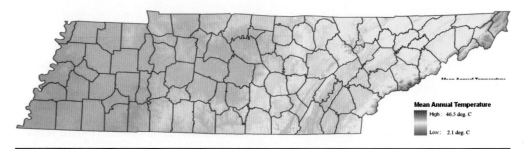

Mean annual temperature in Tennessee July 1971– 2000. Note that the range reflects high and low temperatures for the continental United States and that mean annual temperature in Tennessee varies between 6°C on top of Clingman's Dome and 17°C in the southwest near Memphis. (From: PRISM Group, Oregon State University, Corvallis; http://www.prismclimate.org/)

Mississippi Alluvial Plain

The Mississippi Alluvial Plain is a flat and broad floodplain, 60–90 m (200–300 ft) in elevation that lies between the Mississippi River to west and the Bluff Hills to the east in West Tennessee. This ecoregion is composed of Quaternary alluvial deposits of silt, clay, sand, and gravel. Although most of this ecoregion has been converted to cropland, some southern floodplain forest remains, consisting of oak, tupelo, and bald cypress stands. Streams in this ecoregion are sluggish and meander among many oxbow lakes and cypress swamps. Reelfoot Lake, formed by subsidence during the New Madrid earthquakes of 1811–12, occurs within this ecoregion in northwest Tennessee. Western Cottonmouths, Mississippi Green Watersnakes, Broad-Banded Watersnakes, Diamond-Backed Watersnakes, Plain-Bellied Watersnakes, Western Mudsnakes, Orange-Striped Ribbonsnakes, Midland Smooth Softshells, Mississippi Mud Turtles, False Map Turtles, Mississippi Map Turtles, and Southern Painted Turtles are characteristic reptiles of the Mississippi Alluvial Plain. American Alligators have been reported from southern sections of this ecoregion around Memphis.

Mississippi Valley Loess Plains

The Mississippi Valley Loess Plains ecoregion extends from near the Ohio River in western Kentucky south into Louisiana and is defined by thick loess deposits. Natural vegetation is primarily irregular plains with oak-hickory to oak-hickory-pine forest, but most of this land has been converted for agricultural use in Tennessee. Two subregions of the Mississippi Valley Loess Plains occur in Tennessee, the Bluff Hills and the Loess Plains.

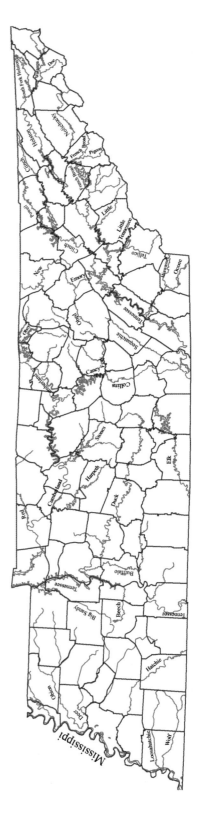

Major rivers and tributaries of Tennessee.

Table 1
Numbers of Species of Reptiles Associated with Ecoregions of Tennessee

Ecoregion	Lizards	Snakes	Turtles	Crocodilians	Total
Mississippi Alluvial Plain	**6**	**25**	**12**	**1**	**44**
Northern Mississippi Alluvial Plain	6	25	12	1	44
Mississippi Valley Loess Plains	**7**	**30**	**11**	**1**	**49**
Bluff Hills	7	26	11	0	44
Loess Plains	7	30	11	1	49
Southeastern Plains	**9**	**32**	**15**	**0**	**56**
Blackland Prairie	8	27	11	0	46
Fall Line Hills	8	28	11	0	47
Flatwoods/Blackland Prairie Margins	8	30	11	0	49
Southeastern Plains and Hills	9	31	12	0	52
Transition Hills	9	27	15	0	51
Interior Plateau	**9**	**28**	**15**	**0**	**52**
Western Highland Rim	9	28	15	0	52
Western Pennyroyal Karst	7	20	9	0	36
Outer Nashville Basin	8	25	11	0	44
Inner Nashville Basin	7	20	11	0	38
Eastern Highland Rim	8	25	10	0	43
Southwestern Appalachians	**8**	**21**	**12**	**0**	**41**
Cumberland Plateau	8	21	8	0	37
Plateau Escarpment	8	20	12	0	40
Sequatchie Valley	8	20	12	0	40
Central Appalachians	**4**	**16**	**6**	**0**	**26**
Cumberland Mountains	4	16	6	0	26
Ridge and Valley	**8**	**21**	**12**	**0**	**41**
Southern Dissected Ridges and Knobs	8	21	4	0	33
Southern Limestone/Dolomite Valleys and Low Rolling Hills	8	21	12	0	41
Southern Sandstone Ridges	8	21	4	0	33
Southern Shale Valleys	8	21	10	0	39
Blue Ridge Mountains	**9**	**21**	**10**	**0**	**40**
Southern Igneous Ridges and Mountains	2	9	3	0	14
Limestone Valleys and Coves	7	21	10	0	38
Southern Sedimentary Ridges	8	21	9	0	38
Southern Metasedimentary Mountains	9	21	9	0	39
Statewide	**9**	**34**	**16**	**1**	**60**

Bluff Hills

The Bluff Hills region consists of disjunct areas of generally dissected and forested terrain over deposits of sand, silt, clay, and lignite that are capped by substantial loess greater than 18 m (60 ft) thick. The Bluff Hills forms a narrow border along the Mississippi Alluvial Plain, which lies to the west. A variety of habitats exist in this subregion, including xeric slopes and ridges, more mesic slopes and ravines, bottomland forest, and small cypress swamps. Upland areas are primarily oak-hickory to mixed oak forest, whereas more mesic ravines and bottomland forest are dominated by maple, hickory, yellow poplar, and beech. Many streams in this subregion are intermittent or ephemeral with sandy to silty substrates; however, higher-gradient streams also occur and have gravel substrates. Timber Rattlesnakes, Southern Copperheads, Southern Black Racers, and Three-Toed Box Turtles are known from this subregion.

Loess Plains

Gently undulating, irregular plains with broad bottomlands and terraces characterize the Loess Plains, with loess deposits up to 15 m (50 ft) thick at elevations of 75–150 m (250–500 ft). This subregion is a very productive agricultural area, and much of the natural mosaic of oak-hickory forest, southern floodplain forest, and bluestem prairie has been replaced by cropland. However, some bottomland forest and cypress-gum swamp forest still remain. Most streams in the subregion have been channelized and are low gradient and murky with sand to silt substrates. Agricultural runoff has degraded water quality in many river systems. Western Cottonmouths, Southern

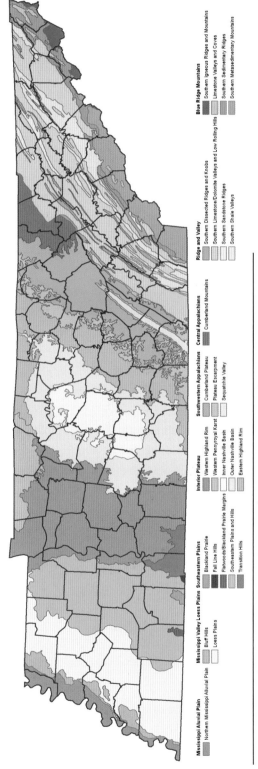

Major ecoregions of Tennessee; see the text for details. (From: Griffith et al. 1998)

(Left) View of the Mississippi River just upstream of the Tennessee border. (Photo by Matthew L. Niemiller) (Right) Reelfoot Lake, a prime habitat for many reptiles, Lake County. (Photo by Brad M. Glorioso)

Copperheads, Rough Earthsnakes, Plain-Bellied Watersnakes, Southern Black Racers, and Three-Toed Box Turtles are known from this subregion. Kirtland's Snakes are known from the Loess Plains in Henry County.

Southeastern Plains

The Southeastern Plains consists of irregular plains, broad areas between streams that contain a mosaic of oak-hickory-pine forest, woodlands, pasture, and croplands. This ecoregion is underlain with Cretaceous- to Tertiary-aged silts, sands, and clays. Elevation and relief are generally greater than the Mississippi Valley Loess Plains to the west but less than the Interior Plateau to the east. Streams in this ecoregion are generally low gradient with sand substrates. Five subregions occur in Tennessee: the Blackland Prairie, Flatwoods/Alluvial Prairie Margins, Southeastern Plains and Hills, Fall Line Hills, and Transition Hills.

Southeastern Plains and Hills

The Southeastern Plains and Hills is the largest subregion of the Southeastern Plains in Tennessee. This subregion is characterized by several bands of sand and clay formations oriented north to south. These formations are Tertiary-aged and have greater relief, with undulating topography to the west and some elevations reaching over 190 m. Natural vegetation is oak-hickory forest transitioning into oak-hickory-pine forest to the south. Streams are located in wide-bottomed areas with broad, rolling terraces and are moderate to low gradient with sandy substrates. Western Cottonmouths, Southern Copperheads, Eastern Hog-Nosed Snakes, Plain-Bellied Watersnakes, and Eastern Six-Lined Racerunners are known from this subregion. Northern Green Anoles, Eastern Fence Lizards, Eastern Slender Glass Lizards, Red Cornsnakes, Western Smooth Earthsnakes, Western Pygmy Rattlesnakes, and Eastern Coachwhips occur in the southern part of the Southeastern Plains and Hills in southwestern Tennessee.

(Left) Bottomland forest, Southeastern Plains and Hills, Hardeman County. (Photo by Stephen Nelson) (Right) Matt and Graham exploring bottomland forest, Southeastern Plains and Hills, Hardeman County. (Photo by Stephen Nelson)

Transition Hills

The Transition Hills subregion is characterized by dissected open hills 120–305 m (400–1,000 ft) in elevation with steep slopes caused by the downcutting of streams into the Mississippian-, Devonian-, and Silurian-aged strata. However, Cretaceous-aged clay, silt, sand, and gravel have been deposited above these older limestones, shales, and chert. This subregion is adjacent to the Western Highland Rim of the Interior Low Plateau in Hardin and Wayne counties. The natural vegetation of most of this subregion is oak-hickory-pine forest. Streams are moderate to low gradient with sand and gravel substrates. Northern Green Anoles, Common Five-Lined Skinks, Northern Pinesnakes, Midland Watersnakes, Mississippi Ring-Necked Snakes, and Eastern Box Turtles occur in this subregion.

Blackland Prairie, Flatwoods/Alluvial Prairie Margins, and Falls Line Hills

The Blackland Prairie, Flatwoods/Alluvial Prairie Margins, and Fall Line Hills subregions have small areas at elevations of 120–210 m (400–700 ft) in southwest Tennessee and extend south into Mississippi. The Blackland Prairie extends into a small portion of southeast McNairy County and is characterized by irregular plains and undulating lowlands. Natural vegetation is oak-hickory forest and blackbelt forest of sweetgum, oak, and cedar with patches of bluestem prairie. Streams in this subregion are low gradient with clay, sand, and silt substrates. The Flatwoods/Alluvial Prairie Margins subregion is characterized by undulating plains and lowlands with sluggish, low-gradient streams with sand substrates. This subregion encompasses a small area in southeast Hardeman County. Natural vegetation is oak-hickory to oak-hickory-pine forest with bottomland hardwood forest. The Fall Line Hills are dissected open hills composed primarily of Cretaceous-aged sand deposits at elevations of 140–210 m (700 ft) in southwest Hardin County. Natural vegetation is oak-hickory-pine forest, and streams are moderate to low gradient with sand substrates. Northern Green Anoles, Eastern Slender Glass Lizards, and Eastern Coachwhips are unique reptiles of these subregions.

Old field habitat in the Interior Plateau, Rutherford County. (Photo by Brian T. Miller)

Interior Plateau

The Interior Plateau ecoregion consists of a diverse landscape of generally hilly topography extending from north Alabama to southern Illinois, Indiana, and Ohio. Within Tennessee, the Interior Plateau is bordered by the Southwestern Appalachians ecoregion to the east and the Southeastern Plains to the west. There are five subregions within the Interior Plateau of Tennessee: the Western Pennyroyal Karst, Western Highland Rim, Eastern Highland Rim, Outer Nashville Basin, and Inner Nashville Basin. The Highland Rim encircles the oval-shaped Nashville Basin (also called the Central Basin) and is 150–180 m (500–600 ft) higher in elevation. The Nashville Basin can be divided into two concentric zones: the relatively flat Inner Nashville Basin and the hillier Outer Nashville Basin.

Western Pennyroyal Karst

The Western Pennyroyal Karst is characterized by relatively flat, weakly dissected terrain with irregular open plains and few perennial streams; however, sinking streams, sinkholes, and depressions are common and subterranean drainages are well developed. The Western Pennyroyal Karst is composed primarily of Middle Mississippian–aged limestones. The natural vegetation of this subregion is oak-hickory forest interspersed with patches of bluestem prairie, but much of this subregion has been cultivated for agriculture or is pastureland. Midwestern Wormsnakes, Southern Black Racers, Prairie Kingsnakes, Midland Watersnakes, Copper-Bellied Watersnakes, Gray Ratsnakes, Midland Brownsnakes, Common Five-Lined Skinks, Eastern Fence Lizards, Snapping Tur-

Abandoned shed, Cannon County. Such buildings provide excellent habitat and hunting ground for many species of snakes and lizards. (Photo by Brian T. Miller)

tles, and Midland Painted Turtles occur in this subregion.

Western Highland Rim

Characterized by dissected, hilly terrain with open hills but heavily forested valleys at 120–300 m (400–1,000 ft) in elevation, the Western Highland Rim is bisected by both the Tennessee and Cumberland rivers. This subregion is composed of Mississippian limestone, chert, and shale covered by cherty, acidic soils. Upland areas have cherty soils, and mixed oak forests predominate, whereas oak-hickory forests are the natural vegetation throughout other areas. However, most of the subregion was deforested in the mid- to late 1800s. Streams are of moderate gradient with sand-gravel substrates.

Northern Scarletsnakes, Red Milksnakes, Smooth Earthsnakes, Northern Pinesnakes, Queensnakes, Eastern Ribbonsnakes, Timber Rattlesnakes, Western Cottonmouths, Western Pygmy Rattlesnakes, Southern Coal Skinks, Northern Map Turtles, and Stripe-Necked Musk Turtles are found in this subregion. Alligator Snapping Turtles, Smooth Softshells, Eastern River Cooters, False Map Turtles, and Ouachita Map Turtles are known from larger rivers.

Eastern Highland Rim

The Eastern Highland Rim averages 305 m (1,000 ft) in elevation (up to 430 m [1,400 ft]) and is only 24 to 32 km (15–20 miles) wide. This subregion is characterized by gently rolling to nearly flat terrain with numerous springs,

Forest on the Eastern Highland Rim, Coffee County. (Photo by Matthew L. Niemiller)

sinkholes, and other karst features. The highest point on the Eastern Highland Rim, Short Mountain in Cannon County (632 m [2,075 ft]), is a remnant of a formerly much broader Cumberland Plateau. Like the Western Highland Rim, the periphery of the Eastern Highland Rim is characterized by steep-sloping valleys that mark the transition into the Outer Nashville Basin. A mosaic of open canopy woodlands and an understory consisting of grasslands called the Barrens is found within the headwaters of the Caney Fork, Duck, and Elk river watersheds. The hardpan soils of this region and low topographic relief keep soils saturated in the winter yet very dry during the summer, inhibiting forest development but fostering a prairie community. This subregion is composed of Mississippian-aged limestones, chert, shale, and dolomite. The natural vegetation is a transition from the oak-hickory forest to mixed mesophytic forest. Streams in this subregion are moderate gradient and nutrient rich with bedrock, cobble, and gravel substrates. Eastern Ribbonsnakes, Eastern Milksnakes, Mole Kingsnakes, Northern Watersnakes, Queensnakes, Northern Red-Bellied Snakes, Broad-Headed Skinks, Common Five-Lined Skinks, Red Cornsnakes, and Northern Map Turtles are found in this subregion.

Cedar glades develop in the Inner Nashville Basin and are home to many species of reptiles. Note the presence of large flat rocks, which provide ideal shelter. Wilson County. (Photo by Brian T. Miller)

Outer Nashville Basin

The Outer Nashville Basin is characterized by rolling hills with numerous knobs slightly higher in elevation than the Inner Nashville Basin. Predominant groundcover includes oak-hickory forest with pasture and cropland. This subregion is underlain primarily by non-cherty Ordovician-aged limestones, although the higher knobs and hills contain more cherty Mississippian-aged limestones and Devonian-aged Chattanooga shale, which are remnants of a once broader Highland Rim. Streams in this subregion are low to moderate in gradient and nutrient rich. Red Cornsnakes, Queensnakes, Northern Copperheads, Midwestern Worm-snakes, Mississippi Ring-Necked Snakes, Tim-ber Rattlesnakes, Eastern Fence Lizards, Eastern Box Turtles, and Northern Map Turtles occur in this subregion.

Inner Nashville Basin

The Inner Nashville Basin is considerably flatter and lower in elevation than the Outer Nashville Basin. Additionally, this subregion has thinner soils and greater karst development. This subregion is underlain by Ordovician-aged limestones. Most of this subregion is cropland and pasture with scattered maple-oak-hickory forests. Cedar glades also are found within the Inner Nashville Basin and contain a unique assemblage of plants and animals adapted to a harsh, xeric environment. The shallow and rocky soils of cedar glades limit forest development and typically are open areas surrounded by eastern red cedar where soils are thicker. Streams are low gradient and nutrient rich with bedrock the predominant substrate. Southeastern Crowned Snakes, Smooth Earthsnakes, Northern Scarletsnakes, Northern Black Racers, Eastern Black Kingsnakes, Eastern Milksnakes, Timber Rattlesnakes, Southeastern Five-Lined Skinks, Eastern Six-Lined Racerunners, Little Brown Skinks, Broad-Headed Skinks, and Eastern Mud Turtles occur in this subregion.

Southwestern Appalachians

The Southwestern Appalachians ecoregion consists of forest and woodland habitats interspersed with some cropland and pasture on low, mountainous terrain. This region extends from Kentucky southwest through Tennessee and into Alabama and northwest Georgia. The eastern edge of this ecoregion, where it meets the Ridge and Valley, is only slightly dissected by eastward-flowing streams. In contrast, the western edge, where it meets the Eastern Highland Rim of the Interior Plateau, is much more rugged and deeply incised. Upland areas are primarily mixed oak–shortleaf pine forest, whereas deeper ravines and escarpment slopes are characterized by mixed mesophytic forest. Three subregions occur within the Southwestern Appalachians: the Cumberland Plateau, Plateau Escarpment, and Sequatchie Valley.

Cumberland Plateau

The Cumberland Plateau is characterized by low hills, undulating uplands, valleys, and ridges about 300 m (1,000 ft) higher in elevation than the Eastern Highland Rim to the west. Elevations range from 360 to 610 m (1,200–2,000 ft), although the Crab Orchard Mountains reach over 900 m (3,000 ft) in elevation. Accordingly, this subregion is slightly cooler and receives more precipitation than surrounding ecoregions. This subregion is underlain by Pennsylvanian-aged conglomerate, sandstone, siltstone, coal, and shale and is covered by acidic, well-drained soils. Although forested, forest composition is highly variable and logging is common. Streams within this subregion are of low to moderate gradient with broader floodplains and more riparian wetlands than in the Plateau Escarpment. Northern Copperheads, Eastern Wormsnakes, Ring-Necked Snakes, Northern Watersnakes, Gray Ratsnakes, Eastern Fence Lizards, Common Five-Lined Skinks, and Eastern Box Turtles occur in this subregion.

Plateau Escarpment

Narrow ridges, cliffs, gorges, and steep, forested slopes characterize the Plateau Escarpment, where local relief can be in excess of 340 m (1,100 ft). This subregion is underlain by Mississippian-aged limestones, sandstone, shale, and siltstone in the lower slopes and western valleys, whereas the upland areas are underlain by Pennsylvanian-aged conglomerate, sandstone, shale, and siltstone. Ravines and gorges are characterized by mixed oak forest on the upper slopes to more mesic forests of beech, tulip-poplar, ash, buckeye, and maple on the middle and lower slopes. Hemlock is common along riparian areas and river birch along floodplain terraces. Streams are of high quality and gradient with numerous riffles and pools and

View from Stone Door on the Cumberland Plateau, Grundy County. (Photo by Brad M. Glorioso)

boulder to bedrock substrates. Several streams have cut down into the Mississippian-aged limestones and have lower gradients, wider valleys, and higher productivity. Northern Copperheads, Timber Rattlesnakes, Midland Watersnakes, Ring-Necked Snakes, Gray Ratsnakes, Common Five-Lined Skinks, Northern Green Anoles, and Eastern Box Turtles are found in this subregion.

Sequatchie Valley

The Sequatchie Valley is an open, rolling valley averaging 6.4 km (4 miles) wide and running from just south of the Crab Orchard Mountains south-southwest for 240 km (150 miles) into northwest Alabama. This subregion is associated with an anticline where erosion has formed a deep valley 180–300 m (600–1,000 ft) in elevation and bounded on either side by the Cumberland Plateau. The eastern section boundary is much narrower and is known as Walden Ridge. The valley floor is nearly 300 m (1,000 ft) lower in elevation than the surrounding highlands. This subregion is underlain

with Mississippian-aged and Ordovician-aged limestones and shale. The Sequatchie Valley is characterized by mixed oak-hickory-maple forest, but there are several areas of cropland and pasture. Streams in this subregion are moderate to low gradient, and several springs are present. Northern Watersnakes, Gray Ratsnakes, Ring-Necked Snakes, Eastern Wormsnakes, Common Five-Lined Skinks, Snapping Turtles, and Eastern Box Turtles occur in this subregion.

Central Appalachians

The Central Appalachians span from northern Tennessee northeast to central Pennsylvania. This ecoregion is characterized as a high plateau of rugged terrain that is highly dissected. Most of this ecoregion is blanketed with Appalachian oak and northern hardwoods forest because of the rugged terrain, cool, moist climate, and infertile soils; however, coal mining is prominent in this ecoregion. The Cumberland Mountains is the only subregion of the Central Appalachians in Tennessee.

Cumberland Mountains

The Cumberland Mountains are characterized as low mountains with long, steep slopes and narrow, winding valleys. This subregion is higher in elevation (360–1,060 m [1,180–3,500 ft]) and much more highly dissected than the adjacent Cumberland Plateau, with local relief up to 610 m (2,000 ft). This subregion is underlain with Pennsylvanian-aged sandstones, shales, siltstones, and coal. This natural vegetation is mixed mesophytic forest consisting of oak, maple, buckeye, beech, and tulip poplar, although the forest composition varies greatly depending on local topography. Streams are high to moderate gradient and clear, with principally bedrock substrates, waterfalls, and many riffles. However, many streams have been negatively impacted by coal mining and logging operations. Northern Copperheads, Timber Rattlesnakes, Gray Ratsnakes, Eastern Garter-snakes, Northern Ring-Necked Snakes, Midland Watersnakes, Common Five-Lined Skinks, and Eastern Box Turtles are found in this subregion.

Ridge and Valley

The Ridge and Valley consists of lower-elevation, roughly parallel ridges and valleys generally running from southwest to northeast between the Cumberland Plateau to the west and

The Cumberland Plateau and Cumberland Mountains provide excellent habitat for reptiles. Rhea County. (Photo by Matthew L. Niemiller)

the Blue Ridge Mountains to the east. This topography is varied in height, width, and geology owing to past tectonic events. This ecoregion is underlain with Paleozoic sandstones, siltstone, shale, chert, limestones, and dolomites that have been folded and faulted. Ridges tend to be composed of more resistant strata, whereas valleys are composed by weaker strata, such as limestones and shales. Dominant natural vegetation is Appalachian oak forest, although only about 50% of this ecoregion is now forested. Streams vary in gradient and substrate composition and are very productive. Springs and caves are also numerous in this ecoregion. Four subregions occur in Tennessee: the Southern Limestone/Dolomite Valleys and Low Rolling Hills, Southern Shale Valleys, Southern Sandstone Ridges, and Southern Dissected Ridges and Knobs.

Southern Limestone/Dolomite Valleys and Low Rolling Hills

Encompassing the largest area of the Ridge and Valley in Tennessee, the Southern Limestone/Dolomite Valleys and Low Rolling Hills subregion is characterized by Ordovician-aged limestones and cherty dolomite forming broad, rolling, fertile valleys and numerous springs and caves. Elevation ranges from 210 to 610 m (2,000 ft), with local relief up to 210 m (700 ft). Natural vegetation is Appalachian oak forest, although bottomland oak and sycamore-ash-elm riparian forests also occur in this subregion. Grassland barren and cedar-pine glades also occur. However, much of this subregion has been converted to cropland and pasture. Streams are moderate to low gradient with sand, gravel, cobble, and bedrock substrates. Northern Cop-

Eastern deciduous forest on the Cumberland Plateau, Cumberland County. (Photo by Matthew L. Niemiller)

perheads, Eastern Wormsnakes, Northern Black Racers, Mole Kingsnakes, Eastern Black Kingsnakes, Eastern Milksnakes, Rough Green-snakes, Gray Ratsnakes, Red Cornsnakes, Queensnakes, Common Five-Lined Skinks, Broad-Headed Skinks, Snapping Turtles, and Eastern Box Turtles are common in this subre-gion. Ouachita Map Turtles, Midland Painted Turtles, Stripe-Necked Musk Turtles, Eastern Spiny Softshells, and Eastern River Cooters are found in larger rivers and streams.

Southern Shale Valleys

The Southern Shale Valleys region is character-ized by lowlands, rolling valleys, and low hills with Ordovician- and Cambrian-aged shale deposits at elevations of 240–460 m (800–1,500 ft). Natural vegetation is Appalachian oak for-est, but much of the land has been converted to pasture and small farms. Streams are mod-erate to low gradient with sand, gravel, cobble, and bedrock substrates. Common reptiles in-clude Northern Watersnakes, Gray Ratsnakes, Northern Black Racers, Eastern Wormsnakes, Northern Ring-Necked Snakes, Northern Cop-perheads, Common Five-Lined Skinks, Snap-ping Turtles, and Eastern Box Turtles.

Southern Sandstone Ridges

The Southern Sandstone Ridges are high, steep-sided, forested ridges with narrow crests rang-ing from 270 to 910 m (900–3,000 ft) in elevation with up to 365 m (1,200 ft) of local relief. These ridges are comprised of Ordovician-, Silurian-, Devonian-, and Mississippian-aged sandstones, shales, siltstones, and conglomerates. Natural

The Hiawassee River is excellent turtle habitat, and many species such as Green Anoles, Northern Copperheads, and Red Cornsnakes occur along the banks. Polk County. (Photo by Stephen Nelson)

vegetation is Appalachian oak forest with some mixed mesophytic forest. Streams are high to moderate gradient with cobble and bedrock substrates. Northern Copperheads, Northern Watersnakes, Gray Ratsnakes, Northern Ring-Necked Snakes, Common Five-Lined Skinks, Eastern Fence Lizards, and Eastern Box Turtles are found in this subregion.

Southern Dissected Ridges and Knobs

The Southern Dissected Ridges and Knobs are characterized by hills, ridges, and knobs that are lower in elevation (240–610 m [800–2,000 ft]) and more dissected than the Southern Sandstone Ridges. The ridges on the east side of the Ridge and Valley ecoregion typically are associated with Ordovician-aged shales, sandstones, and limestones, whereas those in the central and western Ridge and Valley tend to be associated with Cambrian-aged formations. Natural vegetation of higher elevations is chestnut oak-pine forest with Appalachian oak and mixed mesophytic forest at lower elevations. Streams are small in this subregion, with high to moderate gradient and gravel, cobble, bedrock substrates. Gray Ratsnakes, Red Cornsnakes, Eastern Milksnakes, Northern Copperheads, Eastern Fence Lizards, Common Five-Lined Skinks, and Eastern Box Turtles occur in this subregion.

Blue Ridge Mountains

The Blue Ridge Mountains are characterized by narrow mountain ridges that are forested and highly dissected at elevations of 305–2,025 m (1,000–6,600 ft). These mountains are rugged

Hiawassee River, Polk County. (Photo by Brad M. Glorioso)

with high-gradient streams that are cool and clear. This is the wettest ecoregion in Tennessee, with some peaks in the Great Smoky Mountains receiving almost 2 m (80 in) of annual precipitation. Four subregions occur in Tennessee: the Southern Igneous Ridges and Mountains, Southern Sedimentary Ridges, Limestone Valleys and Coves, and Southern Metasedimentary Mountains.

Southern Igneous Ridges and Mountains

The Southern Igneous Ridges and Mountains subregion is characterized by prominent, highly dissected ridges and mountains separated by high gaps and coves with steep slopes. This subregion occurs in northeast Tennessee near the North Carolina border in Unicoi, Carter, and Johnson counties. These ridges and mountains are composed primarily of Precambrian-aged igneous and metamorphic strata. Elevations range from 610 to 1,915 m (2,000–6,300 ft). This area is primarily covered in Appalachian oak, mixed mesophytic, and northern hardwoods forest. Streams are high gradient, cool, and clear with boulder and bedrock substrates. Eastern Gartersnakes, Northern Watersnakes, Eastern Wormsnakes, Northern Ring-Necked Snakes, Eastern Milksnakes, and Eastern Box Turtles occur in this subregion.

Southern Sedimentary Ridges

The Southern Sedimentary Ridges are high-elevation (305–1,370 m [1,000–4,500 ft]), steep-sloped ridges with deep, narrow valleys—primarily composed of Cambrian-aged shale,

A view of the Great Smoky Mountains from within Cades Cove in the Blue Ridge Mountains, Blount County. (Photo by Matthew L. Niemiller)

sandstones, siltstones, quartzite, and conglomerates—and include some of the westernmost foothills areas, such as Bean, Starr, Chilhowee, English, Stone, Bald, and Iron mountains. Natural vegetation is Appalachian oak forest with some mixed mesophytic and northern hardwoods forest. Streams are high gradient with cool, clear water and with boulder and bedrock substrates. Common species of this subregion include Eastern Wormsnakes, Northern Ring-Necked Snakes, Gray Ratsnakes, Eastern Milksnakes, Northern Copperheads, Common Five-Lined Skinks, Eastern Fence Lizards, and Eastern Box Turtles.

Limestone Valleys and Coves

The Limestone Valleys and Coves is the smallest subregion of the Blue Ridge Mountains in Tennessee. This subregion is characterized by small but distinct lowland areas 450—760 m (1,500–2,500 ft) in elevation that have flat to rolling valleys and coves surrounded by high mountains. Cades Cove, Wear Cove, and Tuckaleehee Cove are part of this subregion. Limestone sinks are common. Appalachian oak forest is the dominant natural vegetation, although much of this land has been converted into pasture and cropland. Streams are moderate gradient with boulder and cobble substrates. Northern

Pinesnakes, Northern Watersnakes, Northern Black Racers, Gray Ratsnakes, Eastern Milksnakes, Northern Copperheads, Eastern Slender Glass Lizards, Common Five-Lined Skinks, Broad-Headed Skinks, and Eastern Box Turtles are inhabitants of the Limestone Valleys and Coves subregion. Bog Turtles are found in this subregion in small areas of Johnson County.

Southern Metasedimentary Mountains

The Southern Metasedimentary Mountains are highly dissected, steep-sloped mountains composed primarily of Precambrian-aged metamorphic and sedimentary strata. Elevations range from 305 to 2,010 m (1,000–6,600 ft). Lower and middle elevations have Appalachian oak and northern hardwood forests, while the higher elevations (above 1,675 m [5,500 ft]) have southeastern spruce-fir forest as the dominant natural vegetation. Cool, clear streams are high gradient with boulder and bedrock substrates. Northern Copperheads, Timber Rattlesnakes, Eastern Gartersnakes, Northern Watersnakes, Northern Ring-Necked Snakes, and Eastern Box Turtles can be found at both low and higher elevations. Several reptiles are found only at lower elevations, including Eastern Six-Lined Racerunners, Southeastern Five-Lined Skinks, Northern Green Anoles, Northern Coal Skinks, Northern Pinesnakes, Red Cornsnakes, Eastern Milksnakes, Eastern Wormsnakes, and DeKay's Brownsnakes. Queensnakes, Midland Painted Turtles, Stripe-Necked Musk Turtles, Eastern Musk Turtles, Northern Map Turtles, and Eastern River Cooters are found in larger, low-elevation streams and rivers.

View from near the top of Clingman's Dome, Great Smoky Mountains National Park, Sevier County. (Photo by Stephen Nelson)

3
Reptile Conservation

While significant global declines in amphibian species have been widely reported in the last few decades, comparatively little attention has been given to the decline of many reptile species around the world. Reptiles are extremely vulnerable to human-mediated changes to their environments (anthropogenic habitat modification), both deliberate changes, such as habitat destruction or modification, and unintended changes, such as the introduction of extremely damaging predators, including cats, rats, pigs, and mongoose. Extinction of reptilian species due directly to human causes can be traced back at least 500 years. Many unique species have been lost, particularly on islands. An especially tragic example occurred in the Mascarene Islands (Reunion, Mauritius, and Rodriguez) and serves as a powerful insight into rapid extinction. The herpetofauna of the Mascarenes was decimated after the arrival of humans in the 16th century, as these islands were previously uninhabited. Humans preyed heavily on the giant tortoises of these islands, while introduced rats, cats, and pigs destroyed the eggs, young, and adults of many terrestrial reptiles. To compound the problem, introduced livestock, such as goats and cattle, destroyed much of the available food for tortoises, and introduced plants forever changed the unique vegetation of the islands. As a consequence, the tortoises (five species), several lizards, and one unique boa all became extinct. As another example, the largest gecko known from recent times, the Kawekaweau (*Hoplodactylus delcourti*), succumbed to introduced rats on

New Zealand—the only record of its existence is a single preserved specimen, some bones, and the memory of the animal passed down in the oral traditions of the Maori.

Closer to home, the St. Croix Racer (*Borikenophis sanctaecrucis*) is now extinct on the U.S. Virgin Islands, likely due to the introduction of cats and mongooses. While no reptile species extinctions have been recorded from the contiguous United States, a number of species are threatened or in danger of extinction. The Bog Turtle (*Glyptemys muhlenbergii*) occurs in Tennessee in a few tiny populations and is critically endangered throughout its range. This is a species that could certainly be all but lost in our lifetimes.

Reptiles in Tennessee are vulnerable to a variety of threats, which is why we have included a conservation section for each species account in this book, so that readers might be aware of the diverse threats faced by our state's reptiles. We further hope that Tennesseans who learn about particular threats faced by individual populations and species will participate in mitigation and conservation projects to protect the state's beautiful but vulnerable reptilian fauna. It is our hope that we can improve attitudes about reptiles and bring awareness to the importance of protecting them from extinction. Below are some examples of threats faced by reptiles, both around the world and in the state of Tennessee. We hope our readers will consider these threats and take personal action to assist in the conservation of reptilian biodiversity.

Threats to Reptiles

Habitat Destruction

Habitat destruction is probably the number one cause of reptilian declines both around the world and in Tennessee. The loss of forests, conversion of land to agriculture, draining of wetlands, damming of rivers, and suppression of natural fires all impact local reptilian populations. Many species, such as Pinesnakes (*Pituophis melanoleucus*), are adapted to specific types of pine forest habitat, and alteration of this habitat can lead to the rapid decline or disappearance of a population. Many wetland habitats have been severely altered either by reduction in water quality or changes to natural hydrological patterns. Species, such as River Cooters (*Pseudemys concinna*), prefer clear-flowing water and have likely been impacted by the repeated damming of the Tennessee River.

The removal of native vegetation surrounding the water body can also severely impact aquatic and semiaquatic reptiles. Turtles must all come ashore to lay their eggs, and they require natural banks, soil, and vegetation in order to dig and properly cover their nests. Many turtles in Tennessee exhibit temperature-dependent sex determination, and changes to vegetation cover or soil quality can impact the quality of the nests and influence sex ratios in local populations. Some species in the state, such as Kirtland's Snakes (*Clonophis kirtlandii*) and Bog Turtles (*Glyptemys muhlenbergii*), are so rare and of such limited distribution that any disturbance to their remaining habitat would cause them to rapidly become extirpated in Tennessee.

An all-too-familiar sight: a road killed Yellow-Bellied Kingsnake (*Lampropeltis calligaster*), Coffee County. (Photo by Matthew L. Niemiller)

Road-killed Northern Copperhead (*Agkistrodon contortrix*), Great Smoky Mountains National Park, Blount County. (Photo by Matthew L. Niemiller)

Disturbance

Tennessee has many natural areas that are managed to preserve some aspect of their natural qualities, such as national parks, state parks, state forests, wildlife management areas, and scenic rivers. Overall levels of habitat protection may range from higher levels, such as in the Great Smoky Mountains National Park, to lower levels in regional parks or managed areas, where harvesting of natural resources may occur. While some protection of native habitats is always better than no protection, reptile populations in these less-protected areas are not necessarily safe from human impacts. In wildlife management areas, hunting is allowed nearly year-round and many hunters use firearms to harvest game animals. The combination of easy access to a firearm and a misunderstanding of reptiles can lead to the killing of reptiles, espe-

cially snakes and American Alligators, a situation we have witnessed all too often. It is not uncommon to find rattlesnake carcasses riddled with bullet holes and missing their tails in areas frequented by hunters presumably seeking other prey, such as turkey and deer.

A particularly damaging behavior that we have witnessed in the Inner Nashville Basin of Middle Tennessee is the removal of large, flat rocks from protected cedar glade areas to use in decorative rock walls or other constructions. This might seem a harmless activity, merely picking up rocks in the woods, but these rocks are a vital component of the cedar glade habitat for reptiles. Large flat rocks are the most frequently used hiding place for snakes and lizards when they are not active and are critical to the proper thermoregulation of these species in this unique habitat. We have noted that removing

Some human constructions offer major barriers to reptile species. These Three-Toed Box Turtles (*Terrapene carolina triunguis*) cannot cross railroad tracks except at road crossings and are prevented from dispersing and subject to entrapment or vehicle strikes. Tangipahoa Parish, Louisiana. (Photo by Brad M. Glorioso)

rocks from a glade can lead to the disappearance of reptiles from that glade. Even in supposedly protected areas, reptiles are not safe from human disturbance. Moreover, several cedar glade habitats are severely degraded because of high off-road vehicle (ORV) use. We have even found recently killed snakes and lizards underneath rocks on cedar glades that were run over by ORVs. Such senseless death can be prevented by using ORVs on designated trails.

Malicious Killing

Many people have a mild to strong fear of reptiles, especially snakes, a condition known as herpetophobia (irrational fear of amphibians and reptiles) or ophidiophobia (irrational fear of snakes). While some claim that this fear is innate, we often don't see the appearance of this behavior until children are five or six years old, an age when they can learn this fear from adults they interact with. While a healthy respect of wildlife, especially venomous snakes, is a good thing for both humans and animals, the desire to destroy reptiles out of fear is an extremely misinformed behavior. Nevertheless, the most common interaction a snake can have with a human in Tennessee is to be decapitated with a garden implement, and this is likely a huge source of mortality for snakes. In a lifetime of 10–15 years, a snake in Tennessee is exceedingly likely to encounter a human who would

Road-killed hatchling emydid turtles, Lake County. As hatchlings emerge in the summer, they often must cross roads to travel between the nest site and the water. (Photo by Brad M. Glorioso)

do it harm. Many people feel that killing snakes is a sort of morbid public service or imparts some sort of protection to their families. Others just find it a demented sort of sport or otherwise derive pleasure in destroying wildlife.

Snakes are probably the only vertebrates that elicit such a violent reaction from people. Much of this behavior is probably due to ignorance, as many people cannot tell the difference between a venomous and a nonvenomous snake or mistakenly believe that some species of nonvenomous snakes are indeed venomous. Though only two species of lizards in the Western Hemisphere are venomous (none east of the Rocky Mountains), many people mistakenly believe that the native skinks (genus *Plestiodon*)

are harmful or capable of "stinging." Just as we would not abide someone hacking a songbird to death, it is up to those of us who care about wildlife to reach out and educate people whenever possible to prevent the senseless slaughter of our native fauna.

Hunting

Only one species of reptile may be legally hunted throughout Tennessee (the Snapping Turtle, *Chelydra serpentina*), as long as one possesses a valid fishing license (see chapter 4, **"Finding Reptiles in Tennessee"**). Other nonthreatened turtles can be collected in the Reelfoot Lake Wildlife Management Area in

Road-killed gravid female Pond Slider (*Trachemys scripta*), Lake County. This individual was likely killed on her way to lay eggs. Because aquatic turtles need soft, sandy soil to dig a nest, they often must cross roads to move between the water and a suitable nesting site. This is a common sight in spring on Tennessee's roads. (Photo by Brad M. Glorioso)

Lake and Obion counties. However, some reptiles are routinely hunted illegally—especially the Timber Rattlesnake (*Crotalus horridus*). In the southeastern United States, events known as "rattlesnake roundups" still occur. These morbid celebrations slaughter hundreds or thousands of Eastern Diamondback and Timber Rattlesnakes. These persecutions have led to a drastic decline in populations across the Southeast. Timber Rattlesnakes are killed in Tennessee for the skin and rattle—the former to make tacky clothing accessories and the latter as a vulgar trophy. Entire populations have been wiped out, as very few rattlesnakes that happen across humans survive the encounter.

Timber Rattlesnakes are now quite rare across much of the state, surviving in isolated remnant populations, usually in protected areas. Alligator Snapping Turtles are also in danger from hunting—they are sometimes hooked by fishermen, caught in traps, or encountered in shallow water and are sometimes killed for the meat or to "protect" fish and bird stocks. These are extremely long-lived animals, and the removal of a few female breeders could ultimately doom a population. Because this species is rare in Tennessee, it is likely that without some protection it is in danger of becoming extirpated in our state.

Pet Trade

It is illegal to collect reptiles for pets in Tennessee, yet many people, on finding an interesting reptile in their back yard, decide to keep it. While this act itself is probably not harmful to the population, on a large enough scale, some populations of particularly attractive species, such as Scarlet Kingsnakes (*Lampropeltis elapsoides*), could be greatly reduced. Indeed, Eastern Box Turtles (*Terrapene carolina carolina*) have long been subject to collection, which has likely caused a decline in abundance across the Southeast. While it can be quite educational for children to interact with and appreciate wildlife, we recommend that wild animals not be captured or restrained, and that people make good use of facilities such as zoos or nature centers to appreciate captive animals. One major reason for not bringing wild animals into captivity is that this will cause them stress, which can weaken their immune system and make them vulnerable to disease. They might also pick up a foreign disease if kept with or near exotic animals. If the sick animal is released back into the wild, it will carry this disease to the wild population with potentially serious consequences.

Roads

Roads are a huge source of mortality to reptiles in Tennessee. Snakes, lizards, and turtles must periodically cross roads to move between habitats, and roads are frequently built in areas that separate reptiles from seasonal habitats. Turtles must lay their eggs on land, and many females are killed when crossing roads to reach suitable nesting areas. Indeed, in late spring and early summer the carnage can be huge on busy roads that border aquatic habitats. Snakes are especially vulnerable to motor vehicles, as their long bodies impart a greater chance that a vehicle will run over them. Gray Ratsnakes (*Pantherophis spiloides*) rarely reach the large sizes they are capable of in many areas, likely due in part to the fact that the longer they get, the harder time they will have surviving a trip across a road. A study of road mortality on a population of Eastern Ratsnakes (*Pantherophis alleghaniensis*) found that the loss of more than three females per year due to road mortality increased the probability of extinction for the population to greater than 90%. Many snakes use roads as a source of heat, soaking up the remaining heat from asphalt as the sun goes down. This is obviously a very dangerous behavior for the snakes, and many of them are killed on roads even in protected areas such as the Great Smoky Mountains National Park. Some species of snakes are secretive and extremely vulnerable to road mortality. The most common way to find a Copperhead (*Agkistrodon contortrix*) or Yellow-Bellied Kingsnake (*Lampropeltis calligaster*) is to see them dead on the road. Finding a live individual can be extremely difficult. Copperheads are especially common on roads during warm rainy nights, when they are foraging for amphibians. Some snakes, such as Copperheads and Common Gartersnakes (*Thamnophis sirtalis*), will eat amphibians that have been killed on a road, which usually means that the snake will be killed by the next passing car.

Tragically, some people go out of their way to intentionally strike a reptile with their cars. We have all too frequently watched drivers in front of us endanger themselves, their passengers, and others on the road (including us as we follow them) by swerving off the road or into the median to intentionally strike a turtle or snake. This sort of behavior is irresponsible at best and sadistic at worst.

Introduced Species

Introduced species are taking an enormous toll on reptile populations worldwide, causing extinction and population declines. Rats, cats, mongoose, goats, dogs, and pigs have had a detrimental impact on reptiles, especially on vulnerable island populations. However, Tennessee herpetofauna is not immune to these creatures, and some introduced species can cause tremendous mortality. Cats are popular pets, but owners must realize that they can be incredibly destructive. Pet cats, even if they are well fed, instinctively kill wildlife, whereas cats that are abandoned can establish feral populations that prey heavily on reptiles, small mammals, and songbirds. Fire ants have been introduced to the southeastern United States from South America and continue to spread each year. These ants live in large, densely spaced colonies and have been implicated in the predation of eggs and juvenile reptiles. Several southeastern reptiles, such as Eastern Kingsnakes (*Lampropeltis getula*) and Southern Hog-Nosed Snakes (*Heterodon simus*), have declined precipitously in the last few years, though it remains unclear whether this is directly related to fire ants.

Pollution

The accumulation of toxins from industry can have harmful effects on local reptilian communities. Polychlorinated biphenyls (PCBs) have been shown to bioaccumulate in turtles and American Alligators (*Alligator mississippiensis*), causing sex reversal and poor gonad development, which can cause reproductive failure in heavily affected populations. Coal-ash dumps and the associated toxic heavy metals have been a major news item in Tennessee, and this type of pollution has been shown to affect reptiles in addition to all other wildlife and humans. Northern Watersnakes (*Nerodia sipedon sipedon*) living in coal ash–contaminated habitats were shown to have higher basal metabolic rates, which could mean that they are using too much energy for homeostasis and will have decreased reproduction and growth.

* * *

The global loss and decline of biodiversity demands attention and rectification. This chapter highlights some of the threats and a few examples of obstacles faced by reptile populations both worldwide and here in Tennessee. We encourage readers interested in learning more to consult the many excellent references on reptile conservation and declines, some of which are listed in the **"Recommended Readings, Organizations, and Websites"** section toward the end of this book.

If You Build It . . .

One of my childhood dreams was to landscape my parent's yard to attract species of amphibians and reptiles. My parents were tolerant of my interests and eventually allowed me to bring rock and logs into the yard, creating potential habitat for fence lizards, skinks, and small snakes. Unfortunately, we lived in an area devoid of lizards and only had an occasional gartersnake or box turtle pass through the yard, none lingering in my outdoor sanctuary. Although my attempts at attracting herpetofauna into my parent's yard were unsuccessful, they prodded me to learn more about wildlife habitat, and I quickly realized that the section of lawn my parents had allowed me to landscape was too small to be attractive to most species of amphibians and reptiles and that I lacked a critical component of habitat—an accessible water source. This revelation became important decades later when my wife and I moved into Middle Tennessee.

Fortunately, my wife's tolerance of my interest in amphibians and reptiles equals that of my parents, and she shares my landscaping interests; consequently, our Cannon County property has become the kind of sanctuary for amphibians and reptiles that I could only dream about having around as a child. Admittedly, my current situation is very different. I grew up in the suburbs of St. Louis, Missouri, but I now live in a rural area of Tennessee. Our house is set in the middle of an approximately 20-acre clearing that used to be pasture; much of the original pasture is now old field, but my wife and I "garden" about 3 acres. Our yard is surrounded by hundreds of acres of woods and is bisected by a small creek that flows usually only following heavy rains. Because of the setting, the potential for recruitment of amphibian and reptile species into our yard from surrounding populations is extraordinary. We have about seven garden areas of various sizes that are set aside from lawn, field, or woods by borders formed by either logs or rocks. We mulch the gardens heavily, at least in those years that time allows. During the past quarter century, we have constructed several dry-stacked rock walls, created several small garden ponds and dry "creek" beds, and interspersed scores of small boulders among plants within every garden. Lastly, we laid dozens of sheets of plywood, planks, and corrugated roofing tin to serve as cover boards at the margins of the gardens, lawn, and fields.

Although the interplay of the natural elements (rocks, logs, and plants) is aesthetically pleasing, my secondary goal in constructing the gardens was to attract wildlife, specifically amphibians and reptiles, into the yard, where they easily could be found and observed. The gardens were quickly teeming with invertebrate life, which to my wife's chagrin included black widow spiders and

Spiny Softshell (*Apalone spinifera*). (Photo by Greg Sievert)

scorpions! In addition to invertebrates, the ponds soon attracted amphibians, and the gardens became both acoustically and visually beautiful. Happily, the arrival of the amphibians attracted several species of snakes, including gartersnakes (*Thamnophis*) and watersnakes (*Nerodia*). Overall, I have been very successful in attracting amphibians and reptiles from the surrounding habitats into my yard and garden. To date, we have found 15 species of amphibians and 20 species of reptiles, including Copperheads (*Agkistrodon contortrix*) and Timber Rattlesnakes (*Crotalus horridus*).

It took us many years to find that many species, and the gardens took years to complete (actually, I guess we are still in the process of completing them). To design my first garden, I laid old oak planks end to end in my lawn, in roughly a rectangular pattern. After a few seasons of weathering, the grass beneath the planks died and rotted, and the planks settled into the soil about one-eighth inch. I would occasionally carefully lift the planks to search for creatures hidden beneath ("to see who's home," as my then–young children used to say). The first success I had in my realization of modifying yard to reptile sanctuary occurred with those planks the second or third year we lived on our property. That year I found nests of five Common Five-Lined Skinks (*Plestiodon fasciatus*) with the brooding females in attendance. I was excited to see if my boards would be as productive the following year, but I found only one brooding skink. That was 16 years ago; I have not found any brooding skinks since then. I am not certain why skinks stopped using the boards as nesting sites. However, as we developed more gardens, we were unable to keep weeds at bay by hand pulling them, so we started mulching our gardens with shredded pine bark. Skinks are still common inhabitants of our yard and gardens, and we often find hatchlings. Perhaps the skinks are still brooding eggs in the gardens, but now in the mulch instead of under the planks. Regardless, I had taken what was a mowed lawn and provided habitat that facilitated reptile reproduction. Eureka!

Since then we have found many species courting, breeding, laying eggs, or engaged in other activities that indicated that the gardens are their home. For example, we watched a Box Turtle (*Terrapene carolina*) plop eggs into nest she was digging along the edge of a garden in our front yard, observed a pair of Eastern Fence Lizards (*Sceloporus undulatus*) court and mate on wooden obelisks used to support clematis in one of our gardens, and discovered nests, hatchlings, and neonatal individuals from many reptile species, including milksnakes, fence lizards, red-bellied snakes, brownsnakes, snapping turtles, watersnakes, cornsnakes, kingsnakes, five-lined skinks, ground skinks, and box turtles. Furthermore, we have been attracted to the distress or release call of toads, only to find them being eaten by hog-nosed snakes, and to similar calls of Southern Leopard Frogs being eaten by watersnakes. On several occasions, we have watched as a hog-nosed snake, pestered by a family dog, feigned death! But not all of our encounters have been pleasant.

Unfortunately, we have witnessed many tragedies. Brownsnakes seem to have a preference for the longer, clumped grass that grows along the edges of gardens, swing sets, trees, and other structures in the yard. I have inadvertently killed more of these snakes when mowing the lawn or trimming garden borders than I care to recall. Although brownsnakes are the most commonly encountered species in our yard and gardens, I am sickened each time I notice one sliced by a mower blade, and I have forever set aside motorized trimmers, now keeping hundreds of meters of borders trimmed by hand.

We also were horrified to discover that the mesh we were using to protect our strawberries from birds were death traps for large-bodied snakes. In one weekend the mesh netting trapped and killed at least two racers and one hog-nosed snake! We now allow birds to feast on strawberries—after all, ratsnakes might like eating fledgling birds plump with berries!

Also, we had to be certain that our children were always careful when playing outside. They knew to alert Mom or Dad when they encountered a Copperhead (or any snake when very young), and we never had any close encounters. Our dogs have not been as cooperative as our children. Our border collie, Maria, was a champion snake dog. She had a distinctive bark that she used to alert us (or perhaps our beagle) when she stumbled on a snake of any size or species. Her routine included walking in circles around the snake, slowly decreasing the circle size, hoping to lunge. After we learned Maria's snake bark, we would rush to her aide, hoping to get to her before she lunged. Unfortunately, our presence seemed to boost her confidence and cause her to lunge prematurely. I never did see her bite a snake, although I am certain she must have. However, I did see her get bitten by racers, gartersnakes, and copperheads. Bites from the latter initially prompted a visit to the vet, but eventually we just kept her comfortable and well hydrated for a day or two until the swelling subsided. I often have been asked why I tolerate copperheads (and the rare rattlesnake) in my yard. "You mean you don't kill them?" I find the typical reaction a little confusing. According to data from the Centers for Disease Control, people are much more likely to be bitten by dogs than by venomous snakes in this country, yet we do not advocate killing all stray dogs on sight (thankfully). And as I like to tell my students, horses kill twice as many people each year as do venomous snakes. The lesson is, if you are careful, you can live with wildlife, even venomous spiders, scorpions, and snakes.

Many people reading this might doubt that they can be as successful as I have been in attracting reptiles to their yard. Perhaps, but you never know. For example, Murfreesboro is a relatively large town that has seen considerable growth during the past quarter century. Nonetheless, several wild areas still exist within the city limits, including the downtown section. Several creeks flow through the area, serving as corridors through which wildlife travel. I have friends in the city who have created wonderful wildlife sanctuaries in small yards. Amphibians have somehow found their way into the yards, as have watersnakes. One of my colleagues even brought me a hatchling mud turtle that he unearthed while tending his garden! The purpose of the gardens is not to replace woods, ponds, or streams but to give the reptiles that are rapidly losing habitat a small chance at survival, while giving people countless hours of enjoyment and a plethora of stories to tell.

Brian T. Miller

4
Finding Reptiles in Tennessee

Although observing reptiles at a zoo, aquarium, or pet store can be interesting and satisfying, searching for and discovering the diversity of reptiles that reside in Tennessee in their natural environment is a truly rewarding experience. This is how we became interested in studying reptiles and decided to make a career of it. Even as children, catching turtles, snakes, and lizards proved a meaningful pastime that brought us a deeper understanding of and appreciation for the natural world. Since then, searching for unique and unusual reptiles has taken us to some of the most beautiful natural areas in Tennessee, North America, and the rest of the world. It is this appreciation for wildlife and nature, which we firmly believe must be instilled in children and adults alike, that enables us to understand the value of the natural world and the importance of its conservation. We also believe that reptile appreciation is a grand entry to this end, and below we give advice on observing reptiles in Tennessee to novices and professionals alike.

Preparation

Any foray into the field calls for proper preparation and an understanding of the potential hazards associated with such endeavors. Whether you are a professional herpetologist or just someone with a basic interest in reptiles, prepa-

ration is very important. First and foremost, always make certain that you let someone know where you are going and when you are expected back. In the unfortunate event that something happens to you or your party and you do not return home in a timely fashion, help can be sent to assist you. You should pay attention to the weather and prepare and dress accordingly. Be aware of the potential for severe weather, particularly thunderstorms and heavy rainfall. We have been caught out in the field on several occasions during heavy thunderstorms and had to seek shelter beneath a rock ledge or huddled underneath a tree (but *not* a very tall tree).

Likewise, many streams in Tennessee can rise rapidly after heavy rainfall. Avoid parking or hiking in areas prone to flooding or hiking across such streams when there is a threat of heavy rain. You could easily become stranded on the wrong side of a stream and have to look for alternative routes to cross it again. Several natural areas (e.g., state parks, state natural areas, wildlife management areas) are dissected with trails, making traversing the terrain much easier. We advise searching for reptiles on and around trail systems or along the banks of lakes and rivers to avoid becoming lost or disorientated, especially in dense vegetation. Off-trail searches should only be conducted with those individuals familiar with and experienced in such activities. Be mindful of the

Editors Graham (*left*) and Matt (*right*) photograph the venter of a Florida Red-Bellied Snake (*Storeria occipitomaculata obscura*), McNairy County. (Photo by Stephen K. Nelson)

presence of venomous snakes in the areas you are walking through. You should always scan the ground around any log or rock you intend to flip to avoid an unpleasant and potentially life-threatening encounter with our scaly friends.

Legal and Ethical Issues

Almost all species of reptiles *cannot* be collected in Tennessee without a valid scientific collection permit issued by the Tennessee Wildlife Resources Agency (TWRA). Only Snapping Turtles (*Chelydra serpentina*) may be taken year around and statewide with a valid sport fishing or commercial fishing license by all legal sport fishing methods except archery, spear guns, and dipping. The daily possession limit is five individuals with a minimum carapace length of 30.5 cm

(12 in). Snapping Turtles also may be taken using hoop nets (up to three nets per license holder) in several West Tennessee counties, including Carroll, Chester, Crockett, Decatur, Dyer, Fayette, Gibson, Hardeman, Hardin, Haywood, Henderson, Henry, Houston, Humphreys, Lake, Lauderdale, McNairy, Madison, Obion, Perry, Shelby, Stewart, Tipton, and Weakley. All nets must have at least 7.6-cm (3-in) square mesh and be marked with the name and address of the license holder. All sizes of species of turtles, except those considered threatened, endangered, or deemed in need of management (Table 2), may be taken by any legal commercial fishing method at Reelfoot Wildlife Management Area in Lake and Obion counties. All other reptiles can be collected legally for scientific purposes only via a scientific collection permit issued by the TWRA.

Moreover, collection of reptiles on state parks and state natural areas requires an additional scientific collection permit from the Tennessee Department of Environment and Conservation (TDEC), Division of Natural Heritage. Likewise, collection of reptiles on federal lands requires a proper permit. Several species of reptiles are considered threatened or endangered and are, therefore, protected in Tennessee. These species are listed in Table 2 and should not be taken from the wild unless there is authorization and a very good reason to do so. If you have any questions regarding any restrictions or legality of collecting or observing reptiles in Tennessee, contact your regional TWRA office.

Handling

We do not generally recommend capturing or otherwise harassing wild reptiles, preferring to admire them in situ, but we give advice to those who might wish to physically interact with reptiles and hope that it will be heeded to ensure the well-being of the captured specimen. There are proper and improper ways to capture by hand and handle reptiles, and we discuss some of the most popular techniques below. There are numerous methods that have been developed to capture reptiles, and each person has his or her own favorite techniques to catch particular species. With practice, you likely will develop your own. Remember that you should always wash your hands after handling reptiles, as some individuals can have harmful bacteria on their skin.

Some lizards can be caught be hand by stealthily approaching the lizard from behind while it is at rest on the ground or in a bush or tree. Quickly pounce on the lizard with your hand, making certain not to apply so much force that you injure or kill the individual. Never grab or hold a lizard by the tail, as it might break off in your hands. Lizards can quickly be immobilized by holding their feet between the index finger and thumb or by firmly, but gently, gripping their body with your entire hand such that the head and feet protrude. After capture, lizards should be temporarily placed in a sterile plastic bag, container, or cloth bag. Plastic bags, containers, and cloth bags should be disinfected between uses or thrown away to prevent the spread of infectious diseases. On hot days, bags or containers should be placed in an ice chest or cooler to prevent overheating. Small holes can be punched in the containers, and plastic bags can be periodically opened to allow the flow of fresh air.

Do not attempt to handle venomous snakes unless you have been properly trained on how to handle these species. All nonvenomous snakes can easily be caught by hand, although larger individuals might inflict a somewhat painful bite. To avoid being bitten, nonvenomous snakes can be immobilized by very gently stepping on the snake with your foot or by restraining the snake with a stick and attempting to get the head with the toothy mouth under control. Once the head is under control, you can firmly but gently grab the snake behind the head. Larger snakes should always be supported with two hands to avoid twisting and coiling of the snake, which potentially could lead to injury or death of the snake. NEVER pick up a large-bodied snake by the tail and dangle it, as you might dislocate vertebrae. After a snake is captured, it can be temporarily placed in a bag, container, or snake bag (e.g., pillowcase).

Table 2

Protected and Rare Reptiles in Tennessee (after Withers 2009)

State status is a formal listing by the Tennessee Wildlife Resources Agency. State ranks derive from a ranking system by The Nature Conservancy for estimating the abundance of animals tracked by state Natural Heritage programs. Some species do not have a formal state status but are tracked by Tennessee's state agencies.

State Status	Definition
E	**Endangered:** any species or subspecies of wildlife whose prospects of survival or recruitment within the state are in jeopardy or are likely to become so within the foreseeable future.
T	**Threatened:** any species or subspecies of wildlife that are likely to become endangered within the foreseeable future.
D	**Deemed in Need of Management:** any species or subspecies of nongame wildlife which the executive director of TWRA believes should be investigated in order to develop information relating to populations, distribution, habitat needs, limiting factors, and other biological and ecological data to determine the management measures necessary for their continued ability to sustain themselves successfully.

State Rank	Definition
S1	Extremely rare and *critically imperiled* in the state (often with five or fewer occurrences), or very few remaining individuals, or because of some special condition where the species is particularly vulnerable to extirpation.
S2	Very rare and *imperiled* within the state, 6 to 20 occurrences, or few remaining individuals, or because of some factor(s) making it vulnerable to extirpation.
S3	*Vulnerable*, rare, and uncommon in the nation or state due to a restricted range, relatively few populations (often 80 or fewer), recent and widespread declines, or other factors making it vulnerable to extirpation.
S4	Uncommon but not rare, and *apparently secure* within the state, but with cause for long-term concern due to declines or other factors.

Common Name	Scientific Name	State Status	State Rank
Lizards			
Eastern Slender Glass Lizard	*Ophisaurus attenuatus*	D	S3
Coal Skink	*Plestiodon anthracinus*	D	S1
Snakes			
Kirtland's Snake	*Clonophis kirtlandii*		S1
Mississippi Green Watersnake	*Nerodia cyclopion*	D	S2
Copper-Bellied Watersnake	*Nerodia erythrogaster neglecta*	S2, S3	
Northern Pinesnake	*Pituophis melanoleucus*	T	S3
Western Pygmy Rattlesnake	*Sistrurus miliarius streckeri*	T	S2, S3
Turtles			
Bog Turtle	*Glyptemys muhlenbergii*	T	S1
Alligator Snapping Turtle	*Macrochelys temminckii*	D	S2, S3

Most turtles can be easily handled with two hands, but you should also be mindful of the location of the head and limbs, as all turtle species can bite or scratch. Smaller Alligator Snapping Turtles (*Macrochelys temminckii*) and Snapping Turtles (*Chelydra serpentina*) can be handled safely by grabbing their long tails and making certain that the turtle stays on the ground. Individuals should never be picked up and handled by the tail, as you can dislocate vertebrae and potentially injure the turtle. Instead, you can grab the rear of the turtle with two hands with the tail between your hands or by grabbing the sides of the turtle with two hands. Snapping Turtles have long necks that can reach well up onto the carapace, so handle them with utmost caution. Alligator Snapping Turtles have shorter necks and are safer to handle by the sides of the body. However, very large individuals can be handled by holding the anterior edge of the carapace right behind the head with one hand and the posterior edge of the carapace at the base of the tail with the other hand. Care must be taken to avoid a very serious bite from these species. You should also be mindful of the sharp claws on the limbs. Softshell turtles (*Apalone mutica* and *A. spinifera*) can be quite difficult to hold, and care should be taken so that you do not get bitten (their jaws are sharp and strong). Softshells should be held with two hands away from your body, with one hand used to press against the neck so that the turtle cannot extend its neck out far enough to turn around and bite you. Also pay attention to the sharp claws on the limbs, which can inflict a painful scratch.

Do not attempt! Handling some species of reptiles can be extremely dangerous and should not be attempted unless you are expertly trained. This large Alligator Snapping Turtle (*Macrochelys temminckii*) could easily remove digits with its powerful bite. Savannah River Ecological Laboratory, Aiken County, South Carolina. (Photo by Brad M. Glorioso)

Vince Cobb demonstrating the proper way to hold a snake (*Nerodia cyclopion*), using two hands to support the body and tail. (Photo by Collin Jaeger)

Location and Capture

Reptiles can be captured using a variety of techniques, and the effectiveness of these techniques varies from season to season. Below we describe several of the most common methods employed by herpetologists and wildlife biologists to capture and document reptiles in nature. Several other techniques exist, and we encourage readers seeking additional information on such methods to several references listed under the **"Recommended Readings, Organizations, and Websites"** section.

Visual Encounter Surveys

One of the best approaches to observing reptiles is to simply walk along a stream, edge of a pond, or trail through a forest or field using your eyes and ears to detect lizards, snakes, and turtles. When it is too warm, many reptiles spend daylight hours underneath cover objects, such as rocks, logs, and other debris, or within crevices in rock outcrops, to avoid direct sunlight and the heat of the day. Semiaquatic and terrestrial snakes and lizards can be found underneath rocks, logs, and other debris in a variety of habitats.

Although some people use tools such as potato rakes or snake sticks to overturn rocks and logs, we recommend wearing a pair of gloves and using your hands to overturn cover objects, as lizards and smaller snakes often will escape into burrows quickly once they are exposed. Furthermore, people using tools to overturn rocks are less inclined to properly return them to their original positions. Gloves will protect you from abrasions and from plants that might cause skin irritation, such as poison ivy. However, it is paramount that all cover objects are returned to their original positions in order to preserve the delicate microhabitats found underneath them. Using a potato rake, snake stick, or snake hook can be useful for those who have back problems or in habitats frequented by venomous snakes. Snakes also can be grabbed and restrained with commercially available snake tongs, although we recommend either using padded tongs or seeking training to use regular tongs to avoid injuring the snakes.

Brad "Bones" Glorioso demonstrating an improvised way to hold a Common Five-Lined Skink (*Plestiodon fasciatus*). (Photo by Matthew L. Niemiller)

Night Driving (Road Cruising)

Another highly productive technique for observing snakes at night is driving along roads (also known as road cruising) to detect more mobile species as they are out foraging or on the move, particularly during the warmer summer months. This technique is most productive on warm rainy or humid nights in areas with low vehicle traffic and relatively undisturbed surrounding habitats. Although a single person can perform night driving, it is best if two or more individuals are involved so that one person can drive and the others can spot and capture snakes. This technique can be highly productive. For instance, we have observed as many as seven species of snakes during a single night drive in Tennessee. However, stopping for snakes at night on a road can be very dangerous and drivers should use common sense and the necessary precautions when night driving for reptiles (i.e., do not stop around a curve or at the base or top of a hill).

Turtle Traps

A variety of techniques have been developed to capture aquatic turtles. In larger bodies of water, double-throated hoop nets are particularly effective at catching turtles. These traps are typically baited with dead fish, chicken, or cans of sardines or tuna suspended toward the rear of

Success! A road-cruised Yellow-Bellied Watersnake (*Nerodia erythrogaster flavigaster*) in Middle Tennessee. (Photo by Lisa Powers)

the net. Turtles enter through the mouth of the net in search of the bait but have an extremely difficult time exiting. It is important that these nets be placed such that a portion of the net remains above the water level so the turtles can breathe. Nets should be checked within 12–48 hours after placement. Fyke nets equipped with wings and leaders also can be effective for trapping turtles, particularly more wary species, such as map turtles (genus *Graptemys*). These traps are bag-shaped nets held open by a series of hoops and can be baited or left unbaited. Turtles swim into the leaders and are directed toward the trap. Basking traps can also be used to catch species that routinely bask, such as many of the emydid turtles (family Emydidae). These are floating enclosures with sloping or slippery slides and a basking surface. Turtles are able to climb onto the basking surface but

when they fall into the water in the center of trap, they are unable to get out.

Deep-water crawfish nets are particularly effective at capturing kinosternids (*Kinosternon* and *Sternotherus*) as well as smaller emydid turtles (*Trachemys*, *Graptemys*, and *Chrysemys*) and even Snapping Turtles. Originally designed to catch crayfish, these traps are baited with chicken or other meat and placed in shallow, lentic water, such as ponds, sloughs, and ditches. These circular traps are lowered to the bottom and connected by leads to a float on the surface that can be retrieved with a long pole. When lifted, turtles settle into the pouch in the net as they are carried to shore. Both long-handled and aquarium dipnets are useful for catching young turtles. Long-handled dipnets are typically triangular or rectangular in shape and can be used to sample turtles in a variety of

Old tin and wooden coverboards are great ways to attract and document snakes and lizards. (Photo by David F. Frymire)

habitats. In water bodies with little flow, such as ponds and ditches, dipnets are run through aquatic vegetation or mats of aquatic vegetative debris. Occasionally, smaller aquatic snakes are also caught.

Cover Boards

The use of wood or tin cover boards (or even old carpet) is an inexpensive technique that is commonly used to sample reptiles during herpetological inventories. Many snakes and lizards will seek cover underneath artificial cover objects. Arrays of plywood and tin roofing material can be set out in nearly any terrestrial habitat and periodically flipped and checked for reptiles seeking refuge underneath. When first placed in a new habitat, cover boards are usually left for several weeks before first checking to allow time for the microhabitat underneath cover boards to develop and attract reptiles. Cover boards are then checked every 1–4 weeks throughout the year.

Other Techniques to Locate and Capture Reptiles

Several other methods have proven successful for observing and capturing reptiles. A common but very labor-intensive method is the use of drift fences and pitfall trap arrays. Drift fences can capture species that are secretive, fossorial, or otherwise difficult to capture or observe. Drift fences can be constructed using a variety of materials, including silt fencing, plastic sheeting, aluminum flashing, and sheets of metal, and placed in a variety of habitats. The fencing is staked in the ground such that the bottom of the fencing is buried beneath the surface. Different types of traps can then be placed along the drift to capture reptiles, such as pitfall traps, funnel traps, and box traps. Pitfall traps consist of buckets (typically five-gallon) dug and placed along the drift fence, whereas funnel and box traps are placed along the fencing. As a snake or lizard is moving through a habitat, it reaches the fence and must move along it to get around, at which point it either falls into a pitfall trap or enters a funnel or box trap. Larger snakes can easily climb out of a bucket, so funnel or box traps are often used concurrently with pitfall traps. When in use, traps along a drift fence should be checked daily and shade should be provided during hot weather if traps are exposed to direct sunlight. Drift fences can also be used to effectively catch turtles when placed around aquatic habitats.

Noosing is an effective technique to catch lizards, particularly those in vegetation. Another technique for catching lizards is the use of blowguns with corks to temporarily stun fast-moving lizards, although this is an acquired skill. A common technique used to locate American Alligators at night is spotlighting, in which one walks along the shore of a lake or moves around in a boat and spotlights the water in search of the conspicuous red-orange reflection of American Alligator eyes. Baited snare traps can be used to catch American Alligators, though this technique should be employed only by professionals. Other techniques for capturing and observing turtles include snorkeling, hand capture, fishing with hook and line, seining, and using binoculars to identifying basking turtles from a distance. Softshell turtles (*Apalone mutica* and *A. spinifera*) will bury themselves in sandy habitats in shallow water, and a metal or wooden pole can be used to probe for them. Likewise, probing is an effective method for locating Bog Turtles (*Glyptemys muhlenbergii*) in their muddy habitats. Some breeds of dogs have even been trained to locate Eastern Box Turtles (*Terrapene carolina carolina*) by scent.

It is our hope that this section will aid in the appreciation of reptiles, not the harm or illegal collection of them. Reptiles are an extremely important component of terrestrial and aquatic ecosystems and tend to fair poorly in captivity. Please be responsible when enjoying these animals and be sure that you leave as little a footprint as possible in their environment so that others might share equally in the experience.

Drift fence arrays with funnel traps are often used to capture snakes, lizards, and turtles as they move through the landscape. (Photo by David F. Frymire)

5
How to Use the Species Accounts

We provide an overview of each major group of reptiles (lizards, snakes, turtles, and crocodilians) and an overview of each family preceding the species accounts. Species accounts are arranged alphabetically by scientific name within their respective families. Families are also arranged alphabetically.

Within family accounts, we provide general information on the family. Following the organization of Etnier and Starnes (1993) and Niemiller and Reynolds (2011), general family information includes information on classification, fossil record, diagnostic characters, overall family distribution, biology, and economic importance where available. Taxonomic keys include page number references to easily reference species accounts. Readers should refer to chapter 6, **"How to Use the Taxonomic Keys,"** for assistance in using the keys provided to properly identify Tennessee reptiles.

Species accounts are provided for all reptiles native to Tennessee. Readers should refer to chapter 22, **"Erroneous Species and Species of Possible Occurrence,"** for species that have been reported but are believed to be misidentifications and other species that might be added to the Tennessee's herpetofaunal list. Classification of the species included in this book largely follows *Scientific and Standard English Names of Amphibians and Reptiles of North America North of Mexico* by Crother (2011), with exceptions noted under individual species accounts. Subspecies are not treated separately, as there is often disagreement about the status of subspecific nomenclature. Where appropriate, we mention subspecific variation in individual species accounts, and subspecific variation is included in the taxonomic keys.

Information for each species account comes from a variety of sources, including peer-reviewed journals, government reports, and other available scientific literature, museum records, and individual observations made by the authors of accounts and the editors. We also relied on information gathered and presented in the *Atlas of Reptiles in Tennessee* by Scott and Redmond (2011), *Lizards and Crocodilians of the Southeast* by Gibbons et al. (2009), *Snakes of the Southeast* by Gibbons and Dorcas (2005), *Turtles of the Southeast* by Buhlmann et al. (2008), *Amphibians and Reptiles of Georgia* by Jensen et al. (2008), *Amphibians and Reptiles of Arkansas* by Trauth et al. (2004), *Reptiles and Amphibians of Eastern/ Central North America* by Conant and Collins (1998), *Amphibians and Reptiles of Missouri* by Johnson (2000), *Amphibians and Reptiles of the Carolinas and Virginia* by Beane et al. (2010), and *Reptiles and Amphibians of Alabama* by Mount (1975) . Readers should refer to these works and the **Selected References** section near the back of this book for additional information on a particular species.

A typical species account has the following sections and information.

Scientific and Common Names

The scientific name is composed of the genus name and the unique specific epithet and is italicized (see chapter 6, **"How to Use the Taxonomic Keys"**). Scientific names for several reptile species in Tennessee have changed in recent years as a result of numerous morphological and molecular phylogenetic studies. We include the most recent accepted names in this text in an effort to be consistent with current and future references to these species.

Similarly, common names for a given species can also vary. For this reason, herpetologists maintain standardized lists of common names. Here, we largely follow *Scientific and Standard English Names of Amphibians and Reptiles of North America North of Mexico* by Crother (2011) for common names. For many species, especially those frequently encountered by the general public, there are a number of local and regional colloquial names. We have included the well-known colloquial names under the **Comments** section of a species account.

Description

This section provides information on the size, physical appearance, and coloration of adults, juveniles, and hatchlings. Sizes of adults and juveniles are given in snout-vent length (SVL) or total length. Snout-vent length is measured from the tip of the snout to the posterior margin of the vent and is the preferred measurement, as many reptiles, especially lizards and snakes, might lose a portion of their tail after an attack by a predator. Total length (TL) is measured from the tip of the snout to the tip of the tail. In turtles, standard carapace length is measured from the front to the rear of the carapace, and plastron length is measured from the front to the rear of the plastron. Measurements are presented in metric units. Refer to chapter 1, **"What is a Reptile? Reptile Anatomy and Life History,"** as well as taxonomic keys for diagrams showing how measurements are obtained.

We provide diagnostic characters of each species and discuss potential morphological variation within species and between subspecies if warranted. We have provided many photographs and illustrations to aid in the accurate identification of our state's reptiles. However, accurate identification requires knowledge of the terminology used to describe the many morphological characters that might be unfamiliar to those who are not herpetologists. Readers should therefore refer to **chapter 1** and the **glossary** to better familiarize themselves with these characters. Descriptions of coloration and pattern are for living specimens. In many cases, preservation or death results in the loss of brightly colored pigmentation, but pattern usually remains evident. Coloration and pattern often vary ontogenetically (with age) and between males and females. We have therefore attempted to discuss ontogenetic morphological variation and sexual dimorphism in sufficient detail.

We have included several photographs with each species account to illustrate variation in color, pattern, and other morphological characters. In addition, we have tried to include representative photographs of different life stages for as many species as possible. We strived to provide photographs of individuals actually collected from Tennessee, in spite of the investment of time and energy (and gas) required to locate species that might be rare in our state but abundant elsewhere. Hence most photographs are of Tennessee populations with the county of collection noted. Photographs of specimens collected outside of Tennessee are noted.

Etymology

We provide the derivation of the Latinized scientific name for each species. In many cases, scientific names are based on morphological characters or geography. Other scientific names are based on the last name of an individual, or individuals, often as a tribute to him, her, or them. We also provide a pronunciation guide in brackets next to names with nonphonetic or nonintuitive pronunciation. We do not adhere to traditional pronunciation jargon or shorthand; instead we provide a phonetic close approximation to the most commonly used pronunciation of the name. We hope that this will make these names more accessible, though the pronunciations we offer might differ depending on whom you ask and whether that individual actually speaks Latin or Greek (we trust the latter individuals will forgive us).

Similar Species

Many species, particularly snakes, skinks, and some turtles, are quite similar in appearance, especially to novice naturalists. We have therefore included this section to quickly alert readers to similar looking species that can be easily confused with that particular species. We provide diagnostic characters for distinguishing

between these species when appropriate. In some cases, geography can be used to discern a species identity, so it is important to keep track of where a particular specimen was observed. However, some individuals can be nearly impossible to identify to species because of hybridization between two or more species. We have therefore tried to provide detailed and accurate information on documented cases of hybridization involving a species and where it is known to occur.

Distribution

A range map is illustrated for each species with delineation of subspecific ranges when warranted. We have not provided a generalized overall range map depicting the range of the species in North America. However, readers can easily find this information online or in several other texts listed in the **Selected References** section. Range maps show the distribution of the species in light red with confirmed county records depicted in darker red. Individual subspecies are illustrated in additional light and dark colors, with dark colors always indicating confirmed county records. Readers should note that for counties within which the range of the species only partially occurs, the darker red shading for county records only appears for that the portion of the range within the county and not the entire county itself. This differs from many other state herpetological texts, which highlight the entire county instead of just a portion of the county. Ranges and county records were determined from a number of sources, including museum collections, the scientific literature, and personal observations, although not all county records have a museum voucher associated with them. The depicted range within the state represents an estimated distribution generated from known locality data and information on habitat preferences and availability, physiography, and hydrology. Questionable records are discussed in the written account. Likewise, information on physiography or river drainages that cannot be easily determined from the range maps is discussed when appropriate.

Habitat

We provide information for each species on the general habitat and, for some species, microhabitat preferences. Readers should refer to chapter 2, **"Physiography, Climate, and Habitats of Tennessee,"** for a discussion of major habitats found in the state. We also pro-

vide discussion on the breeding habitat of each species if this information is available.

Natural History

In this section we provide a brief discussion on the life history and ecology of each species. We present information on reproduction and development, including courtship, breeding, and hatching. We also present information on the ecology and behavior of each species, including daily and seasonal activity patterns, diet, predators, and defense. We relied primarily on information available from Tennessee but expanded our review to include information obtained in adjacent states when such data were not available for Tennessee populations.

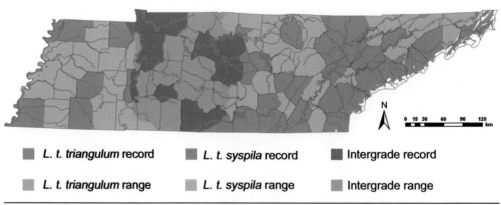

Example Range Map (*Lampropeltis triangulum*)

■ *L. t. triangulum* record	■ *L. t. syspila* record	■ Intergrade record
■ *L. t. triangulum* range	■ *L. t. syspila* range	■ Intergrade range

Two subspecies of milksnake are recognized in Tennessee: the Eastern Milksnake (*Lampropeltis triangulum triangulum*), whose range appears in orange, and the Red Milksnake (*L. t. syspila*), whose range appears in blue. The purple zone indicates an area of intergradation in which characteristics of both subspecies may be found and range boundaries between the two are unclear. County records are in darker colors and ranges are in lighter colors.

Conservation Status

We comment on the status of each species referring to the relative abundance or rarity within its range and within Tennessee. We provide state and federal listings along with potential causes of decline or threats to particular species when applicable. Readers should refer to chapter 3, **"Reptile Conservation,"** for more information on major threats facing many reptile species in Tennessee and worldwide.

Comments

This section contains other anecdotal information that is not included in other sections of a species account, such as common colloquial names, taxonomic uncertainty, and vocalizations (crocodilians).

An Eastern Wormsnake (*Carphophis amoenus*) consuming a beetle larva, Polk County. (Photo by Stephen Nelson)

6
How to Use the Taxonomic Keys

Standardized taxonomy, or the classification and naming of living organisms, is an important pursuit in modern science. Taxonomy allows us to give names to groups of organisms in a way that reduces confusion about what organism is being referred to, a situation that frequently arises when using common names. Common names are limited in their utility, as they tend to vary from place to place and often encompass many colloquial names. For instance, Mudsnakes (*Farancia abacura*) are referred to as hoop snakes, horn snakes, and stinging snakes depending on whom you ask and where you might be asking. Hence taxonomists give each species a scientific name, also known as a Latin binomial, that places the organism within a hierarchical framework of Linnaean taxonomy. These names include two or three parts—first the name of the genus to which the organism belongs, in this case *Farancia,* followed by the specific epithet *abacura.* The species name is occasionally followed by an additional name, which is known as the subspecies name. Subspecies, though widely used, are frequently in debate, as they usually just describe regional variation within a species. The Latin binomial for the Mudsnake tells us more than just the proper name of the species—it gives us an idea of its evolutionary relationship to other organisms on the planet in a hierarchical manner. The species is nested within the genus, which is in turn nested within the successively more inclusive groups of family, order, class, phylum, kingdom, and domain. So for the Mudsnake, we have **domain** Eukaryota, **kingdom** Animalia, **phylum** Chordata, **class** Reptilia, **order** Squamata, **family** Colubridae, **genus** *Farancia,* **species** *Farancia abacura.* Thus from the name we can deduce that Mudsnakes are a species of North American colubrid snake (Colubridae), sharing ancestry with another North American aquatic snake (*Farancia erytrogramma*).

Taxonomic keys are scientific tools used to identify unknown specimens. Most keys are dichotomous in that for each step there are two options to choose from. You may think of them as a sort of choose-your-own-adventure book, where at each step you are asked a question about the characteristics of the organism and then proceed according to the instructions given by the option you select. The goal is to eventually reach the correct name of the organism that you are "keying" based on a series of steps that list the characteristics of that organism. One uses a key by starting at **1** and selecting one of the two options: **1a** or **1b**. At the end of the boldface arrow (**→**) is a number, instructing you to go to that number in the key to select from two additional options. For instance, a lizard with two pairs of limbs (**1b . . .** go to **2**), frontal scale not V shaped (**2b . . .** go to **4**), one postmental scale present (**4a . . .** go to **5**), and from West Tennessee will key to a **Southern Coal Skink** (*Plestiodon anthracinus pluvialis*). If identifying a reptile is difficult based on the description and photographs, use of the key should allow positive identification in most cases. For more information on reptile taxonomy, we refer the reader to the **"Recommended Readings, Organizations, and Websites"** section.

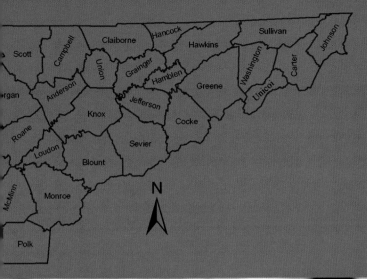

Part 2
Lizards

Green Anole (*Anolis carolinensis*), Rhea County. (Photo by Matthew L. Niemiller)

7

Key to the Lizards of Tennessee

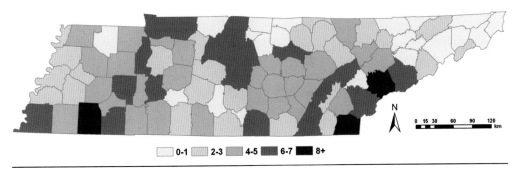

Diversity of lizard species in Tennessee based on county records.

1a. Legs absent—**Eastern Slender Glass Lizard (*Ophisaurus attenuatus longicaudus*) (Family Anguidae)**

1b. Four legs present ➔ **2**

2a. Body scales smooth, typically shiny, and overlapping without a granular, keeled, or spinous appearance ➔ **3 (Family Scincidae)**

2b. Body scales not smooth but keeled, granular, or spinous in appearance ➔ **8**

3a. Frontonasal scale in contact with rostral scale; frontal scale V shaped; supranasal scales absent; two thin, dark dorsolateral stripes present—**Little Brown Skink (*Scincella lateralis*)**

3b. Frontonasal scale separated from rostral scale by a pair of supranasal scales; frontal scale not V shaped; light-colored longitudintal stripes running the length of body usually present ➔ **4**

4a. One postmental scale present; broad dark lateral stripe bordered by two light-colored thin stripes present; four longitudinal stripes on dorsum ➔ **5**

4b. Two postmental scales present; five longitudinal stripes on dorsum (often obscured in older adults) ➔ **6**

5a. From West Tennessee (Benton, Henderson, and Humphreys counties)—**Southern Coal Skink (*Plestiodon anthracinus pluvialis*)**

5b. From the Blue Ridge Mountains of East Tennessee—**Northern Coal Skink (*Plestiodon anthracinus anthracinus*)**

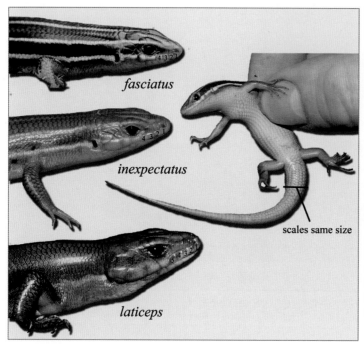

fasciatus

inexpectatus

scales same size

laticeps

Skinks in Tennessee are best differentiated by scale counts. Common Five-Lined Skinks (*Plestiodon fasciatus*) and Southeastern Five-Lined Skinks (*P. inexpectatus*) generally have four labial scales and two postlabial scales, while Broad-Headed Skinks (*P. laticeps*) have five labial scales and no postlabial scales. Common Five-Lined Skinks can be differentiated from Southeastern Five-Lined Skinks by having an enlarged row of scales underneath the tail, a characteristic they share with Broad-Headed Skinks. Southeastern Five-Lined Skinks have scales the same size underneath the tail.

6a.   Midventral scale row on tail similar in size to scales in adjacent rows—**Southeastern Five-Lined Skink (*Plestiodon inexpectatus*)**

6b.   Midventral scale row on tail much wider than scales in adjacent rows ➔ **7**

7a.   Four preorbital supralabial scales present; posteriormost supralabial scale separated from temporal scale by one or two postlabial scales—**Common Five-Lined Skink (*Plestiodon fasciatus*)**

7b.   Five preorbital supralabial scales present; posteriormost supralabial scale not separated from temporal scale or separated by a single postlabial scale—**Broad-Headed Skink (*Plestiodon laticeps*)**

8a.   Body scales strongly keeled and overlapping; dorsum usually with a paired series of dark, wavy bands—**Eastern Fence Lizard (*Sceloporus undulatus*) (Family Phrynosomatidae)**

8b.   Body scales granular and nonoverlapping ➔ **9**

9a.   Scales on head small and numerous; toes flattened and expanded; belly scales granular; longitudinal stripes absent—**Northern Green Anole (*Anolis carolinensis carolinensis*) (Family Dactyloidae)**

9b.   Scales on head larger and fewer; toes not flattened or expanded; belly with eight rows of large transverse scales; six light, longitudinal lines present—**Eastern Six-Lined Racerunner (*Aspidoscelis sexlineatus sexlineatus*) (Family Teiidae)**

8
Family Anguidae (Glass and Alligator Lizards)

Glass and Alligator Lizards are typically slender, elongate lizards that have long tails and either lack limbs entirely or possess short limbs. The legless species superficially resemble snakes, but unlike snakes they have external ears, moveable eyelids, and small scales on the venter. Nonetheless, the superficial resemblance to snakes is acknowledged in the genus name *Ophisaurus* [oh-feh-SOAR-us], which is formed by combining the Greek words *ophis* (serpent) and *sauros* (lizard). Thus this name literally translates as "serpent-lizard" and explains why some people refer to these creatures as "snake-lizards." Glass and Alligator Lizards are distributed on nearly every continent (they are absent from Australia and Antarctica) and include around 95 species in eight genera. *Ophisaurus* is a wide-ranging genus, with several species in Asia and southern Europe, and four species recognized in eastern North America, one of which occurs in Tennessee. Fossil anguids are known from the Paleocene of North America. Some species can reach nearly 1 m (3.2 ft) TL, with much of the length in the fragile and easily breakable tail. Some individuals will purposefully break off their tail when attacked or captured by a predator—a behavior known as tail autotomy. However, some individuals will autotomize their tail even without being handled; hence the common name Glass Lizards.

Eastern Slender Glass Lizard adult, Aiken County, South Carolina. (Photo by Greg Sievert)

Slender Glass Lizard

Ophisaurus attenuatus

Description: Slender Glass Lizards are elongate, legless lizards with adults 55–100 cm (21.6– 39.4 in) in TL but reaching 118 cm (46.5 in). The tail is extremely long, comprising nearly two-thirds the total length. The ground coloration of the dorsum and sides is tan, brown, golden brown, or bronze. A dark, narrow, longitudinal middorsal stripe, which is sometimes broken into a series of dashes, extends the length of the dorsum, including the tail. This stripe typically is more prominent toward the head and becomes fainter toward the tail. The stripe also fades with age and increasing body size. A lateral fold is present along the body from the neck to the vent, and dark longitudinal stripes run below this fold and under the tail. The venter is white to cream colored. Males grow larger than females and typically have a speckled "salt-and-pepper" dorsal patterning. Some adults develop distinctive pale but dark-bordered crossbars on the dorsum and tail. Hatchlings and juveniles resemble adults but typically lack the darker crossbars often present in older individuals. The stripes also are bolder and darker than those of adults. Some hatchlings might only have a single dark longitudinal stripe, rather than several thinner stripes, extending along the sides of the body. Eyelids and external ear openings are present on all individuals. Two subspecies are recognized, but only the Eastern Slender Glass Lizard (*O. a. longicaudus*) occurs in Tennessee.

Etymology: The specific epithet, *attenuatus* [ah-ten-you-AY-tus], is Latin for "thin" or "slender" and refers to the tail shape. The subspecific epithet, *longicaudus* [lawn-ge-CAW-dus], is from the Latin *longus* (long) and *caudus* (tail), referring to the extreme length of the tail relative to the body.

Similar Species: Slender Glass Lizards cannot be confused with other lizard species in Tennessee, as all others have four limbs. However, this species might be confused with snakes, which also lack limbs. Eastern Slender Glass Lizards can easily be distinguished from all snakes in Tennessee by possessing eyelids, obvious external ear openings, and small ventral scales.

Distribution: Slender Glass Lizards are found throughout much of the southeastern United States from southeastern Virginia southward to Florida then west to eastern Texas and north into Nebraska, Missouri, Illinois, and northwestern Indiana. This species is apparently absent from the Mississippi River Valley. The Eastern Slender Glass Lizard occurs east of the Mississippi River, while the Western Slender Glass Lizard (*O. a. attenuatus*) occurs west of the river. In Tennessee, Eastern Slender Glass Lizards likely occur throughout much of the state but are absent from northeastern Tennessee and the higher elevations of the Cumberland Mountains and Blue Ridge Mountains, although this species has been documented from Cades Cove in the Great Smoky Mountains National Park. Records are scattered throughout most of the state and none exist for northwestern Tennessee. These records are likely incomplete as this species is quite secretive and rarely encountered in our state.

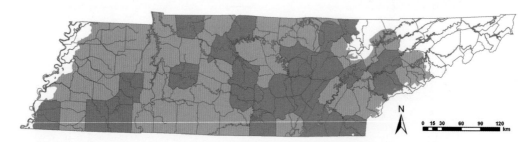

Left: Eastern Slender Glass Lizard female with eggs, Coffee County. (Photo by Rufus Darden) *Right:* Eastern Slender Glass Lizard adult, Montgomery County. (Photo by A. Floyd Scott)

Habitat: Slender Glass Lizards are found in a variety of habitats with well-drained soils and numerous open areas, including open fields, upland oak-pine forest, barrens, and even vacant lots and clear-cut areas. Rather than burrow, this species will hide in or under low vegetation throughout much of the day when they are active on the surface. Slender Glass Lizards will enter burrows underground during hibernation. Individuals are usually seen on the move in the early morning or late afternoon during summer, particular on sunny days after rainfall the previous day or night. Most individuals are observed basking on or crossing roads.

Natural History: Little is known about the natural history of Slender Glass Lizards in Tennessee. In adjacent states, Slender Glass Lizards are seasonally active from late March into November. Mating probably occurs from late April into early June. Females lay 4–19 eggs in shallow nests under logs, vegetation, and other cover. A female will remain with her nest until eggs hatch, which takes about eight weeks. Sexual maturity is reached in 3–4 years, and some individuals are known to live at least nine years.

Slender Glass Lizards are active predators, feeding on a variety of prey, including grasshoppers, ants, beetles, arachnids, caterpillars, snails, lizards, small snakes, young mice, and eggs of ground-nesting birds. Predators include birds of prey, predatory mammals, and several lizard-eating snakes, such as Coachwhips (*Masticophis flagellum*), Racers (*Coluber constrictor*), and Kingsnakes (*Lampropeltis getula*). As a defense mechanism, Slender Glass Lizards are quick to autotomize their tails when grasped by a predator or human. Indeed, photographing this species intact can be tricky, as any restraint will cause them to drop their tail. They sometimes will even break off their tails when not physically touched. Although the tail will regenerate, it typically will not grow to the original length or color, often appearing stumpy or conical. Slender Glass Lizards seldom bite when captured.

Conservation Status: Slender Glass Lizards are uncommonly encountered, and little is known about the status of the species in Tennessee. As such, this species is listed "Deemed in Need of Management" in the state. Threats to this species include loss of habitat and death by motor vehicles when crossing or basking on roads.

Comments: Slender Glass Lizards are occasionally referred to as "legless lizards," "glass snakes," and "horn snakes." The last two colloquial names refer to how quickly this species will break off its tail when threatened or handled. The tail often does not regenerate fully and appears stumpy or hornlike, hence the name "horn snake."

Matthew L. Niemiller and
R. Graham Reynolds

9

Family Phrynosomatidae (Fence and Spiny Lizards)

The Fence and Spiny Lizards are small (usually less than 10 cm [4 in] SVL), robust lizards covered with keeled, overlapping scales. This family is distributed throughout much of North America, from southern Canada south through the United States and Mexico and into Central America. Some species within this family are found on Caribbean islands. Many species are adapted to the hot, xeric conditions of rocky deserts; however, some species, including the only member of this family found in Tennessee, inhabit mesic forests. Around 135 species are described in 10 genera. Fossils are known from the Eocene of North America. Most species are oviparous, but some, particularly those species found at higher elevations, give birth to live young. This family includes the horned lizards (genus *Phrynosoma*), which are extremely spiny and exhibit a unique defensive behavior in which they squirt blood at predators from sinuses in their orbits of the eyes. Our lone species is a member of the genus *Sceloporus* [ske-lop-POUR-us], from the Greek words *skelos* (leg) and *porus* (pore) in reference to the obvious femoral pores, and is commonly found throughout the state in a variety of habitats.

Adult male Eastern Fence Lizard, Wilson County. (Photo by Matthew L. Niemiller)

Eastern Fence Lizard

Sceloporus undulatus

Description: Eastern Fence Lizards are medium-sized spiny lizards 10–18 cm (4–7 in) in TL. They have overlapping scales that are strongly keeled and pointed. The dorsum usually is shades of gray, grayish brown, or brown. Coloration varies between the sexes. Adult females are gray with a series of dark gray or black, wavy bands that form a ladder-like pattern on the back of the body and tail, and small pale blue spots on the sides of the belly and on the throat. Adult males are usually brown and, because they lack the dark ladder-like pattern of females, are uniform in color. The venter is white to pale gray with scattered darker flecks in both sexes. However, adult males have an elongated hyacinth blue to turquoise blotch on each side of the venter and a similarly colored collar at the base of the throat. The blue blotches often are encircled with black and generally are not visible unless being displayed during courtship or combat. Females grow to larger sizes than males. Hatchlings are 4–6.5 cm (2.6 in) in TL and are patterned like females but are duller in overall coloration. No subspecies are recognized.

Etymology: The specific epithet, *undulatus* [un-due-LAY-tus], is Latin for "having a wave-like pattern."

Similar Species: Eastern Fence Lizards cannot be confused with any other lizard species in Tennessee. All skinks have glossy, smooth scales, whereas Eastern Fence Lizards appear spiny and have rough scales.

Distribution: Eastern Fence Lizards are widely distributed in the eastern United States and can be found from Delaware to Florida and west to New Mexico and Arizona. In Tennessee, this species is common and found nearly statewide, absent only from the highest elevations of the Blue Ridge Mountains.

Habitat: Eastern Fence Lizards are commonly found in open pine, deciduous and mixed pine-hardwood forests, cedar glades, woodlots, old fields, rock and brush piles, urban areas with downed trees or wood fences, and around homes and barns. This species is frequently found in xeric habitats, but it can also be found in mesic areas. Eastern Fence Lizards are primarily arboreal and their coloration and patterning makes them nearly invisible on the trunk or branch of a tree. Individuals encountered on the ground will, when approached, dash for the nearest tree or other tall structure, including a person's leg!

Natural History: Eastern Fence Lizards are diurnal and active from early March into October in Tennessee. As evening approaches, these lizards take refuge in secluded and protected sites, such as within rock crevices or tree cavities, underneath loose bark, or beneath rocks. This species is one of the first lizards to emerge from hibernation during spring. Fence lizards have a very complex social system and observing interactions between lizards can be quite interesting. Males are highly territorial and actively defend their territory. Males can often be seen displaying to other rival males via head shuddering and pushups. Courtship, which also involves male pushup displays, and mating occur during spring. Females lay 6–10 eggs (up to 16) during May or June in rotting logs or loose

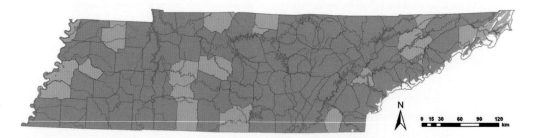

Top left: Excavated nest of the Eastern Fence Lizard, Cannon County. (Photo by Brian T. Miller) *Top right:* Eastern Fence Lizard, juvenile, Wilson County. (Photo by Matthew L. Niemiller) *Bottom right:* Venter of a male Eastern Fence Lizard, Wilson County. (Photo by Matthew L. Niemiller)

soil. Larger females lay more eggs than smaller females, and they often lay two clutches of eggs during a single year. Eggs hatch in about 9–11 weeks. Sexual maturity is reached within two years. Eastern Fence Lizards prey on a variety of invertebrates, including beetles, flies, stink bugs, leafhoppers, ants, moths, grasshoppers, roaches, spiders, millipedes, and snails. They often observe their prey while perched in a tree, shrub or rock, then scamper down the perch site to grab the invertebrate. Documented predators include several species of snakes, such as Racers (*Coluber constrictor*), Gray Ratsnakes (*Pantherophis spiloides*), Milksnakes (*Lampropeltis triangulum*), Yellow-Bellied Kingsnakes (*L. calligaster*), and Copperheads (*Agkistrodon contortrix*), as well as birds, predatory mammals, and domestic cats and dogs. Eastern Fence Lizards rely on a quick sprint or camouflage for defense. They often remain motionless even when captured.

Conservation Status: Eastern Fence Lizards are common across the state and considered secure throughout most of their range.

Comments: Most Tennesseans who venture outdoors encounter Eastern Fence Lizards. Many of us who have attempted to capture these lizards note their tendency to not only take refuge in trees but also try to remain on the opposite side of the trunk from the pursuer. A common regional name of this species is "swift," an acknowledgment of the escape behavior.

Lisa Powers and
Matthew L. Niemiller

10
Family Dactyloidae (Anoles)

Anoles are an extremely speciose and well-studied family of lizards. Most species are relatively small (generally less than 12 cm [4.7 in] in SVL) and adapted to an arboreal existence. More than 650 species are recognized in 13 genera distributed throughout South America, Central America, the Caribbean, and the southeastern United States. A single species of the genus *Anolis* [AH-no-lis] occurs in Tennessee, a name derived from the Antillean word *anoli,* which was used to describe these lizards. Fossil anoles are known from the Tertiary period of Mexico and the Dominican Republic. Anoles typically have pointed snouts, expanded toe pads, extensible dewlaps, powerful jaws, and small granular scales that cover the body. Males typically are highly territorial and possess brightly colored dewlaps, which they use in defensive and courtship displays. The dewlap is an extendable flap of skin along the gular region, which is often brightly colored. Many species can slowly change skin color, a trait that has led to the colloquial name of "chameleon," although true chameleon lizards are only found in Africa and Madagascar. Anoles are also unique in their ability to continually lay eggs one at a time rather than during a confined reproductive season. Anoles are very popular in the pet trade around the world and have been widely introduced. At least nine anole species from the Caribbean have been introduced in southern and central Florida. One species, the Brown Anole

(*Anolis sagrei*), has become firmly established and is now displacing native Green Anole (*A. carolinensis carolinensis*) populations in parts of Florida. The family name is from the Greek dactylos (finger) and –oides (like or resembling).

Green Anole

Anolis carolinensis

Description: Green Anoles are small- to medium-sized lizards with adults 12–20 cm (4.7–7.9 in) in TL. Males reach 7.5 cm (3 in) SVL and are larger than females, which reach about 5.5 cm (2.2 in) SVL. Scales on the body are very small, keeled, and non-overlapping. Noted for their ability to change color, individuals may be bright green, brown, or mottled green-brown, depending on temperature, state of agitation, and light conditions. Most females, and rarely males, possess a white to cream-colored mid-dorsal stripe. Both sexes possess a reddish or pinkish dewlap, or "throat fan" which is much larger in males than in females. Males extend the dewlap during courtship displays and both sexes do so during aggressive encounters. The belly and chin are white to grayish white. The head is relatively long and pointed, and a ridge is present between each eye and nostril. The ventral surface of each toe includes an area of expanded scales (the toe pad) that are adhesive and allow the lizard to climb smooth surfaces.

Green Anole adult female, Blount County. (Photo by Arthur C. Echternacht)

The tail is long and accounts for more than 60 percent the total length. Hatchlings are 50–65 mm (1.9–2.6 in) TL and resemble adults, but they have proportionally wider heads and a shorter tails. Two subspecies are recognized, but only the Northern Green Anole (*A. c. carolinensis*) occurs in Tennessee.

Etymology: The specific and subspecific epithet, *carolinensis* [ka-ro-len-EN-sis], refers to the Carolinas, the area from which the originally described specimens were presumably collected.

Similar Species: Green Anoles are unlikely to be confused with any other lizard species na-

tive to Tennessee. No other lizards in the state have the ability to change color and pattern as quickly and dramatically as Green Anoles. Additionally, all other species lack a dewlap and expanded toe pads, none exhibit a head or body shape like that of Green Anoles, and none are ever uniformly bright green.

Distribution: Green Anoles are widely distributed in the southeastern United States from North Carolina to the Florida Keys, westward to southeastern Oklahoma and central Texas. There is an isolated population is Tamaulipas, Mexico. Green Anoles also have been accidentally or deliberately introduced to all of the ma-

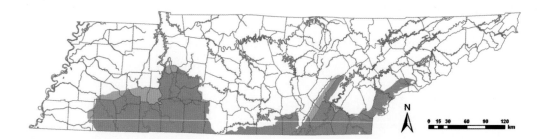

Top left: Green Anole adult, Rhea County. (Photo by Matthew L. Niemiller) *Top right:* Green Anole adult male, Blount County. (Photo by Arthur C. Echternacht) *Bottom right:* Green Anole adult basking on a sunny winter day, Blount County. Note the icicle in the foreground. (Photo by Arthur C. Echternacht)

jor Hawaiian Islands and elsewhere outside of the continental United States. In Tennessee, this species ranges across the southern tier of counties from Polk to Hardeman County, with more northerly populations reported in Hickman, Perry, and Lewis counties in the Western Highland Rim and in Grundy, Rhea, Monroe, Blount, and Sevier counties in the Cumberland Plateau, Ridge and Valley, and Blue Ridge Mountains. Only in North Carolina does this species, and genus, occur farther north than in Tennessee.

Habitat: The habitat of Green Anoles is characterized by its structure. In Tennessee, at the northern extent of its range, the preferred habitat occupied in late spring, summer, and fall is often riparian (i.e., along the banks of a river or stream), which supports trees with an understory of woody shrubs and broadleaf herbaceous plants. They are essentially an "edge species," occupying habitat along the margins of forested areas. Hatchlings are frequently found on the stems and leaves of herbaceous plants, such as White Crown-Beard (*Verbesina virginica*). As Green Anoles grow, they shift first to woody shrubs and then, as adults, to trees. Adults seem to favor smaller trees with a trunk diameter in the 50–200 cm (20–80 in) range, although they can also be found in the canopies of much larger trees. In most areas where they occur in Tennessee, the trees are hardwoods, but they also utilize pines and cedars. This is not intuitively obvious; it might be thought that the sap exuding from the trunks of these softwoods would render the lizard's adhesive toe pads useless. As winter approaches, the lizards move to areas, usually on south-facing slopes or rock faces, where they overwinter in dry rock crevices, holes in the ground, or, less frequently, under rocks.

Natural History: Green Anoles are diurnal and can be found active every month of the year in Tennessee. In late autumn, Green Anoles move from their summer territories to the winter habitat described above. They do not hibernate and emerge on any clear winter day when the sun warms the openings of the crevices or holes in which they take refuge from the cold, even if the air temperature a meter or so above the crevice or hole is below freezing. The movement from summer to winter habitat is accompanied by a loss of territoriality and aggregations of several anoles may occupy the same winter refuge without aggressive interaction. During the late spring, summer, and autumn, both male and female Green Anoles are territorial. That is, they defend a part of the habitat in which they live against other anoles of the same sex. The territories of males are larger and may contain the territories of more than one female. Not all males possess territories and those that don't are considered "floaters" and must move furtively through the territories of more dominant males if they are to mate.

Courtship occurs throughout the late spring and summer, and its onset is dependent on both day length and temperature. This activity involves male pushup and dewlap displays directed toward females. Females normally lay one egg at a time, from alternate ovaries, at approximately 7–14 day intervals. There is no nest or any other form of parental care. Generally, eggs may be found where they have been abandoned under leaves on the ground where moisture is adequate and the temperature not too high. If it has been very dry, females may hold an egg and, because the egg in the opposite oviduct continues to develop, lay two eggs instead of one. The incubation period is temperature dependent but probably averages about 30 days. Those that survive their first winter reach sexual maturity in their second summer. Although Green Anoles in captivity, without predation or winter mortality, can live up to a reported seven years, in the wild they live an average of less than two years. An anole that survives into its third summer is a rarity. Green Anoles prey mainly on insects, spiders, and other small invertebrates. They are preyed upon by several species of snakes, such as Racers (*Coluber constrictor*), Broad-Headed Skinks (*Plestiodon laticeps*), birds, such as kestrels and shrikes, and small carnivorous mammals. Like all other lizard species in Tennessee, Green Anoles can autotomize their tails to escape predators and regrow a new one.

Conservation Status: Limited as they are by suitable winter habitat, Green Anoles are locally abundant, but populations may be widely separated. Largely because of the discovery of previously unknown populations, the species is no longer considered under threat in Tennessee.

Comments: Because of its obvious ability to change colors, the Green Anole is often incorrectly called the American Chameleon.

Anoles are native only to the Western Hemisphere (North, Central, and South America and the West Indies). True chameleons occur naturally only in the Eastern Hemisphere, especially Madagascar and Africa. The two families are not closely related and differ considerably in body form and behavior. The ability to change color is much better developed in the true chameleons.

A second species of anole, the Brown Anole (*Anolis sagrei*), is occasionally found in Tennessee (See **chapter 22**). This nonnative species was introduced in South Florida during the 1940s and has expanded its range northward into at least southern Georgia and westward into Texas. This species colonizes new areas by "hitchhiking," often in ornamental plants transported from nurseries in Florida to commercial greenhouses farther north and west. How far the species will expand its range is unknown. Brown Anoles differ from Green Anoles in their stockier build, gray to brown coloration with more patterning, the lack of any green coloration, and orange dewlap. Their ability to change colors is limited to lightening or darkening. Especially in disturbed areas, established populations of Brown Anoles can displace or extirpate populations of Green Anoles.

Arthur C. Echternacht

11
Family Scincidae (Skinks)

Skinks are one of the most diverse and speciose families of lizards. More than 1,200 species of skinks have been described, which accounts for roughly 30% of all known lizard species (exclusive of the snakes). Skinks are found in temperate and tropical regions and on nearly every continent (none are found in Antarctica). Most skinks, including those in North America, are medium-sized, with body lengths up to 12 cm (4.7 in), and long tails that can be autotomized and regenerated. Most species are covered with smooth and overlapping scales on both the dorsum and venter. The smooth scales give skinks a shiny or glossy appearance. Some species are limbless. The fossil record of skinks is poor, but fossils are known from the Late Cretaceous and Early Paleocene of North America. The young of many species have brightly colored tails that fade with age. Skinks occupy a variety of habitats, with some species being fossorial and others almost entirely arboreal. Nearly 45% of species give birth to live young, although all species in Tennessee lay eggs. Tennessee is home to five species in two genera, *Plestiodon* [plee-stee-OH-don] (formerly *Eumeces*), probably from the Greek *plesios* (near) *odon* (teeth), and *Scincella* [sin-SELL-ah], from the Latin *scincus* (skink) and *ella* (little). Skinks are the most commonly observed lizards in our state, frequently found in residential and urban areas in backyards, abandoned lots, parks, and gardens, where they especially favor rock walls, stone steps, and other places with both hiding and basking areas.

Coal Skink

Plestiodon anthracinus

Description: Coal Skinks are medium-sized lizards with most adults reaching 125–180 mm (5–7 in) TL or 50–70 mm (2–2.75 in) SVL. The scales are overlapping, smooth, and glossy in appearance. A conspicuous, broad dark brown to black stripe is present on each flank and is bordered above and below by a thin pale yellow, cream, or white stripe. The dark, broad stripe is up to four scales wide, whereas the thin, light stripes are 1–1.5 scales wide. In total, there are two broad stripes and four light stripes, all of which usually extend from the head past the hind limbs onto the tail. The dorsum typically is a monochromatic greenish gray to brown but occasionally has faint, darker longitudinal lines. The venter is gray to bluish gray and lacks any appreciable markings. Six or seven labial scales are present on each side of both the upper and lower lips. A single postmental scale is present on the undersurface of the lower jaw. Males and females are similar in appearance; however, mature males develop conspicuous orange to red coloration on the lips and side of the head that may extend back to the insertion of the fore limbs. Hatchlings are 46–51 mm (1.8–2.0 in) TL and 20–25 mm, (0.8–1.0 in) SVL. Hatchlings and

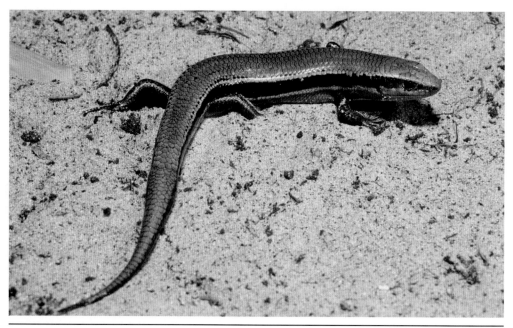

Northern Coal Skink adult, Blount County. (Photo by Dick Bartlett)

juveniles are considerably darker than adults and may lack markings entirely. In some populations, the tail of hatchlings is blue to violet, but this color fades with age. In other populations, the tail of hatchlings is jet black. Two subspecies are recognized and both occur in Tennessee: the Northern Coal Skink (*P. a. anthracinus*) and the Southern Coal Skink (*P. a. pluvialis*). The Northern Coal Skink has a continuous pale stripe through the posterior supralabial scales and typically has 25 or fewer scale rows around the midbody. The Southern Coal Skink lacks a continuous stripe through the posterior supralabial scales and typically has 26 or more scale rows around the midbody. Rather, the posterior supralabial scales have a pale spot in the center with darker edges. Most of the body of hatchling and juvenile Southern Coal Skinks are jet black, although the pale labial spots might be present and the snout and lips might be reddish in color. Hatchling and juvenile Northern Coal Skinks resemble adults but have blue tails.

Etymology: The specific and subspecific epithet, *anthracinus* [an-thrah-SIGN-us], is from

Two subspecies of Coal Skinks are recognized in Tennessee: the Northern Coal Skink (*Plestiodon anthracinus anthracinus*), whose range appears in orange, and the Southern Coal Skink (*P. a. pluvialis*), whose range appears in blue.

Top left: Underside of the head of a Northern Coal Skink, Blount County. (Photo by Dick Bartlett) *Top right:* Northern Coal Skink adult, Sevier County. (Photo by Dick Bartlett) *Bottom right:* Southern Coal Skink, Calloway County, Kentucky. (Photo by David F. Frymire)

the Greek *anthrakos* (coal), in reference to the dark lateral stripe of adults of this species. The subspecific epithet, *pluvialis* [plew-vee-AL-is], is Latin for "pertaining to rain," probably in reference to the mesic habitats that this species prefers.

Similar Species: Coal Skinks most closely resemble other skinks, particularly Common Five-Lined Skinks (*P. fasciatus*), Southeastern Five-Lined Skinks (*P. inexpectatus*), and Broad-Headed Skinks (*P. laticeps*). Coal Skinks have a single postmental scale, whereas other skink species have two. In addition, Coal Skinks have just four pale stripes that extend the length of the body, whereas the other species have five stripes. In Tennessee, Coal Skinks are the only skinks with a dark stripe running along each side of the body that is more than two body scales wide. Juvenile Coal Skinks lack the two pale lines on the head found in juveniles of the other three similar skink species.

Distribution: Coal Skinks have a patchy distribution from southern New York south to the Florida Panhandle then west into eastern Texas and northwestern Louisiana and north to southern Missouri. The distribution west of the Mississippi River is more continuous than that east of the river. The distribution in Tennessee is not well known, and populations in the state are disjunct. Coal Skinks have only been reported from six counties. In West Tennessee, they are known from along the Tennessee River in Benton and Humphreys counties and from Henderson County. The Northern Coal Skink is represented by two records: one near Cades Cove in Great Smoky Mountain National Park in Blount County and another around Sugarlands in Great Smoky Mountain National Park in Sevier County. Presumed intergrades between the two subspecies have been reported from southern Polk County in southeast Tennessee.

Habitat: Coal Skinks prefer more mesic habitats than other species of *Plestiodon* in Tennessee, occurring in heavily shaded and moist hillsides in deciduous to mixed hardwood-pine forests. Individuals are found often near rocky areas and bluffs in close proximity to water, such as springs and streams. Adults and juveniles can be found underneath rocks and logs in such areas. Coal Skinks are considerably less arboreal and more fossorial than other species of *Plestiodon* in Tennessee, and individuals are often found underground. Coal Skinks also can occasionally be found basking at the edges of road cuts and other forest openings.

Natural History: Little is known about the natural history of Coal Skinks in Tennessee. In southern portions of their range, Coal Skinks court and mate in February and March, and this likely occurs from March to early May in Tennessee. Females typically lay 4–10 eggs underneath rocks, logs, and other cover from April to June and will brood a clutch until hatching 4–5 weeks later. Sexual maturity is reached in two years. Coal Skinks are generalist predators that eat a variety of small invertebrates that inhabit leaf litter on the forest floor, including beetles, crickets, spiders, and earthworms. They actively seek out their prey, rustling through leaf litter and rapidly flicking their tongue to pick up chemical trails. A variety of predators prey on adults and juveniles, including mammals, birds of prey, larger lizards, and forest-dwelling snakes. Coal Skinks will jump into water or hide underground when threatened. Like all lizards in Tennessee, Coal Skinks also will autotomize their tails when attacked.

Conservation Status: Perhaps because of their secretive existence and limited distribution, Coal Skinks are seldom encountered and thought to be rather uncommon throughout their range. In Tennessee, the species is "Deemed in Need of Management" and considered "Critically Imperiled" because of presumed rarity. Populations in East Tennessee occur on federal land and are afforded some protection. Conservation strategies that help protect mesic forests likely would benefit this species.

Comments: Skinks are sometimes referred to as "scorpions," probably because they are erroneously considered by some to be venomous.

Matthew L. Niemiller

Common Five-Lined Skink female with eggs, Cannon County. (Photo by Brian T. Miller)

Common Five-Lined Skink

Plestiodon fasciatus

Description: Common Five-Lined Skinks are medium-sized lizards, with adults 125–215 mm (4.9–8.5 in) TL and 55–80 mm (2.2–3.1 in) SVL. The scales are overlapping, smooth, and glossy in appearance. Juveniles are dark brown to black, with five conspicuous white to yellow stripes running the length of the body and onto a bright blue to violet tail. The stripes include one thin middorsal stripe and two pairs of broad lateral stripes. In adult females, these stripes often fade and the tail turns gray in color. In adult males, the stripes also fade, the dorsum becomes brown or tan, and the tail turns gray with age. Some individuals can lose all hints of striping. Adult males also develop conspicuous orange to red to pink coloration on the head during the breeding season. Additionally, the jaws of males enlarge slightly during the breeding season such that the head appears more arrow-shaped. Males attain a larger size than females. The venter is gray to bluish gray and lacks any appreciable markings. Under the base of the tail, scales in the middle row are enlarged

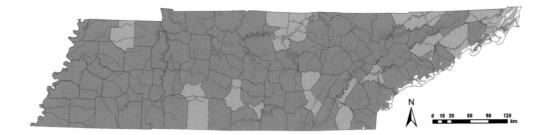

Left: Common Five-Lined Skink juvenile, Rutherford County. (Photo by Matthew L. Niemiller) *Right:* Common Five-Lined Skink male, Lake County. (Photo by Brad M. Glorioso)

such that the middle scale row is much wider than adjacent scale rows. Usually seven labial scales (four before the eye) and two enlarged postlabial scales are present. Two postmental scales are present on the undersurface of the lower jaw. The stripes of hatchlings and juveniles are considerably brighter than adults and often orange to reddish orange (rather than yellow) on the head. The tail is bright blue to violet, but this color fades with age (see above). Hatchlings are 50–65 mm (1.9–2.6 in) TL. No subspecies are recognized.

Etymology: The specific epithet, *fasciatus* [fah-see-AY-tus], is Latin for "banded" or "striped," in reference to the dorsal stripes of juveniles, females, and young males.

Similar Species: Common Five-Lined Skinks most closely resemble other skinks, particularly Southeastern Five-Lined Skinks (*P. inexpectatus*), Broad-Headed Skinks (*P. laticeps*), and Coal Skinks (*P. anthracinus*). Coal Skinks have a single postmental scale and just four prominent stripes, whereas Common Five-Lined Skinks have two scales and five stripes. Common Five-Lined Skinks can be distinguished from Southeastern Five-Lined Skinks by examining the scale rows under the tail (subcaudal scales). In Common Five-Lined Skinks the middle row of scales is much wider than adjacent rows, whereas all scale rows are similar in size in Southeastern Five-Lined Skinks. Broad-Headed Skinks usually have five labial scales preceding the scale that contacts the eye and no large postlabial scales, whereas Common Five-Lined Skinks have just four labial scales before the scale that contacts the eye and two postlabial scales.

Distribution: Common Five-Lined Skinks occur throughout much of the eastern United States from the Great Lakes and mid-Atlantic regions southward throughout Florida, the Gulf Coast, and eastern Texas. This species occurs throughout much of Tennessee, with the exception of the very highest peaks of the Blue Ridge Mountains. Common Five-Lined Skinks are without a doubt the most commonly encountered lizard in our state.

Habitat: Common Five-Lined Skinks occur in many habitats, including swamps, bottomland forests, cedar glades, and upland pine and hardwood forests. Individuals also are common in urban and residential environments, such as abandoned lots, back yards, and gardens. Common Five-Lined Skinks often are found basking or active around fallen trees, rotting logs, rock piles, and other debris that can be used as cover. Adults will often climb trees when disturbed or threatened. Adults and juveniles can found under rocks, bark, logs, and other cover and debris when not active.

Natural History: Common Five-Lined Skinks become active during warmer weather in March. Courtship and mating occurs from April into June, with oviposition occurring from May into June. Both males and females often can be observed in close proximity to each other during the breeding season. Females typically lay 4–15 eggs in rotten logs, stumps, or vegetation piles, or underneath rocks, logs, boards or other cover, and will brood them until hatching 3–8 weeks later. Females will not actively forage while brooding but have been known to cannibalize eggs. Sexual maturity typically is reached within two years. Adults and juveniles actively forage and bask during the day, feeding on a variety of invertebrates, such as insects, arachnids, and other arthropods, other lizards, and even small mice. When threatened, Common Five-Lined Skinks will quickly run to cover or climb vegetation. The bright blue to violet coloration of the tail in juveniles is thought by some researchers to warn potential predators that the skink is distasteful and unpalatable. However, no studies have confirmed this hypothesis, and as with many lizard species in Tennessee, Common Five-Lined Skinks will autotomize their tails when attacked. The autotomized tail wriggles wildly; hence the blue color of the wriggling tail might attract predator attention to the tail and away from the young skink. Predators include snakes, dogs, cats, small mammals, and birds. Although harmless, this species is quick to bite when captured.

Conservation Status: Common Five-lined Skinks have a wide distribution and are abundant in suitable habitat. Consequently, this species is thought to be secure and is not listed as a species of conservation concern in Tennessee. As with most lizard species in Tennessee, the primary threat to Common Five-Lined Skink populations is loss of habitat associated with urbanization and other land-use practices (e.g., mining and timbering). However, little information exists on the demography and vulnerability of this species in Tennessee.

Comments: Skinks are sometimes referred to as "scorpions," probably because they are erroneously considered by some to be venomous.

Matthew L. Niemiller and
R. Graham Reynolds

Southeastern Five-Lined Skink male, Wilson County. (Photo by Brad M. Glorioso)

Southeastern Five-Lined Skink

Plestiodon inexpectatus

Description: Southeastern Five-Lined Skinks are medium-sized lizards with adults 140–210 mm (5.5–8.3 in) TL and 55–85 (2.2–3.3 in) SVL. The scales are overlapping, smooth, and glossy in appearance. Adult females and juveniles are dark brown to black, with five conspicuous white to yellow stripes running the length of the body onto the tail. The stripes include one thin middorsal stripe and two pairs of broad lateral stripes. In adult males, the middle stripe typically fades, and the dorsum becomes brown or tan with age. The lateral stripes also often fade, and the tail becomes grayish. Some individuals can lose all hints of striping. Adult males also develop conspicuous orange to red to pink coloration on the head during the breeding season. Additionally, the jaws of males enlarge during the breeding season such that the head appears arrow-shaped. The venter is gray to bluish gray and lacks any appreciable markings. All subcaudal scales are similar in size. Usually seven

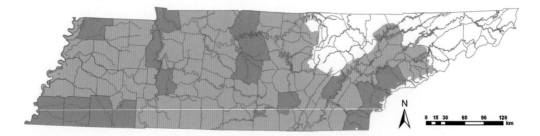

Top left: Southeastern Five-Lined Skink female, Wilson County. (Photo by Matthew L. Niemiller) *Top right:* Underside of Southeastern Five-Lined Skink tail showing scales that are all the same size. (Photo by Matthew L. Niemiller) *Bottom right:* Southeastern Five-Lined Skink male, Wilson County. (Photo by Matthew L. Niemiller)

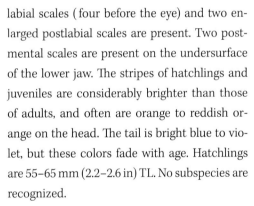

labial scales (four before the eye) and two enlarged postlabial scales are present. Two postmental scales are present on the undersurface of the lower jaw. The stripes of hatchlings and juveniles are considerably brighter than those of adults, and often are orange to reddish orange on the head. The tail is bright blue to violet, but these colors fade with age. Hatchlings are 55–65 mm (2.2–2.6 in) TL. No subspecies are recognized.

Etymology: The specific epithet, *inexpectatus* [in-ex-peck-TAY-tus], is Latin for "unexpected," probably in reference to the discovery of this third sympatric species of similar-looking skink.

Similar Species: Southeastern Five-Lined Skinks most closely resemble other skinks, particularly Common Five-Lined Skinks (*P. fasciatus*), Broad-Headed Skinks (*P. laticeps*), and Coal Skinks (*P. anthracinus*). Coal Skinks have a single postmental scale and just four prominent stripes, whereas Southeastern Five-Lined Skinks have two postmental scales and five stripes. Southeastern Five-Lined Skinks can be distinguished from Common Five-Lined Skinks and Broad-Headed Skinks by examining the scale rows under the tail. In Common Five-Lined Skinks and Broad-Headed Skinks the middle row of subcaudal scales is much wider than adjacent rows, whereas all subcaudal scale rows are similar in size in Southeastern Five-Lined Skinks. Broad-Headed Skinks also usually have five labial scales preceding the scale that contacts the eye and no large postlabial scales, whereas both the Common Five-Lined Skink and Southeastern Five-Lined Skink have just four labial scales before the scale that contacts the eye and two postlabial scales.

Distribution: Southeastern Five-Lined Skinks occur throughout much of the eastern United States from Missouri, southern Illinois, Indiana, Ohio, Maryland, Virginia, and Delaware southward into northern Florida, the Gulf Coast, and eastern Texas. The distribution is patchy but poorly documented in Tennessee. The species likely occurs statewide, with the exception of northeastern Tennessee and the higher elevations of the Cumberland and Blue Ridge Mountains. Most records are from southwestern Tennessee, the Inner Nashville Basin, and Blue Ridge Mountains of southeastern Tennessee.

Habitat: Southeastern Five-Lined Skinks occur in a variety of habitats, including swamps, bottomland forests, cedar glades, and upland pine and hardwood forests. Individuals also can be found in more urban and residential environments, such as abandoned lots and back yards. This species prefers more xeric habitats than other *Plestiodon* species in Tennessee. Southeastern Five-Lined Skinks often are found around fallen trees, rotting logs, rock piles, and other debris than can be used as cover. Adults are less likely to climb trees than other *Plestiodon* species but can still be observed actively climbing when disturbed or threatened. Adults and juveniles can found under rocks, bark, logs, and other cover debris when not active.

Natural History: Little is known about the natural history of Southeastern Five-Lined Skinks in Tennessee. In other parts of their range, courtship and mating occur in early spring, with oviposition occurring from early June through August. Both males and females often can be observed in close proximity to each other during the breeding season. Females typically lay 3–10 eggs in rotten logs, stumps, or vegetation piles, and she will brood them until hatching 3–8 weeks later. Sexual maturity is reached within two years. Adults and juveniles actively forage and bask during the day, feeding on a variety of invertebrates, such as insects, arachnids, and other arthropods. When threatened, Southeastern Five-Lined Skinks quickly run to cover or climb vegetation. The bright blue to violet coloration of the tail in juveniles is thought by some researchers to warn potential predators that the skink is distasteful and unpalatable. However, no studies have confirmed this hypothesis. Documented predators include several snakes (e.g., Scarlet Kingsnakes and Scarletsnakes), dogs, cats, other mammals, and birds. Like all lizards in Tennessee, Southeastern Five-Lined Skinks also will autotomize their tails when attacked.

Conservation Status: Like most lizards in Tennessee, the primary threat to Southeastern Five-Lined Skink populations is loss of habitat associated with urbanization and other land-use practices (e.g., mining and timbering). However, little information exists on the demography and vulnerability of this species in Tennessee.

Comments: Skinks are sometimes referred to as "scorpions," probably because they are erroneously considered by some to be venomous.

Matthew L. Niemiller and
R. Graham Reynolds

Broad-Headed Skink male, Rutherford County. (Photo by Brad M. Glorioso)

Broad-Headed Skink

Plestiodon laticeps

Description: Broad-Headed Skinks are the largest skinks in our state, with adults 165–325 mm (6.5–12.8 in) TL and 75–120 mm (2.9–4.7 in) SVL. The scales are overlapping, smooth, and glossy in appearance. Hatchlings and juveniles are dark brown to black, with five or seven white to yellow stripes running the length of the body and onto the tail. The stripes are bright in hatchlings but usually fade with age. Males have broad jaws, giving them the appearance of having an enlarged head relative to the body. Because of their enlarged jaws, males are distinguished easily from females, which have smaller jaws. Furthermore, males often have bright orange or red heads during the breeding season, which contrasts greatly with the brown to olive color of the body. Females are usually uniformly brown, with younger individuals retaining faint stripes associated with juvenile patterning. The tail often turns gray with age in both sexes. The venter is an immaculate (unpatterned) gray to

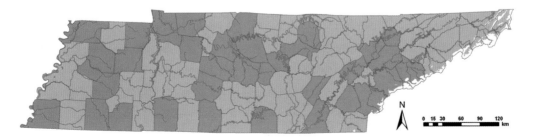

N

0 15 30 60 90 120
km

Left: Head of a Broad-Headed Skink, Cannon County. (Photo by Brian T. Miller) *Right:* Broad-Headed Skink juvenile, Hardeman County. (Photo by Matthew L. Niemiller)

bluish gray at all ages. Under the base of the tail, scales in the middle row are enlarged such that the middle row is much wider than adjacent scale rows. Usually there are five labial scales before the eye and no postlabial scales at the posterior of the jaw. Juveniles are extremely similar in appearance to juvenile skinks of other species. Their dorsum is black with either five or seven yellow stripes running down the back and sides, and their tail is bright blue to violet. Hatchlings are 55–85 mm (2.2–3.2 in) TL. No subspecies are recognized.

Etymology: The specific epithet, *laticeps* [LAH-teh-cepts], is from the Latin words *latus* (broad) and *ceps* (head), in reference to the large jaws of males, which make the head look larger relative to the body.

Similar Species: Male Broad-Headed Skinks are fairly distinct, especially in the spring, when males have orange- to red-colored heads. Females and especially juveniles are easily confused with other skink species, particularly Common Five-Lined Skinks (*P. fasciatus*) and Southeastern Five-Lined Skinks (*P. inexpectatus*). Broad-Headed Skinks can be best identified by a combination of scale patterns not found in other skinks in our state. Broad-Headed Skinks have both an enlarged middle

scale row on the undersides of their tails and five labial scales before the eye with no postlabial scales. Other skinks in our state have either enlarged subcaudal scales and four labial scales with two postlabial scales (Common Five-Lined Skinks), four labial and two postlabial scales and no enlarged caudal scales (Southeastern Five-Lined Skinks), or a single postmental scale (Coal Skinks).

Distribution: Broad-Headed Skinks are widely distributed across the eastern United States from southern Ohio, Indiana, and Illinois to the Gulf Coast and northern peninsular Florida. This species occurs nearly statewide in Tennessee with the exception of the very highest peaks of the Blue Ridge Mountains. However, records are scattered and lacking for several counties.

Habitat: Broad-Headed Skinks tolerate a wide range of moisture levels and are more tolerant of xeric habitats than other species of skinks. They especially prefer habitats with access to basking areas in direct sunlight. Forested sunlit patches on tree trunks or fallen logs with loose bark are favored areas, but individuals can also be found underneath logs, or on fallen debris along road cuts. Broad-Headed Skinks are the most arboreal skinks in Tennessee and frequently seek refuge in cavities and woodpecker excavations high in trees. This species also is frequently encountered in residential areas, such as abandoned lots, back yards, and gardens. Adults and juveniles can found under rocks, bark, logs, and other cover debris when not active.

Natural History: Broad-Headed Skinks emerge from communal underground hibernation sites in March or April, or as soon as daytime temperatures allow basking and diurnal activity. Breeding occurs between April and June, when males establish and defend territories that include both a female and a nesting site. Females typically lay 6–19 eggs in decomposing logs in midsummer, and they remain with their nests until the eggs hatch. Females are known to move their eggs to a new location if conditions become unfavorable. Hatchlings emerge during late summer, usually in 4–8 weeks after the eggs were laid. Individuals reach sexual maturity in about two years. Broad-Headed Skinks feed on a variety of invertebrates that are found in leaf litter or underneath bark, including crickets, wasps, bees, spiders, and snails. They locate this prey largely by use of tongue flicks while actively foraging. They also eat other skinks and small vertebrates. Black Racers (*Coluber constrictor*) and Coachwhips (*Masticophis flagellum*) are major predators of this species, and cats are especially proficient at killing skinks, often dragging them to their owners' doormats. Like all lizards in Tennessee, Broad-Headed Skinks will autotomize their tails when attacked. Broad-Headed Skinks also will bite when captured, but the bite is not particularly painful, nor does it usually break the skin.

Conservation Status: Broad-Headed Skinks are documented from scattered areas in Tennessee, but this species is widespread and locally abundant; hence the species is not of conservation concern.

Comments: Broad-Headed Skinks are sometimes referred to as "scorpions," probably because they are erroneously considered by some to be venomous and because males can give a sharp nip if grabbed.

R. Graham Reynolds and
Matthew L. Niemiller

Little Brown Skink adult, Wilson County. (Photo by Matthew L. Niemiller)

Little Brown Skink

Scincella lateralis

Description: Little Brown Skinks are small (7.5–14.5 cm [3–6 in] TL), slender lizards with short legs, and shiny, smooth, and overlapping dorsal scales. Coloration of the dorsum is coppery, reddish brown, or dark brown, often with darker flecks scattered along the midline. The head is often more reddish brown in color than the rest of the dorsum. Two dark dorso-lateral stripes, which begin at the nares and extend through the eyes to the tip of the tail, define the border between dorsal and ventral coloration. The venter is white to pale yellow, often with scattered dark flecking. The tail is long and can easily autotomize when captured. Regenerated tails are usually a lighter brown compared to the overall body coloration. A portion of each lower eyelid is transparent or translucent; this translucent patch acts as a window that allows individuals to see when the eyes are closed. The venter of males is more yellow than that of females, particularly during the breeding season. Hatchlings are around 4.4 cm (1.7 in) TL and resemble adults. No subspecies are recognized.

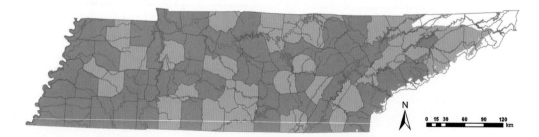

Etymology: The specific epithet, *lateralis* [lah-ter-AL-iss], is Latin meaning "of the side" and refers to the stripe pattern on the flanks.

Similar Species: No other lizards in Tennessee resemble the Little Brown Skink. Other skinks in Tennessee (genus *Plestiodon*) have pale longitudinal stripes running the length of the body. This species is sometimes confused with some of the many salamanders found in the state, but salamanders do not have scales or claws on their feet.

Distribution: Little Brown Skinks are found throughout much of the southeastern United States from eastern Oklahoma, northeastern Texas, and Louisiana eastward into Kentucky and Tennessee and northward along the Piedmont and Atlantic Coastal Plain into Maryland, Delaware, and New Jersey. Little Brown Skinks likely occur throughout Tennessee, except for the higher elevations of the Blue Ridge Mountains. However, records for this species are scattered and the exact distribution in northeast Tennessee is not well known.

Habitat: Little Brown Skinks are primarily fossorial and live in a variety of habitats with substrates of deep leaf litter or loose soil. They are often abundant in deciduous or mixed deciduous/coniferous forests, and along the edges of forested streams and ponds but can also be found in drier habitats, such as cedar glades, open fields, and grasslands, including lawns. Adults and juveniles can be found under logs, rocks, and aged building materials, such as old bards and sheets of tin, and within leaf litter. Adults are often seen moving in the typical stop and start motion of skinks, as they forage or bask in light patches that reach the forest floor. Hikers often inadvertently flush Little Brown

Little Brown Skink adult, Obion County. (Photo by Matthew L. Niemiller)

Skinks, and these startled skinks make a rustling sound as they scramble away from the trail and seek refuge in leaf litter.

Natural History: Little Brown Skinks are active during the warmer months of spring, early summer, and autumn. However, this species is also active on warm days during winter. Active during the day, Little Brown Skinks more often are heard scurrying through leaf litter than seen. The reproductive biology of Little Brown Skinks is not well studied, but mating occurs over a broad period from February through August. Females lay small clutches of 1–7 (usually 2–3) eggs in moist soil, decaying logs, or under rocks from late March through August. Females may lay more than one clutch a year. Unlike other skink species in Tennessee, female Little Brown Skinks do not remain with their eggs and are known to communally nest. Eggs hatch in 4–8 weeks. Sexual maturity typically is reached within a year. Little Brown Skinks seldom climb, more often they forage through leaf litter and grass for small invertebrates, including beetles, wood roaches, ants, leafhoppers, butterfly larvae and adults, spiders, millipedes, earthworms, and isopods. Predators include lizard-eating snakes such as Racers (*Coluber constrictor*), Eastern Black Kingsnakes (*Lampropeltis nigra*), Scarlet Kingsnakes (*L. elapsoides*), and Copperheads (*Agkistrodon contortrix*), predatory mammals, and birds such as Eastern Bluebirds and Barred Owls. Cats and dogs also will prey on Little Brown Skinks.

Conservation Status: Little Brown Skink populations are thought to be secure and are not listed as a species of conservation concern in Tennessee.

Comments: Little Brown Skinks are also known as "ground skinks" and "brown-backed skinks."

Lisa Powers, Matthew L. Niemiller, and
Brian T. Miller

12

Family Teiidae (Whiptail Lizards)

Whiptail Lizards are a large and diverse group (approximately 240 species in nine genera) restricted to the Western Hemisphere. The family includes the medium-sized (20–40 cm [8–16 in] TL) Whiptails and Racerunners (genus *Aspidoscelis*) of North America, as well as some unusual species, such as the giant Caiman Lizards (*Dracaena* sp.) and Tegus (*Tupinambis* sp.) of South America. Fossils are known from the Miocene and Pleistocene of South America. North American Whiptails are generally medium-sized diurnal species with small granular dorsal scales and plate-like ventral scales. They are best characterized by their speed and ability, indeed preference, to withstand extremely high temperatures, and they remain active even in the hottest parts of the day by moving in and out of shade. They have very long tails, often 100–200% of the body length, and move in a rapid, sometimes jerky motion of stops and starts, often scratching leaf litter in search of insects. All species are oviparous. Several species are capable of parthenogenic reproduction and consist of populations composed entirely of females that reproduce in the absence of males. The genus *Aspidoscelis* [ass-peh-doh-SELL-iss], from the Greek *aspis* (shield) and *scelis* (leg), in reference to the large scutes on the anterior surface of each limb, was formerly subsumed within the larger genus *Cnemidophorus*, the latter being now restricted to Central and South America and the Caribbean. *Aspidoscelis* contains 28 species north of Mexico, and includes the lone species that occurs in Tennessee.

Six-Lined Racerunner

Aspidoscelis sexlineata

Description: Six-Lined Racerunners are medium-sized, long-tailed, slender lizards with a pointed snout and six thin white to yellow to pale green stripes extending from the head back of the eye onto the base of the tail. The body length of adults is usually 65–85 mm (2.6–3.2 in), but the tail is longer than the body so that TL approaches 275 mm (10.8 in) in old adults that have never experienced tail loss. These lizards look velvety and feel soft and smooth because of numerous tiny, non-overlapping granular scales on the dorsum and sides of the body. The dorsum typically is greenish brown to brown to black. The belly is covered by a series of rectangular scutes that are aligned in eight rows. Scutes also adorn the top of the head and the anterior surface each limb. Furthermore, the tail is covered with scutes, and the posterior edges of these enlarged scales are slightly elevated from adjacent posterior scales, giving the tail a roughened texture and providing stark contrast to the velvety feel of the body. A series of small pores, known as femoral pores, are located on the posterior edge of the thighs. These pores are more conspicuous in males than in

Six-Lined Racerunner adult, Wilson County. (Photo by Brad M. Glorioso)

females, and males are more colorful than females. The head, back, and sides of adult males are washed vivid green, and the anterior half of their belly is pale blue. The stripes are often indistinct in older males, whereas the stripes of females are bright and clearly defined throughout life. Although the belly of females is occasionally pale blue, their backs and sides lack the vivid green washing that is characteristic of males; rather, the backs and sides of females typically are pale brown to tan. Hatchlings are relatively small (28–32 mm [1.1–1.2 in] SVL and <100 mm [3.9 in] TL) and have a blue tail, which fades to tan as they age. Except for the tail, the coloration of hatchlings and juveniles resembles that of females. Three subspecies are recognized, but only the Eastern Six-Lined Racerunner (*Aspidoscelis s. sexlineata*) occurs in Tennessee.

Etymology: The specific and subspecific epithet, *sexlineata* [sex-len-nee-AH-ta], is from the Latin words *sex* (six) and *lineata* (line) and refers to the six conspicuous stripes that extend from the head to the tail.

Similar Species: Six-Lined Racerunners most closely resemble skinks in the genus

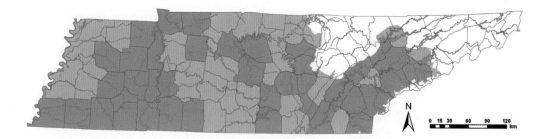

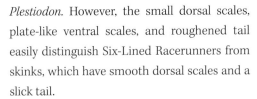

Top left: Six-Lined Racerunner subadult, Wilson County. (Photo by Matthew L. Niemiller) *Top right:* Venter of a Six-Lined Racerunner, Wilson County. (Photo by Brad M. Glorioso) *Bottom right:* Femoral pores on a Six-Lined Racerunner adult, Wilson County. (Photo by Matthew L. Niemiller)

Plestiodon. However, the small dorsal scales, plate-like ventral scales, and roughened tail easily distinguish Six-Lined Racerunners from skinks, which have smooth dorsal scales and a slick tail.

Distribution: Six-Lined Racerunners are found primarily east of the Mississippi River in the southeastern United States; however, populations also exist west of the Mississippi River in Louisiana, eastern Texas, southeastern Oklahoma, and southeastern Missouri. Several of these more western populations possibly allow for intergradation between the Prairie Racerunner (*A. s. viridis*) and the Eastern Six-Lined Racerunner. The Six-Lined Racerunner is locally abundant, with scattered occurrences throughout much of Tennessee. However, records are lacking for several large areas of the state, including much of the Coastal Plain of northwest Tennessee, the Cumberland Plateau of Middle and East Tennessee, the Ridge and Valley of

northeast Tennessee, and much of the Nashville Basin. The absence of the species from these areas may be real, or may be an artifact of limited collection data.

Habitat: Six-Lined Racerunners are found in association with open, sunny habitats that are prone to high daytime temperatures, including cedar glades, prairies, clearings in forests, and rail, road, and power-line rights-of-way. This species is terrestrial when active during the heat of the day, but it becomes fossorial, using its feet and pointed snout to dig into loose soil to seek the refuge of burrows, during cool periods of the day and at night. Racerunners are most easily captured in their burrows during the cool early morning.

Natural History: Little is known about the specifics of the natural history of Six-Lined Racerunners in Tennessee. Mating occurs during spring, shortly after adults exit hibernation burrows, and females begin to lay eggs a few weeks after breeding. Courtship is not elaborate but does involve the male biting the neck of a receptive female, and then wrapping his tail around hers to facilitate coitus. Eggs are laid in burrows excavated by the female, which are occasionally found beneath rocks or logs. Females abandon the eggs after the clutch is complete (1–6 eggs); consequently, they often lay a second clutch of eggs during a single breeding season. Hatching occurs 8–9 weeks after eggs are laid. Six-Lined Racerunners are dietary generalists and feed on a variety of small invertebrates, including insects, spiders, and centipedes. They are sight feeders that also rely on olfaction to locate prey. Six-Lined Racerunners are commonly observed during the heat of the day scampering around and flicking their forked tongue in search of insect odor trails. Predators include snakes, birds, and predatory mammals. Six-Lined Racerunners are well known for their running speed and have been listed for decades in the Guinness Book of World Records as the fastest lizard in the world, clocked at speeds of 18 mph. They are the only group of lizards I have chased that, apparently confident in their escape, pause to feed while being pursued. As with all species of lizards in Tennessee, Six-lined Racerunners will readily autotomize their tail to escape predation.

Conservation Status: Six-Lined Racerunners are thought to be secure and currently are not listed as species of conservation concern in Tennessee. Indeed, unlike many species of amphibians and reptiles that suffer from habitat fragmentation, Six-Lined Racerunners are found often in the cleared and otherwise disturbed habitats associated with highways, railways, and power lines. Because ample sunlight reaches the ground in these artificially maintained open areas, they are relatively hot compared to adjacent forested areas and potentially allow dispersal of the species throughout Tennessee.

Brian T. Miller

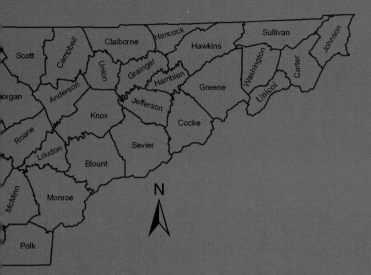

Part 3
Snakes

Western Cottonmouth (*Agkistrodon piscivorus*) in duck-weed. (Photo by David F. Frymire)

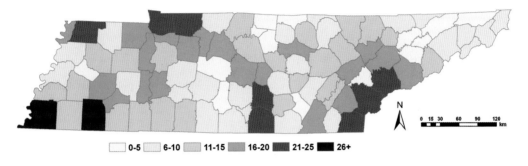

Diversity of snake species in Tennessee based on county records.

1a.   Loreal pit between nostril and eye present; eye pupil elliptical and vertically orientiated; single row of subcaudal scales ➜ **2 (Family Viperidae)**

1b.   Loreal pit between nostril and eye absent; eye pupil round; double row of subcaudal scales ➜ **6 (Family Colubridae)**

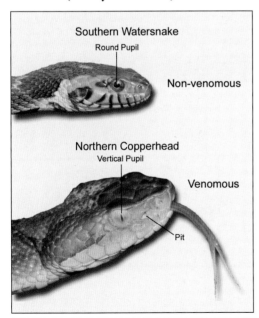
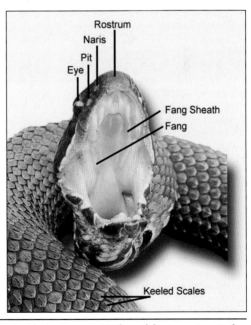

Left: Venomous snakes in Tennessee are readily distinguished by having a vertical pupil, heat-sensing pits below the eyes, *and* a triangular-shaped head. Note that most harmless watersnakes will attempt to impersonate a venomous snake by flattening their heads when disturbed. *Right:* Some key features used to identify a venomous Cottonmouth include pits below the eyes, keeled scales, and large fangs. This individual is exhibiting a threat posture typical of this species in which the head is thrown back and the mouth exposed. (Photo by Brad M. Glorioso)

2a.   Tail without rattle or button; scales weakly keeled ➔ **3**

2b.   Tail with rattle or button; scales distinctly keeled ➔ **5**

3a.   Loreal scale absent; dorsum without hourglass-shaped crossbands but with dark brown or black bands; head typically dark with a broad dark stripe through and behind eye; pale labial stripe present—**Western Cottonmouth (*Agkistrodon piscivorus leucostoma*)**

3b.   Loreal scale present; dorsum with hourglass-shaped crossbands typically on an orange-brown or tan background color; head typically not dark; a narrow dark stripe through and eye ➔ **4**

4a.   Dark hourglass crossbands complete across dorsum—**Northern Copperhead (*Agkistrodon contortrix mokasen*)**

4b.   Dark hourglass crossbands often broken and offset middorsally; from West Tennessee—**Southern Copperhead (*Agkistrodon contortrix contortrix*)**

5a.   Several small scales between supraocular scales on dorsal surface of head; large rattle; total length of adults exceeding 64 cm (25 in)—**Timber Rattlesnake (*Crotalus horridus*)**

5b.   Large scales present between supraocular scales on dorsal surface of head; small rattle; maximum total length of 64 cm (25 in)—**Western Pygmy Rattlesnake (*Sistrurus miliarius streckeri*)**

6a.   Rostral scale upturned and keeled—**Eastern Hog-Nosed Snake (*Heterodon platirhinos*)**

6b.   Rostral scale not upturned and keeled ➔ **7**

7a.   Anal plate divided ➔ **8**

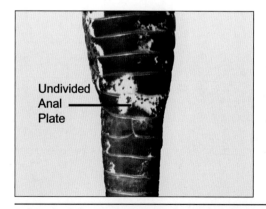 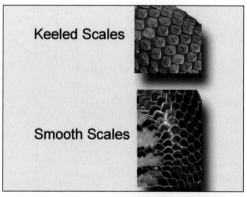

Left: Undivided anal plate on a Western Cottonmouth (*Agkistrodon piscivorus*). Note that venomous snakes in Tennessee have a single row of scales on the underside of the tail, though this individual has one pair of scales immediately below the anal plate. (Photo by Lisa Powers) *Right:* Body scales in snakes can be smooth or keeled, and this is often a useful characteristic to use in identification. (Photos by Brad M. Glorioso)

7b.   Anal plate not divided ➔ **34**

8a.   Dorsal scales strongly or moderately keeled ➔ **9**

8b.   Dorsal scales smooth or very weakly keeled (see *Virginia*) ➔ **24**

9a.   Either loreal scale or preocular scale absent ➔ **10**

9b.   Both loreal scale and preocular scale present ➔ **13**

10a.   Prefrontal scale in contact with eye—**Rough Earthsnake (*Virginia striatula*) (in part)**

10b.   Prefrontal scale not in contact with eye ➔ **11**

11a.   Two preocular scales present; six upper labial scales; usually 15 dorsal scale rows—**Northern Red-Bellied Snake (*Storeria occipitomaculata occipitomaculata*)**

11b.   One preocular scale present; seven upper labial scales; usually 17 dorsal scale rows ➔ **12**

12a.   Two rows of paired dark spots present on dorsum not connected by transverse lines—**Northern Brownsnake (*Storeria dekayi dekayi*)**

12b.   Two rows of paired dark spots on dorsum connected by transverse lines—**Midland Brownsnake (*Storeria dekayi wrightorum*)**

13a.   Seventeen dorsal scale rows; dorsum uniformly pigmented green in life; venter uniformly pigmented yellow in life—**Northern Rough Greensnake (*Opheodrys aestivus aestivus*)**

13b.   Nineteen or more dorsal scale rows ➔ **14**

14a.   Dorsal scales weakly keeled seldom extending to tips of scales; ventral scales 190 or more ➔ **15**

14b.   Dorsal scales strongly keeled often extending to tips of scales; ventral scales 159 or fewer ➔ **16**

15a.   V-shaped mark on head present; postocular stripe extending onto neck; dorsal markings reddish to orange in life—**Red Cornsnake (*Pantherophis guttatus*)**

15b.   V-shaped mark on head absent; postocular stripe stopping at posterior margin of mouth—**Gray Ratsnake (*Pantherophis spiloides*)**

16a.   Nineteen dorsal scale rows ➔ **17**

16b.   Twenty-one or more dorsal scale rows ➔ **18**

17a.   Seven or more upper labial scales; usually two preocular scales present; dorsum unpatterned; venter cream to yellow with a double row of midventral dark spots; pale yellow stripe on dorsal scales rows 1 and 2—**Queensnake (*Regina septemvittata*)**

17b. Six or fewer upper labial scales; dorsum patterned; one preocular scale present; venter pink, reddish, or orange with a row of dark spots along each side—**Kirtland's Snake (*Clonophis kirtlandii*)**

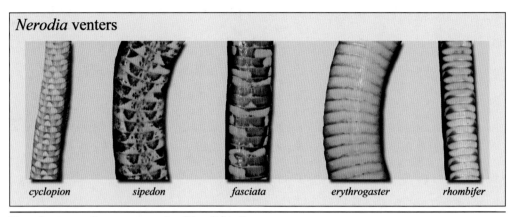

Nerodia venters

cyclopion sipedon fasciata erythrogaster rhombifer

Ventral patterns of some watersnakes (*Nerodia*) in Tennessee.

18a. Subocular scales present—**Mississippi Green Watersnake (*Nerodia cyclopion*)**

18b. Subocular scales absent ➔ **19**

19a. Venter typically immaculate or with some dark markings present only on the anterior edges of ventral scales ➔ **20**

19b. Venter heavily marked with conspicous markings ➔ **21**

20a. Venter plain yellow and occasionally washed with orange; from south-central and south-western Tennessee—**Yellow-Bellied Watersnake (*Nerodia erythrogaster flavigaster*)**

20b. Venter plain red or orange-red; from lower Cumberland River basin—**Copper-Bellied Watersnake (*Nerodia erythrogaster neglecta*)**

21a. Dorsal pattern of spots in alternating median and lateral rows that are widely separated but connected at corners forming a diamond-backed appearance; venter typically yellow with dark spots or half moons; 25 or more dorsal scale rows—**Diamond-Backed Watersnake (*Nerodia rhombifer*)**

21b. Dorsal pattern variable with crossbands or blotches but not forming a diamond-backed appearance; 23 or fewer dorsal scale rows➔ **22**

22a. Dark line present frome eye to angle of jaw; dorsal pattern of dark crossbands that are much wider than intervening paler spaces; venter with squarish red to black markings—**Broad-Banded Watersnake (*Nerodia fasciata confluens*)**

22b.   Dorsal pattern of dark crossbands and blotches with crossbands usually not extending the length of the body; venter with a double row of red to black half moons or crescents

23a.   Dark crossbands and blotches on dorsum from the neck to above the cloaca usually 30 or less and separated by more than one scale row—**Midland Watersnake (*Nerodia sipedon pleuralis*)**

23b.   Dark crossbands and blotches on dorsum from the neck to above the cloaca usually 30 or more and separated by one scale row or less—**Northern Watersnake (*Nerodia sipedon sipedon*)**

24a.   Dorsum uniformly dark and shiny; venter with alternating red and black bands with black bands continuous with dorsum; tail tip with pointed terminal scale—**Western Mudsnake (*Farancia abacura reinwardtii*)**

24b.   Venter without alternating red and black bands ➔ **25**

25a.   Either loreal or preocular scale absent ➔ **26**

25b.   Both loreal scale and at least one preocular scale present ➔ **30**

26a.   Thirteen dorsal scale rows ➔ **27**

26b.   Fifteen or more dorsal scale rows ➔ **28**

27a.   Prefrontal and internasal scales separate—**Eastern Wormsnake (*Carphophis amoenus amoenus*)**

27b.   Prefrontal and internasal scales fused—**Midwestern Wormsnake (*Carphophis amoenus helenae*)**

28a.   Seven or more upper labial scales; prefrontal scale not in contact with eye; head black with light band across head followed by black band 3–5 scales wide—**Southeastern Crowned Snake (*Tantilla coronata*)**

28b.   Six upper labial scales present; prefrontal scale in contact with eye ➔ **29**

29a.   Scales very weakly keeled; 17 dorsal scale rows—**Western Smooth Earthsnake (*Virginia valeriae elegans*)**

29b.   Scales smooth or with very weak keel on posterior half of body; 15 dorsal scale rows—**Eastern Smooth Earthsnake (*Virginia valeriae valeriae*)**

30a.   Light-colored ring present on neck; one anterior temporal scale present ➔ **31**

30b.   Light-colored ring absent on neck; two or three anterior temporal scales present ➔ **32**

31a.   Venter uniform yellow or with a row or partial row of irregular dark spots arranged along midline; neck ring never interupted—**Northern Ring-Necked Snake (*Diadophis punctatus edwardsii*)**

31b.   Venter with irregular dark spots arranged along midline either attached or in separate pairs; neck ring narrow and often interupted—**Mississippi Ring-Necked Snake (*Diadophis punctatus stictogenys*)**

32a.   Thirteen or fewer dorsal scale rows present at posterior of body before tail; venter typically pale orange—**Eastern Coachwhip (*Masticophis flagellum flagellum*)**

32b.   Fifteen or fewer dorsal scale rows present at posterior end of body before tail; venter lacking orange coloration ➜ **33**

33a.   From eastern and south-central Tennessee; enlarged basal hemipenial spine of males less than 3x the length of preceeding spine in same row—**Northern Black Racer (*Coluber constrictor constrictor*)**

33b.   From western and north-central Tennessee; enlarged basal hemipenial spine of males 3x or more the length of preceeding spine in same row—**Southern Black Racer (*Coluber constrictor priapus*)**

34a.   Dorsal scales strongly or weakly keeled ➜ **35**

34b.   Dorsal scales smooth ➜ **39**

35a.   Twenty-seven or more dorsal scale rows; four prefrontal scales present—**Northern Pinesnake (*Pituophis melanoleucus melanoleucus*)**

35b.   Twenty-six or fewer dorsal scale rows; two prefrontal scales present ➜ **36**

36a.   Seven or fewer lower labial scales present—**Rough Earthsnake (*Virginia striatula*) (in part)**

36b.   Eight or more lower labial scales present ➜ **37**

37a.   Lateral light stripe present on dorsal scale rows 2 and 3; venter with one or two rows of dark spots present along each side; supralabials with narrow black bars—**Eastern Gartersnake (*Thamnophis sirtalis sirtalis*)**

37b.   Lateral light stripe present on dorsal scale rows 3 and 4; venter immaculate; supralabial lacking dark bars ➜ **38**

38a.   Brightly colored parietal spots on top of head large and contacting each other; from West Tennessee—**Orange-Striped Ribbonsnake (*Thamnophis proximus proximus*)**

38b.   Bright-colored parietal spots on top of head absent or faint and not contacting each other if present—**Eastern Ribbonsnake (*Thamnophis sauritus sauritus*)**

39a.   Snout pointed with enlarged rostral scale; venter without dark markings—**Northern Scarletsnake (*Cemophora coccinea copei*)**

39b.   Snout rounded without enlarged rostral scale; venter with at leat some dark markings ➜ **40**

40a.   Dorsal pattern with reddish, gray, tan, or brown blotches or crossbands ➜ **41**

40b.   Dorsal pattern without blotches or crossbands; dorsum black with greatly reduced white to yellow chainlike pattern—**Eastern Black Kingsnake (*Lampropeltis nigra*)**

41a.   Twenty-one or fewer dorsal scale rows ➜ **42**

41b.   Twenty-one or more dorsal scale rows ➜ **44**

42a.   Nineteen dorsal scale rows; red, black, and yellow bands continuing completely onto venter—**Scarlet Kingsnake (*Lampropeltis elapsoides*)**

42b.   Twenty-one dorsal scale rows; venter with irregularly checkerboarded black and white ➜ **43**

43a.   Middorsal blotches distinctly red and large extending well onto the sides of the body; lateral blotches reduced or absent; from West Tennessee—**Red Milksnake (*Lampropeltis triangulum syspila*)**

43b.   Middorsal blotches red to gray to brown and smaller; lateral blotches typically present; from Middle and East Tennessee—**Eastern Milksnake (*Lampropeltis triangulum triangulum*)**

44a.   Dorsal blotches smaller numbering up to 71; 21–23 dorsal scale rows present—**Mole Kingsnake (*Lampropeltis calligaster rhombomaculata*)**

44b.   Dorsal blotches larger numbering 50–65; 25–27 dorsal scale rows present—**Prairie Kingsnake (*Lampropeltis calligaster calligaster*)**

14
Family Colubridae (Colubrid Snakes)

The Colubridae is a "hodgepodge" family that includes more than two-thirds (>1,700) of all described species of snakes. The family includes both nonvenomous and venomous species, which are found in tropical and temperate areas on nearly every continent (none are found in Antarctica). Seven distinct lineages within the family are recognized as subfamilies. However, many species are placed into this family because they cannot be placed in another family with confidence. Because of this practice, the Colubridae is paraphyletic (all species within the family do not share a common ancestor), and the interrelationships of the subfamilies are poorly understood. Not surprisingly, the Colubridae has long been the subject of debate by systematists, and some recognize the subfamilies as families. We realize that the taxonomy is likely to change but treat the Colubridae as a large family composed of several subfamilies until their interrelationships are resolved.

The colubrids are morphologically diverse. Some species are small, reaching maximum TL of only 19 cm (7.5 in), whereas other species are large, exceeding 3.7 m (12 ft) TL. Furthermore, some species are slender bodied, never attaining girths as large as your little finger, whereas other species are robust and might have girths as large as your forearm! Colubrids inhabit a variety of habitats, including deserts, grasslands, forests, cedar glades, lakes, rivers, streams, and brackish marshes. Some species are fossorial

and have adaptations that allow them to burrow through soil, whereas other species are arboreal and have adaptations that allow them to move efficiently through trees or shrubs. Thirty colubrid species from 18 genera associated with three subfamilies occur in Tennessee. These subfamilies include the Colubrinae (Harmless Egg-Laying Snakes), Natricinae (Harmless Live-Bearing Snakes), and Xenodontinae (Harmless Rear-Fanged Snakes).

The Natricinae is the largest subfamily in Tennessee, with 14 species in six genera (*Clonophis*, *Nerodia*, *Regina*, *Storeria*, *Thamnophis*, and *Virginia*). Many of these snakes are either aquatic (*Nerodia* and *Regina*) or commonly found around water (*Clonophis* and *Thamnophis*). All have keeled scales and give birth to live young (ovoviviparous). The Colubrinae is the second largest subfamily, with 12 species in eight genera (*Cemophora*, *Coluber*, *Lampropeltis*, *Masticophis*, *Opheodrys*, *Pantherophis*, *Pituophis*, and *Tantilla*). This subfamily comprises a group of species with diverse natural histories. For example, one species, *Tantilla coronata*, is fossorial and rear-fanged; it uses mild venom to subdue prey. Another species, *Pantherophis guttatus*, is found under rock or logs, on the ground, or up in trees; its diet consists primarily of rodents, which it subdues by constriction. All species in this subfamily are oviparous. The smallest subfamily is the Xenodontidae, with four species in four genera (*Carphophis*, *Diadophis*, *Farancia*, and *Heterodon*). This subfamily also is highly diverse in body form and natural history. Adults of some species are less than 8 cm (3 in) TL, but adults of other species exceed TL of more than 1 m (3.3 ft). Furthermore, some species are primarily fossorial (e.g., *Carphophis amoenus*), while others are aquatic (e.g., *Farancia abacura*). All species in this subfamily are also oviparous.

Regardless of subfamily, all colubrid snakes in Tennessee are harmless. Nonetheless, individuals from several species are often killed because they are mistaken for venomous species in the family Viperidae. As might be expected from species that occupy such a diversity of habitats, there is considerable diversity in diet among the colubrids. Some species are dietary specialists, such as *Cemophora coccinea*, which prefers eggs of lizards and snakes, or *Regina septemvittata*, which feeds almost exclusively on crayfish that have recently shed their exoskeletons. Other species are dietary generalists, such as *Coluber constrictor* and *Pantherophis spiloides*, which eat a variety of prey, including insects, amphibians, birds, and mammals.

The family name, Colubridae [koh-LOO-breh-dee], is from the Latin *coluber* (serpent). The etymology of the 18 genera is as follows. The genus name *Carphophis* [kar-FOH-fis] is from the Greek *karphos* (straw) and *ophis* (snake), perhaps in reference to the color or shape of some preserved specimens. The genus name *Cemophora* [seh-MOF-or-ra] is from the Greek *kemos* (muzzle) and *phoreus* (bearer), which refer to the enlarged rostral scale on the snout (muzzle). The genus name *Clonophis* [kloh-NOH-fis] is from the Greek *klonos* (writhing motion) and *ophis* (snake), in reference to the defensive behavior of this species. The genus

name *Coluber* [koh-LOO-ber] is Latin for serpent. The genus name *Diadophis* [die-ah-DOH-fis] is from the Latin *diadema* (crowned) and *ophis* (snake), in reference to the ring around the neck. The genus name *Farancia* [fah-RAN-see-ah] is an unknown reference, although Gotch (1986) and Beltz (2006) suggest that J. E. Gray, who described the genus in 1842, was known for inventing meaningless scientific names. The genus name *Heterodon* [het-er-ROH-don] is from the Greek *heteros* (different) and *odontos* (tooth), which describe the enlarged postmaxillary teeth in this genus. The genus name *Lampropeltis* is from the Greek *lampros* (shiny) and *pelta* (shield), which describe the smooth scales of this species. The genus name *Masticophis* [mast-ti-KOH-fis] is from the Greek *mastix* (whip) and *ophis* (snake), which describes the slender body and braided patterning of the tail. The genus name *Nerodia* [neh-RO-dee-ah] honors the Greek sea nymph Nereis, who was considered to possess great wisdom and foresight. The genus name *Opheodrys* [oh-FEE-oh-drees] is from the Greek *opheos* (a snake) and *drymos* (forest), which describe the arboreal nature of the first species described for this group. The genus name *Pantherophis* [pan-ther-ROH-fis] is from the Greek *pan* (bread), *thero* (wild animal found in the summer), and *ophis* (snake), which refer to the cross-sectional shape of the body in members of this group (like a loaf of bread) and the preferred activity period. The genus name *Pituophis* [pit-you-OH-fis] is from the Greek *pitys* (pine) and *ophis* (snake), which refer to the preferred habitat of the first described species. The genus name *Regina* [reh-JIE-nah], is Latin for "queen," but the nature of this name is unclear. The genus name *Storeria* [stoh-REH-ree-ah] honors David H. Storer, who was a prominent mid-19th-century New England natural historian. The genus name *Tantilla* [tan-TILL-ah] is from the Latin *tantalum,* meaning "a small thing." The genus name *Thamnophis* [tham-NO-fis] is from the Greek *thanes* (shrub) and *ophis* (snake), which likely refer to a commonly used

basking and hunting habitat for this group. The origin of the genus name *Virginia* [vir-GIN-nee-ah] is obscure. Presumably the name honors the region (present day Washington, D.C.) from where early individuals were collected in the 18th century.

Eastern Wormsnake adult, Blount County. (Photo by Matthew L. Niemiller)

Eastern Wormsnake

Carphophis amoenus

Description: Eastern Wormsnakes are small, fossorial snakes with adults typically 18–28 cm (7–11 in) TL but occasionally reaching 34 cm (13.4 in). Superficially, this species resembles an earthworm in coloration and size. Body scales are smooth and opalescent, but the anal plate is divided. The dorsum typically is purplish brown, brown, or dark gray and sharply contrasts with the ventral coloration, which is immaculate white or pink. The ventral coloration extends

onto the first 1–2 dorsal scale rows on the sides. Individuals that have recently shed their skin appear iridescent. The head is small and pointed, and the eyes are tiny. The tail is short and ends in a small spine. Females are larger than males but also have a shorter tail. Hatchlings are 7.5–10 cm (3–4 in) TL. Hatchlings and juveniles resemble adults but are often darker in coloration. Two subspecies are recognized and both occur in Tennessee: the Eastern Wormsnake (*C. a. amoenus*) and Midwestern Wormsnake (*C. a. helenae*). The two subspecies are nearly identical except that in the Midwestern

Wormsnake the prefrontal scales are fused with the corresponding internasal scales.

Etymology: The specific and subspecific epithet *amoenus* [ah-MEAN-nus] is Latin for "pleasing," which perhaps refers to the soft pink and iridescent coloration of many individuals. The subspecific epithet *helenae* [heh-LEN-nay] honors Helen Tennison, who collected the types (specimens from which the taxon was described) of this subspecies.

Similar Species: Eastern Wormsnakes resemble other small woodland snakes that are unpatterned, including Smooth Earthsnakes (*Virginia valeriae*), Rough Earthsnakes (*V. striatula*), Red-Bellied Snakes (*Storeria occipitomaculata*), and Southeastern Crowned Snakes (*Tantilla coronata*). Unlike Eastern Wormsnakes, all of these species lack an iridescent sheen to their scales and lack a spine at the tip of tail. Both species of *Virginia* are not strongly bicolored like Eastern Wormsnakes. The scales of Red-Bellied Snakes are keeled. Southeastern Crowned Snakes have black on the head and neck that is separated by a light-colored band.

Distribution: Eastern Wormsnakes are found from southern New England southwestward into Alabama, Mississippi, and southeastern Louisiana then north into southern Illinois, southern Indiana, and southern Ohio. Eastern Wormsnakes are found statewide in Tennessee, except for the highest elevations of the Blue Ridge Mountains (above 1,200 m). In general, Eastern Wormsnakes occur in the Ridge and Valley and Blue Ridge Mountains of eastern Tennessee, whereas Midwestern Wormsnakes occur throughout the rest of the state to the west. However, intergrades between the two subspecies are known, and some individuals that are morphologically Midwestern Wormsnakes (with fused scales) have been found in the range of the Eastern Wormsnake and vice versa.

Habitat: Eastern Wormsnakes spend much of their lives underground in a variety of habitats, from mesic hardwood forests to more xeric cedar glades in central Tennessee where they are particularly abundant. Indeed, we have found as many as 30 or more individuals in the latter habitat in just a few hours of searching. Adults and juveniles can be found under rocks, rotting logs, and other debris, especially during spring. During extremely cold or hot weather, Eastern Wormsnakes retreat deep underground. Individuals are rarely found on the surface, especially during the day. However, we have found Eastern Wormsnakes crossing roads on rainy nights.

Natural History: Many aspects of the life history of Eastern Wormsnakes are poorly

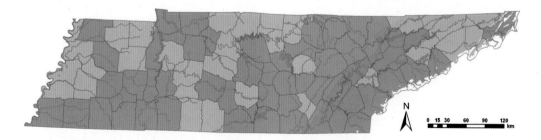

Two subspecies of wormsnakes are recognized in Tennessee: the Eastern Wormsnake (*Carphophis amoenus amoenus*), whose range appears in orange, and the Midwestern Wormsnake (*C. a. helenae*), whose range appears in blue.

Top left: Midwestern Wormsnake adult, Wilson County. (Photo by Matthew L. Niemiller) *Top right:* Venter of a Midwestern Wormsnake, Wilson County. (Photo by Matthew L. Niemiller) *Bottom right:* Head of a Midwestern Wormsnake, Wilson County. (Photo by Matthew L. Niemiller)

understood because of the secretive nature of this species. They have several morphological adaptations for a fossorial existence, including a compact skull, tiny eyes, smooth scales, and spiked tail, the latter of which is used for "thrust creeping" through tunnels. Females lay 1–12 eggs under rocks and rotting logs from June into July. Eggs hatch 8–12 weeks later in August or September. Sexual maturity is reached in about 3 years. Their diet also belies their fossorial nature, as Eastern Wormsnakes feed primarily on earthworms; however, they also eat insect larvae, slugs, and other soft-bodied invertebrates. Known predators include large invertebrates, birds (e.g., American Robins, Barn Owls, and Thrushes), opossums, and snakes, such as Copperheads (*Agkistrodon contortrix*), Racers (*Coluber constrictor*), Black Kingsnakes (*Lampropeltis nigra*), and Milksnakes (*L. triangulum*). When handled, Eastern Wormsnakes rarely bite and are harmless to humans. They will release a foul-smelling musk from the vent, and often press their tail spine against their captor's hand.

Conservation Status: Although secretive, Eastern Wormsnakes are common throughout Tennessee and are not a species of conservation concern.

Matthew L. Niemiller and
R. Graham Reynolds

Black, Red, and Yellow Make for a Happy Fellow

One benefit of attending Middle Tennessee State University (MTSU) for my master's degree was that Murfreesboro is located within a couple hours' drive of several different ecoregions and their associated habitats. Consequently, I didn't have to travel far from campus to explore a different ecological setting with its own distinct assortment of amphibians and reptiles. A one- to one-and-a-half-hour drive to the east allowed me to enter the Cumberland Plateau; a similar drive to the west took me to the Coastal Plain ecoregions. The cedar glades and forests found in several state-owned and -managed properties were only a 30-minute drive up the road. Cedar glades, unique and highly imperiled ecosystems found primarily within the Inner Nashville Basin of Middle Tennessee, support a diverse assemblage of specialized flora and fauna, including several endemic species of plants. These naturally treeless areas have

shallow soils overlaying limestone, which is often exposed in many areas. Cedar glades are tough places to make a living for many species, particularly during summer, when ground temperatures in open glades well exceed 120°F and little moisture is retained near the surface.

Despite the extreme conditions, many species of snakes and lizards make their homes in and around cedar glades in Middle Tennessee. During my tenure at MTSU, I made several trips a year to cedar glades to photograph their reptilian denizens. Several species, such as Eastern Six-Lined Racerunners (*Aspidoscelis s. sexlineata*), Little Brown Skinks (*Scincella lateralis*), and Midwestern Wormsnakes (*Carphophis amoenus helenae*), come to the surface from subterranean haunts after a few straight days of 70°F temperatures and sunshine in early spring. However, activity really increases from April into May, when most larger-bodied snakes come out of their winter shelters and are active on the surface looking for mates. At this time, Racers (*Coluber constrictor*), Eastern Black Kingsnakes (*Lampropeltis nigra*), and Eastern Milksnakes (*L. t. triangulum*) are common, as are Eastern Fence Lizards (*Sceloporus undulatus*), Southeastern Five-Lined Skinks (*Plestiodon inexpectatus*), and several species of small woodland snakes, such as Mississippi Ring-Necked Snakes (*Diadophis punctatus stictogenys*), Southeastern Crowned Snakes (*Tantilla coronata*), and Smooth Earthsnakes (*Virginia valeriae*).

One snake species that I really wanted to find was the Northern Scarletsnake (*Cemophora coccinea copei*). Scarletsnakes are relatively small serpents that can easily be confused with Milksnakes by having alternating

---

Northern Scarletsnake (*Cemophora coccinea*), Cedars of Lebanon State Forest, Wilson County. (Photo by Matthew L. Niemiller)

red, black, and whitish yellow bands or saddles. This coloration also resembles that found in the Harlequin Coral Snake (*Micrurus fulvius*), a venomous species found in the Southeast, though not in Tennessee, and both harmless species may be Batesian mimics of the deadly venomous species. You might be familiar with one of the several mnemonics to distinguish between the harmless and venomous snakes in the Southeast: "Red on yellow will kill a fellow, but red on black is a friend of Jack"—in reference to the order of the red, black, and yellow bands. Scarletsnakes, although broadly distributed throughout the southeastern United States, are uncommon in Tennessee, and few individuals have been found in Middle Tennessee. Brad "Bones" Glorioso and I once thought we had found a road-killed individual near Reelfoot Lake in northwestern Tennessee, but upon closer examination it was determined that our initial identification was mistaken. Instead we had discovered an aberrantly patterned Red Milksnake (*L. t. syspila*) that lacked the characteristic black-and-white checkerboard patterning of the belly. Both of us were disappointed to find such a beautiful snake dead on the road.

Because the Scarletsnake is a documented reptile egg eater, I always thought the cedar glades in Middle Tennessee, with their abundance of other lizards and snakes, would offer the best chances of finding the rare snake. However, several hundreds of hours of searching by myself, Bones, and Richie Wyckoff failed to produce even a glimpse of this elusive snake . . . until one day in early May.

It was a beautiful and warm morning when Bones, his girlfriend (and now wife) Melita, and I drove just up the road to Cedars of Lebanon State Forest. We routinely visited the several very productive glades there in order to take photos of their herpetofauna. This day started off like so many other trips to the cedar glades. Eastern Fence Lizards and Eastern Six-Lined Racerunners were abundant underneath rocks lying on top of the shallow soil. We caught a racer here and an Eastern Black Kingsnake there. Even an Eastern Box Turtle became a model for a day in front of our cameras. All in all, it was quite a productive morning.

Just after noon, Bones and I decided to check out a series of large, flat rocks along a road cut nearby that had produced several larger-bodied snakes in the past, including racers, kingsnakes, milksnakes, and even Northern Copperheads (*Agkistrodon contortrix mokasen*). As in previous trips, we were not disappointed and captured several larger snakes, including the largest Copperhead we had ever seen in Tennessee. As we worked our way along the road cut, we came upon a cluster of several large, flat rocks—so large that it required two people to check them (one person to lift the rock and the other to search and grab anything hiding underneath). We had found several milksnakes under this set of rocks in the past, and they almost always produced a snake or two.

I lifted the first two rocks nearest me, but only abandoned rodent burrows awaited our eyes. Bones then lifted the next slab—again, no luck. Bones began lifting the final rock of the cluster and we both immediately saw the flash of red and black attempting to escape down into a rodent burrow. "Snake!" we both yelled. Before I could even blink, Bones had grasped the banded snake by the back end of its body. We both knew what species it was, but neither of us said anything. Capture was still in doubt. I sprang into action and grabbed the rock slab so Bones could use two hands to secure his grasp on the snake and safely extract it from the burrow. After a few tense moments, he had it. Before our eyes was a beautiful adult scarletsnake, representing a "lifer" for both us. All I could say was "Holy ——, it's a scarletsnake!!!"

This was, by far, our most significant and memorable find at the cedar glades.

Although we could have continued searching, the day was over for us. We got back in the car and headed back to Murfreesboro. We were anxious to show Brian Miller, my advisor and Bones' committee member at the time, our catch. Brian had only found two *Cemophora* in his twenty years of living in Middle Tennessee and none this size. We kept the snake for a couple of days to take photographs and to show our find to our herpetology class (we had wildlife collection permits). As an added treat, in addition to being an excellent photographic subject, it regurgitated the shells of several lizard eggs it had consumed during its last meal. A few days later, we returned to the exact rock where we had captured it, released the snake, and watched as it slithered on its way down the burrow underneath. We positioned the rock exactly how we had found it, so that others also might one day partake in such a memorable experience.

Matthew L. Niemiller

Scarletsnake adult, Wilson County. (Photo by Matthew L. Niemiller)

Scarletsnake

Cemophora coccinea

Description: Scarletsnakes are relatively small, slender snakes, with adults 35–50 cm (14–20 in) TL, occasionally reaching 83 cm (33 in). Dorsal scales are smooth and arranged in 19 rows. The anal plate is undivided. The body is marked with red, black, and whitish to yellow bands that do not encircle the body. The reddish coloration forms broad saddles bordered by black pigment and separated by the lighter whitish to yellowish pigmentation. Small dark spotting or flecking often is present in the whitish to yellowish banding. The venter is whitish to yellowish and is immaculate. The head is narrow with a pointed, slightly upturned, enlarged snout that is red in color, and the hazel and black eyes are relatively small. Six supralabial scales are present. Juveniles and hatchlings resemble adults in coloration. Hatchlings are 11–18 cm (4.3–7.0 in) TL. Three subspecies are recognized, but only the Northern Scarletsnake (*C. c. copei*) occurs in Tennessee.

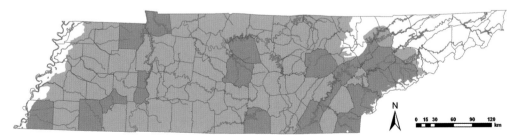

Etymology: The specific epithet, *coccinea* [kon-SIN-nee-ah], is from the Latin *coccineus* for "scarlet" or "crimson." The subspecific epithet, *copei* [koh-PEE-i], honors Edward Drinker Cope, a prolific naturalist and scientist.

Similar Species: Scarletsnakes most closely resemble Scarlet Kingsnakes (*Lampropeltis elapsoides*) and Milksnakes (*L. triangulum*), as both of these species have red, black, and whitish to yellowish bands or saddles. The red, black, and yellow bands completely encircle the body in the Scarlet Kingsnake but do not in the Scarletsnake, which has an immaculate venter. Milksnakes in Tennessee are readily identified as usually having a black and white checkered venter.

Distribution: Scarletsnakes are found throughout much of the southeastern United States from eastern Oklahoma, northeastern Texas, and Louisiana eastward into Kentucky and Tennessee and northward along the Piedmont and Atlantic Coastal Plain into Maryland, Delaware, and New Jersey. They are absent from the higher elevations of the Appalachian Mountains. They are likely found throughout Tennessee, except for the Ridge and Valley of northeastern Tennessee and the higher elevations of the Blue Ridge Mountains. However, records for this species are scattered and the exact distribution in Tennessee is not well known.

Habitat: Scarletsnakes are found in association with pine, hardwood, and mixed pine-hardwood forests with sandy to loamy soils and abundant leaf litter and other organic debris. This species is primarily fossorial and, therefore, extremely secretive. The pointed snout is used to burrow or dig into loose soil. Scarletsnakes are active on the surface at night only

Venter of an adult Scarletsnake, Wilson County. (Photo by Brad M. Glorioso)

from late spring through early autumn. Adults and juveniles are found underneath rocks, logs, and other ground cover objects. Adults are occasionally found crossing roads at night, and some individuals have been excavated when fields are plowed.

Natural History: Little is known about the reproductive biology of Scarletsnakes. Mating likely occurs during spring with oviposition occurring shortly thereafter. Around five elongate eggs (range of 2–9) are laid during the summer in underground retreats, such as a burrow, or under rocks or other cover objects. Eggs hatch in 8–12 weeks in late summer or early autumn. Females may produce more than one clutch during a single breeding season. Scarletsnakes feed primarily on the eggs of other snakes and lizards but also will prey on small snakes, lizards, and amphibians. Live prey is killed by constriction. They possess enlarged posterior teeth that are used to slit open eggs. Scarletsnakes constrict larger eggs to squeeze out the contents but often will ingest smaller eggs whole. Predators include other snakes, such as Black Kingsnakes (*Lampropeltis nigra*), as well as large frogs, birds, and mammals. The color and pattern of the Scarletsnake is thought to deter predation as it mimics that of the venomous Eastern Coral Snake (*Micrurus fulvius fulvius*), which does not presently occur in Tennessee. Scarletsnakes rarely will bite when handled.

Conservation Status: Although secretive and rarely seen, Scarletsnakes are thought to be secure and are not listed as a species of conservation concern in Tennessee. However, the status of this species is difficult to assess because of its reclusive nature. Habitat fragmentation and road mortality are thought to be significant threats to this species.

Matthew L. Niemiller

Kirtland's Snake adult, Ripley County, Indiana. (Photo by Todd Pierson)

Kirtland's Snake

Clonophis kirtlandii

Description: Kirtland's Snakes are small, robust snakes with most adults 30–50 cm (12–20 in) TL but reaching at least 62 cm (24 in). Body scales are keeled and the anal plate is divided. The dorsum is reddish brown to gray in color with four rows of rounded darker gray to black blotches or spots running the length of the body. In total, 46–57 blotches or spots are in each row; however, this patterning can be indistinct in some individuals. The venter is pink to red to orange in color with a prominent row of black, round spots along each side on each ventral scale. The head is small and darker than the rest of the body, whereas the labial scales, chin, and throat are white to cream to yellow in color. Females are slightly larger than males. Neonates are 11–16 cm (4.3–6.3 in) TL at birth. Both young and juveniles are darker than adults, with a largely unpatterned dorsum and a deep red venter. No subspecies are recognized.

Etymology: The specific epithet, *kirtlandii* [kert-LAN-dee-i], honors Jared Kirtland, a naturalist and physician from Ohio who founded the Kirtland Society of Natural Sciences.

Similar Species: Kirtland's Snakes are most similar in appearance to Brownsnakes (*Storeria dekayi*) and Red-Bellied Snakes (*S. occipitomaculata*). Brownsnakes occasionally can have a pinkish venter with a row of small dark spots on each side of the ventral scales. The venter is more vibrant in color in Kirtland's Snakes with the darker spots being more prominent. Likewise, adult Brownsnakes do not have four rows of darker blotches or spots on the dorsum. Red-Bellied Snakes have similar ventral coloration as Kirtland's Snakes but lack the row of round black spots along each side of the ventral scales. Red-Bellied Snakes also have three lighter colored spots or blotches around the neck that are lacking in Kirtland's Snakes.

Distribution: The main range of Kirtland's Snake is from western Pennsylvania west through Ohio, Indiana, Illinois, and into Missouri and northern Kentucky. This species was recently found in extreme northwestern Henry County in 2006, but its true distribution in the state is unknown. Only three individuals are known to have been found in Tennessee, including one that was found dead on a road.

Kirtland's Snake adult, Henry County. (Photo by David F. Frymire)

Habitat: Kirtland's Snakes inhabit a variety of mesic to moist habitats throughout their range, including the margins of ponds, lakes, streams, ditches, and swamps, bogs, open grassy fields, meadows, forests, and other habitats near sources of water. This species is mostly confined to smaller patches of suitable habitat, and, surprisingly, larger populations are associated with more urban areas, including city parks and vacant lots. Kirtland's Snakes use crayfish burrows, particularly those constructed by chimney crayfish, for shelter, hibernation, and aestivation. When active on the surface, Kirtland's Snakes can be found under cover, such as rocks, logs, boards, and within crayfish burrows.

Natural History: Although nothing is known of the Tennessee population, Kirtland's Snakes have been studied elsewhere in their range. This is a reclusive species that is seldom seen active during the day and is thought to be mostly nocturnal. Most individuals are observed in April, when they are reproductively active. Courtship and mating occur in April and May. Females give birth to 4–15 young from late July into September. Kirtland's Snakes feed predominantly on earthworms and slugs. Predators include several other snake species, birds, and mammals. When threatened, this species will flatten its body and become rigid, but individuals are not quick to bite when handled.

Conservation Status: The status of the Tennessee population(s) has not been assessed, although it is thought that Kirtland's Snakes are extremely rare in Tennessee. Documented threats to Kirtland's Snakes include habitat loss and degradation, collection for the pet trade, and road mortality.

Matthew L. Niemiller and
R. Graham Reynolds

Southern Black Racer adult, Wilson County. (Photo by Matthew L. Niemiller)

Racer

Coluber constrictor

Description: Racers are medium-sized snakes 100–150 cm (39 to 59 in) TL and reaching at least 185 cm (73 in). Adult racers are like no other snake in Tennessee. They are long and slender, round in cross-section, have a narrow head, and a divided anal scale. The entire body is covered by dark gray or bluish skin with smooth, dark gray to jet-black scales that exhibit a satiny sheen. In bright light, and especially before shedding, the scales may shine a beautiful iridescence due to crystals of guanine (an amino acid) reflecting all the colors of the rainbow. The scales of the chin, neck, and lower jaw (including the infralabial scales) are always white or cream colored. The tip of the rostrum

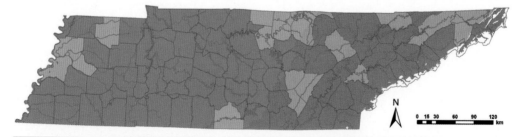

Two subspecies of racers are recognized in Tennessee: the Northern Black Racer (*Coluber constrictor constrictor*), whose range appears in orange, and the Southern Black Racer (*C. c. priapus*), whose range appears in blue.

is often brownish, and a broad black postocular band, or "mask" may be present in some (see below). The ventral scales are often a bit paler than the dorsal and lateral scales. Hatchlings are 25–30 cm (10–12 in) TL. Both hatchlings and juveniles are very colorful, with a series of oval, dark brown, rust, or reddish blotches running down the dorsal midline and smaller alternating blotches on both sides and the ventral scales; all of which are superimposed on a pale gray to bluish background. The blotches are more distinct anteriorly than posteriorly; often the tail in hatchlings is uniformly colored. This blotched pattern fades with age, and is generally absent by the time individuals reach 70 cm (28 in) TL. Racers are not sexually dimorphic. Eleven subspecies are currently recognized in eastern North America, only two of which occur in Tennessee. The Northern Black Racer (*C. c. constrictor*) is described above. The Southern Black Racer (*C. c. priapus*) has a larger area of white on the chin and neck, red or orange eyes, and hatchlings with reddish spots. Without molecular methods, Southern Black Racers can only reliably be distinguished by a fairly esoteric, internal characteristic—the spine at the base of the hemipene is three or more times as long as the next most proximal spine on the hemipene, whereas in Northern Black Racers the basal hemipenial spine is less than three times as long as the next most proximal spine. Southern Black Racers in Tennessee occur west of the Cumberland Mountains. The Black-Masked Racer (*C. c. latrunculus*) is known from eastern Arkansas and northern Mississippi and could potentially occur in extreme southwest Tennessee. Black-Masked Racers are grayish, bluish, or olive in dorsal color with similarly colored, but paler, ventral scales and an obvious broad black postocular band, or "mask."

Etymology: The specific and subspecific epithet, *constrictor,* is from the Latin terms *con* (with) and *strictus* (tight or drawn together). Ironically, racers do not constrict their prey. Instead, they "pin" their prey against the ground with one section of their body. The subspecific epithet, *priapus* [pre-AH-pus], is Latin for the god of procreation and is a reference to the difference in hemipenal morphology in this subspecies. The subspecific epithet, *latrunculus* [la-truhn-KEW-lus], is Latin for "bandit" and refers to the dark mask of this subspecies.

Similar Species: With their plain black body and head, adult racers will be difficult to confuse with any other snake in Tennessee. Even the darkest, most melanistic of Gray Ratsnakes (*Pantherophis spiloides*) and Common Gartersnakes (*Thamnophis sirtalis*) will have some patterning, pale skin, a "loaf of bread" cross-sectional shape, and, most important, keeled scales. Coachwhips (*Masticophis flagellum*) transition from black to brown from anterior to posterior and do not have a white chin. Hatchlings and juvenile racers are easy to confuse with many other Tennessee snakes of the same size, as they have a distinct dorsal pattern and coloration, which resembles young Gray Ratsnakes and Red Cornsnakes (*P. guttatus*). However, both of these species typically have a checkerboard pattern of white and black on the ventral scales.

Distribution: This is one of the most widespread species of snake in North America (arguably second only to Common Gartersnakes) with a coast-to-coast distribution that includes Mexico and Canada (the latter only barely). Racers occur statewide in Tennessee. The distribution of the two subspecies in Tennessee is not known with certainty. Northern Black Racers generally occur east of the western margin

Top left: Venter of an adult Southern Black Racer, Wilson County. (Photo by Matthew L. Niemiller) *Top right:* Southern Black Racer juvenile, Hardeman County. (Photo by Matthew L. Niemiller) *Bottom right:* Venter of a juvenile Southern Black Racer, Hardeman County. (Photo by Matthew L. Niemiller)

of the Cumberland Plateau, whereas Southern Black Racers occur to the west.

Habitat: Racers are nearly ubiquitous in Tennessee and can be found in nearly every terrestrial habitat type but are most abundant in ecotonal situations, such as the broad transition between open oak or pine forest and wetland habitats or the transition between natural and disturbed habitats such as old-field or wide fencerows in agricultural areas. They appear to be less abundant in mature forests of any type (oak, pine, or mixed), particularly where the canopy is thick and light penetration is minimal. Racers are most often seen as a fleeting black tail across a trail or road, although careful observers will spot them "periscoping," or raising their heads up in tall grass to have a look around. Adults and juveniles are also found underneath rocks and other cover.

Natural History: True to their name, racers are active, incredibly fast snakes, and even with minimal irregularities in the substrate they can easily "outrun" any human. When approached, they usually disappear silently, after which the pursuer usually gives up, but almost invariably the snake will be found about one meter up in a small shrub less than 20 m (66 ft) away. Adults will also flee up into trees, such as cedars, climbing as high as 10 m (33 ft). Racers are active during the warmer months and will hibernate in animal burrows and in crevices under logs and stumps. Racers are active throughout the day, remaining active in the afternoon heat long after other reptiles have retired to the shade or hiding places. Racers move relatively large distances in short periods of time and reach densities of up to three snakes per hectare.

They mate during spring and females nest during midsummer, depositing 4–36 eggs (the average clutch is about 12 in Georgia but over 20 in Virginia) in all sorts of natural crevices (under logs and rocks) and anthropogenic substrates (under plywood and discarded mattresses, and in sawdust piles). Hatchings emerge about two months later, grow rapidly, and reach sexual maturity in under two years. Racers consume a vast array of live prey items, including terrestrial arthropods, frogs, toads, salamanders, lizards, other snakes, songbirds, and small mammals; if they can catch it, subdue it, and swallow it, it is a potential prey item. Racers are themselves vulnerable to the diurnal vertebrate predators that can catch them, including feral cats and dogs in and around human habitation. They are frequently killed crossing roads. Although they will first try to escape from a predator or potential captor, racers aggressively bite and vibrate their tails when cornered.

Conservation Status: This widespread, abundant species is stable, affected only by the usual habitat alteration, fragmentation, road mortality, chemical pollution, and other disturbances that impact all terrestrial animals. In fact, given its abundance in open, disturbed habitats, it may be doing equally well (but not likely better) in altered systems, especially when the disturbance results in an increase of their prey base (e.g., house mice and certain lizards).

Comments: Racers are one of many "black snakes" that, when described after the fact using this vernacular, have no hope of being accurately identified, given the other slender, black, shiny snakes in Tennessee. This may even be one of the many fictional "hoop snakes" of American mythology, a species thought to chase people by putting its tail in its mouth and roll along like a hula hoop (this tall tale was told to me twice in the Big South Fork National River and Recreation Area). On a more serious note, the taxonomy of this widespread species is not considered controversial but should probably be evaluated using molecular genetics techniques, given its widespread distribution and geographic variation. Widespread species are incredibly valuable for studying the process of speciation in action.

Todd Campbell

Northern Ring-Necked Snake adult, Unicoi County. (Photo by Matthew L. Niemiller)

Ring-Necked Snake

Diadophis punctatus

Description: Ring-Necked Snakes are small woodland snakes with adults 25–38 cm (10–15 in) TL but occasionally reaching 70 cm (27.6 in) TL. This species derives its common name from the conspicuous yellow to cream-colored band around the neck. Body scales are smooth and the anal plate is divided. The dorsum is uniformly colored brown, bluish gray, or bluish black color, whereas the venter is a contrasting yellow or orange color. The venter can be immaculate or contain either a longitudinal row or partial row of small, black half-moon-shaped spots aligned in the center of the ventral scales or a more irregular arrangement of spots clustered along the midline of the venter. Hatchlings are 9–14 cm (3.5–5.5 in) TL. Hatchlings and juveniles resemble adults, although the

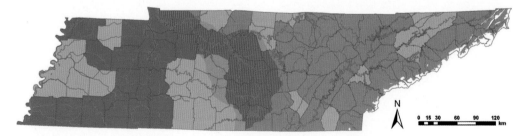

Two subspecies of Ring-Necked Snakes are recognized in Tennessee: the Northern Ring-Necked Snake (*Diadophis punctatus edwardsii*), whose range appears in orange, and the Mississippi Ring-Necked Snake (*D. p. stictogenys*), whose range appears in blue. A contact zone between the subspecies occurs in Middle Tennessee, shown in purple.

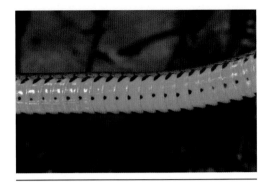

Top left: Venter of a Northern Ring-Necked Snake, Sevier County. (Photo by Matthew L. Niemiller) *Top right:* Mississippi Ring-Necked Snake adult, Wilson County. (Photo by Matthew L. Niemiller) *Bottom right:* Venter of a Mississippi Ring-Necked Snake, Stewart County. (Photo by Matthew L. Niemiller)

dorsum is often darker in younger individuals. Twelve subspecies are currently recognized and two are found in Tennessee: the Northern Ring-Necked Snake (*D. p. edwardsii*) and the Mississippi Ring-Necked Snake (*D. p. stictogenys*). The Northern Ring-Necked Snake grows to a much larger size up to 70 cm (27.6 in) TL, has a broader, complete neck ring, and has an immaculate yellow venter or occasionally a row or partial row of small black spots along the midline of the venter. The Mississippi Ring-Necked Snake does not grow as large, reaching only 48 cm (19 in) TL, has a narrow neck ring that is often incomplete, and has a venter with irregular black spots that are clustered along midline of the venter in attached or separated pairs. However, intergrades between these two subspecies exist where their respective ranges meet in Tennessee. Northern Ring-Necked Snakes might also intergrade with Southern Ring-Necked Snakes (*D. p. punctatus*) in southeastern Tennessee. Southern Ring-Necked Snakes typically have an incomplete neck ring and have a venter

with large, half-moon-shaped black spots on each ventral scale arranged along the midline.

Etymology: The specific epithet, *punctatus* [punk-TAY-tus], is Latin for "dotted" and refers to the black spotting present on the venter of some subspecies and individuals. The subspecific epithet, *edwardsii* [ed-ward-SEE-i], honors George Edwards, an ornithologist and naturalist who collected the type specimen. The subspecific epithet, *stictogenys* [stick-toe-GEN-nees], is from the Greek *stiktos* (dotted) and *genys* (cheek) and refers to the ventral pattern in some individuals.

Similar Species: Ring-Necked Snakes can be confused with Southeastern Crowned Snakes (*Tantilla coronata*) and juvenile Midland Brownsnakes (*Storeria dekayi wrightorum*). Southeastern Crowned Snakes have a pale neck ring but have a black head and black band around the neck. Also, this species has a tan dorsum and a cream-colored, rather than yellow or orange venter. Juvenile Midland Brownsnakes have a pale neckband but have keeled rather than smooth dorsal scales, and their venter is cream colored rather than yellow or orange.

Distribution: Ring-Necked Snakes have a broad distribution that extends from southeastern Canada and New England southwestward throughout the Great Lakes, Ohio Valley, Gulf Coast through Florida, and southern Great Plains. In Tennessee, they occur statewide. In general, Northern Ring-Necked Snakes occur in the eastern half of the state, whereas Mississippi Ring-Necked Snakes occur in the western half. However, there is broad zone of intergradation in central Tennessee and the actual distributions of each subspecies have not been accurately defined.

Habitat: Ring-Necked Snakes occur in a variety of habitats in Tennessee, such as mesic hardwood forests, xeric cedar glades, floodplain forests, and swamps. Most individuals are encountered under rocks, logs, bark slabs, and other cover. However, individuals can be found active on the surface on warm nights, particularly during rainy weather. We often find them in close proximity to streams in forest areas where one can also find several species of salamanders. Ring-Necked Snakes also are common on the cedar glades of the Nashville Basin during spring before conditions become too warm and dry.

Natural History: Ring-Necked Snakes can be found every month of the year as long as surface conditions are not too cold or too dry. Individuals will hibernate underground when surface conditions are intolerable during winter. Mating occurs both in the spring and in autumn, and females are capable of storing sperm during the winter months. Females lay 2–7 eggs under rocks and rotting logs in late spring. Eggs hatch 7–8 weeks later in August. Sexual maturity is reached in about 3 years. Ring-Necked Snakes feed primarily on earthworms and salamanders, but they will also eat insect larvae, small frogs, small lizards, and even small snakes. Ring-Necked Snakes have enlarged post maxillary teeth and release a mild toxin that mixes with saliva and can paralyze prey while the snake manipulates the prey in its mouth. Furthermore, individuals occasionally constrict struggling prey. Several predators feed on them, including birds, mammals, snakes, American Bullfrogs, and toads. When threatened, they first try to flee. If escape is not an option, some individuals will display the undersurface of the bright-colored tail, others will become motionless, as if dead. When handled, Ring-Necked Snakes rarely bite and are harmless to humans, although some people have reported a burning sensation when bitten. This species will release a foul-smelling musk from the vent when handled or alarmed.

Conservation Status: Ring-Necked Snakes are common throughout Tennessee and are not a species of conservation concern.

Comments: Recent molecular analyses suggest that *D. punctatus* is comprised of several independently evolving lineages that do not coincide with the distributions of currently recognized subspecies. Although several of these lineages may be recognized in the future, we refrain acknowledging any at this time until more thorough systematic studies are conducted.

Matthew L. Niemiller and
R. Graham Reynolds

Mudsnake subadult, St. Martin Parish, Louisiana. (Photo by Matthew L. Niemiller)

Mudsnake

Farancia abacura

Description: Mudsnakes are large, heavy-bodied aquatic snakes with adults 100–140 cm (39–55 in) TL, but they can attain TL of at least 207 cm (81 in). Dorsal body scales are smooth, shiny, and iridescent. The anal plate typically is divided. The dorsum is predominantly black with red to pink triangular or rounded blotches extending up from the venter. The ventral color-ation consists of alternating red to pink bands and black bands that give the belly a checkerboard appearance. The head is rounded and the eyes are small with red irises. The chin can be yellowish in color with the upper labial scales distinctly marked with black blotches. The tail is relatively short and ends in a moderately sharp spine at the tip. Females are larger than males and have longer tails. Males typically have fewer black markings on the underside of the tail. Hatchlings are 16–27 cm (6.3–10.6 in) TL and

Top left: Head of a Mudsnake adult, Lake County. (Photo by Matthew L. Niemiller) *Top right:* Venter of a Mudsnake, Lake County. (Photo by Brad M. Glorioso) *Bottom right:* Mudsnake hatching, Graves County, Kentucky. (Photo by David F. Frymire)

resemble adults. However, both hatchlings and juveniles have more pronounced red triangular or rounded blotches on the sides that occasionally coalesce forming bands on the dorsum, especially toward the head. Two subspecies are recognized, but only the Western Mudsnake (*F. a. reinwardtii*) occurs in Tennessee.

Etymology: The specific epithet, *abacura* [ah-bah-KUR-ah], is from the Greek *abakos* (abacus) and *oura* (tail), which refers to the subquadrate (nearly square) blotches on the tail that resemble the subquadrate shapes of the beads of an abacus. The subspecific epithet, *reinwardtii* [rine-ward-DEE-i], honors Caspar Reinwardt, a naturalist with the Leyden Museum in the Netherlands.

Similar Species: Western Mudsnakes cannot be confused with any other snake species in Tennessee. The combination of smooth, shiny scales, black and red ventral coloration, and small eyes with red irises readily distinguish Mudsnakes from other snakes.

Distribution: Mudsnakes occur through the much of the Atlantic Coastal Plain, Gulf Coastal Plain, and Mississippi River Valley in the southeastern United States. In Tennessee, Western Mudsnakes can be found in the Mississippi Alluvial Plain, Mississippi Valley Loess Plains, and southern Southeastern Plains in West Tennessee. This species is common around Reelfoot Lake in northwestern Tennessee.

Habitat: Mudsnakes inhabit a variety of aquatic habitats, including cypress swamps, marshes, streams, ditches, oxbow lakes, ponds, and other flooded wetlands. Such habitats typically have little current, have abundant aquatic vegetation or organic debris, and are inhabited by Three-Toed Amphiumas (*Amphiuma tridactylum*) and Western Lesser Sirens (*Siren intermedia*), which are the primary prey of Mudsnakes. However, this species is uncommonly observed because of its secretive and burrowing habits. Mudsnakes can be spotted at night moving in shallow, vegetated water, or found under leaf litter, rocks, logs, and other debris on land near aquatic habitats. Although predominantly aquatic, adults and juveniles can be found active on land both during the day and at night during warmer weather, although Mudsnakes become primarily nocturnal during the summer months.

Natural History: Little is known about the reproductive biology of Mudsnakes. Mating occurs from late April through June. Females lay 4–104 eggs in nests that they construct in or under rotting logs, rotting vegetation, or even in nests of the American Alligator (*Alligator mississippiensis*). Females likely reproduce every year. Unlike most snakes, female Mudsnakes remain with their eggs until hatching, which occurs in August or September. Hatchlings typically remain in or around the nest until dispersing the following spring. Sexual maturity is reached within two years. Home ranges are presumably small. During the winter months, Mudsnakes hibernate under logs, stumps, and other debris, or within burrows near aquatic habitats. Adult Mudsnakes feed primarily on large, aquatic salamanders (i.e., the Three-Toed Amphiuma and Western Lesser Siren). Presumably, Mudsnakes use the spine on the tail to assist in holding onto and subduing these slippery amphibians. Juveniles eat a variety of fishes and amphibians, including frogs, tadpoles, and salamanders. Predators include raccoons, birds of prey, large fish, alligators, and other snakes, such as Racers (*Coluber constrictor*) and Western Cottonmouths (*Agkistrodon piscivorus leucostoma*). Juveniles also are prey for American Bullfrogs (*Lithobates catesbeianus*). Mudsnakes are docile and rarely bite when handled. When threatened, some individuals will curl and elevate their tail displaying the red coloration underneath; this flash of bright colorations may startle a potential predator. Mudsnakes also often will coil into a ball to protect their heads and will use their tail spine to poke against a person's skin, or they will occasionally cock their heads to one side and "false strike" by thrashing the head toward the threat with the mouth closed. Additionally, when molested Mudsnakes produce a foul-smelling musk that is thought to discourage predation.

Conservation Status: Although secretive and rarely encountered, Western Mudsnakes are thought to be secure and are not listed as a species of conservation concern in Tennessee. However, the status of this species is difficult to assess because of its reclusive nature. Loss of wetlands and surrounding nesting habitat, aquatic habitat degradation, and road mortality from motor vehicles are thought to be significant threats to populations.

Comments: Mudsnakes are locally known as "hoop snakes" because of their tendency to lie in a loose coil and "horn snakes" or "stinging snakes" because of the spine on the tail, which is harmless and too blunt to puncture skin.

Matthew L. Niemiller

Eastern Hog-Nosed Snake adult, Franklin County. (Photo by Matthew L. Niemiller)

Eastern Hog-Nosed Snake

Heterodon platirhinos

Description: Eastern Hog-Nosed Snakes are medium-sized, robust snakes with most adults 50–84 cm (20–33 in) TL but reaching at least 116 cm (46 in). Body scales are keeled and the anal plate is divided. Dorsal coloration can be extremely variable, although most adults in Tennessee are dark gray to black with little patterning. However, dorsal background coloration can vary from pale gray to orange or yellow with spotted or crossband patterns from white to gray or black. The venter is mottled or immaculate and usually similar in color, although paler, than the dorsal background color. The editors have found a particularly striking specimen in Stoddard County, Missouri (about 75 km [47 miles] from the Tennessee border), an example of the bright colorations that are oc-

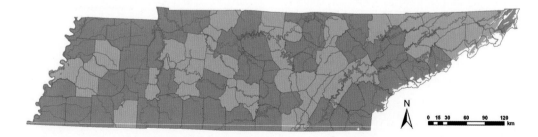

Top left: Eastern Hog-Nosed Snake hatchlings, Franklin County. (Photo by Brian T. Miller) *Top right:* Venter of an Eastern Hog-Nosed Snake, Franklin County. (Photo by Matthew L. Niemiller) *Bottom right:* Eastern Hog-Nosed Snake adult, Butler County, Missouri. (Photo by R. Graham Reynolds)

casionally seen in this species (see photo). The head is large and the snout appears upturned with a large highly keeled rostral scale at the terminus of the snout. Females are generally larger than males. Hatchlings are 130–250 mm (9.8 in) TL and have contrasting and conspicuous patterning, which might or might not fade as they age. The venter of some juveniles can be almost black. This species has an enlarged pair of rear teeth, which are used to puncture inflated toads and other anurans. No subspecies are recognized.

Etymology: The specific epithet, *platirhinos* [pla-tea-RINE-nos], is from the Greek *plati,* meaning "broad," and *rhinos,* meaning "nose," which refer to the enlarged and upturned rostral scale of the snout.

Similar Species: Eastern Hog-Nosed Snakes might be mistaken for Racers (*Coluber constrictor*), as the coloration can be similar. However, Eastern Hog-Nosed Snakes have keeled scales, upturned snout, and a stout body, which readily distinguishes them from the more slender, smooth-scaled Racer. Although they do not occur in Tennessee, Southern Hog-Nosed Snakes (*H. simus*) are very similar and can be distinguished by their more upturned snout and having the underside of the tail the same color, not paler, as the venter.

Distribution: Eastern Hog-Nosed Snakes range widely across the central and eastern United States, from Kansas to Cape Cod, Massachusetts and the Florida peninsula, although they appear to be restricted to certain areas within their range and are not found in all habitat types. Populations are highly scattered and likely fragmented. In Tennessee, documented populations are also scattered, but individuals can be quite secretive, making it difficult to ascertain the exact distribution of this species. Nonetheless, Eastern Hog-Nosed Snakes likely occur statewide, except for the highest elevations (> 760 m) of the Blue Ridge Mountains. Areas with many records include southwest Tennessee, Land Between the Lakes, and Great Smoky Mountains National Park, although these records are likely due to more intense sampling effort in these areas.

Habitat: Eastern Hog-Nosed Snakes are completely diurnal and tend to inhabit more xeric areas with loose soil, as they prefer to burrow. This species can be locally common in areas of pine forest with loose soil, and they will readily use root holes, rotten stumps, and rubbish piles for cover and overwintering sites. Individuals are most frequently encountered while traveling during the day, either crossing trails or roads in warmer weather. They occasionally can be found in thicker deciduous forest or near water, although they seem to be far less common in these types of habitats.

Natural History: Eastern Hog-Nosed Snakes are one of the most interesting species in our state, as they have a complex and remarkable defensive display. When approached, this species will begin by spreading its head and neck like a cobra and waving its head around while hissing. They occasionally will "strike" by lunging toward the threat with the mouth closed, although they almost never actually bite. If this fails to deter the threat, the snake will roll onto its back, defecate on itself, open its mouth, hang out its tongue, and otherwise pretend to be quite dead. Often Eastern Hog-Nosed Snakes will even regurgitate their last meal. A snake in this condition, called thanatosis, is limp when picked up, and if it is rolled right side up, it will promptly roll over onto its back again. These snakes are ever watchful of the predator's presence. Indeed, an amusing response can be observed by walking slowly away from the snake feigning death–their eyes follow you as you depart. When you are sufficiently far away, the snake will quickly right itself and slither off. Eastern Hog-Nosed Snakes are most frequently encountered in late spring and early autumn, and mating takes place in early spring. Females lay 4–60 eggs from late May into August that hatch in 8–10 weeks in July into September. Sexual maturity is reached in two years. Eastern Hog-Nosed Snakes are diurnal predators of amphibians, especially toads and ambystomatid salamanders, and are able to overcome the bufotoxins in the skin of toads. Hog-Nosed Snakes have a few adaptations to assist in feeding on toads, as they will excavate toads that have buried themselves with their upturned rostrum, which acts like a spade, and they can puncture toads, which often inflate their bodies in defense, with two large rear teeth. Eastern Hog-Nosed Snakes also will consume other small vertebrates and invertebrates, and juveniles likely rely heavily on arthropods. Predators include Black Kingsnakes (*Lampropeltis nigra*), birds of prey, and predatory mammals.

Conservation Status: The status of the Tennessee populations has not been assessed, although populations appear highly fragmented. Threats to Eastern Hog-Nosed Snakes include habitat loss and degradation, collection for the pet trade, and road mortality. Eastern Hog-Nosed Snakes are slow moving and diurnal; hence they are highly vulnerable to being run over by motor vehicles.

Comments: Colloquial names for Eastern Hog-Nosed Snakes are plentiful and varied, mostly describing an aspect of their elaborate defensive display. Some examples are "spreading adder," "puffing adder," "possum snake," "spread-head moccasin," "blowing viper," and "black adder." They are frequently thought to be venomous, although this is not the case, and this species will usually never bite even if provoked.

R. Graham Reynolds and
Matthew L. Niemiller

Yellow-Bellied Kingsnake adult, Warren County. (Photo by Matthew L. Niemiller)

Yellow-Bellied Kingsnake

Lampropeltis calligaster

Description: Yellow-Bellied Kingsnakes are medium-sized, slender snakes with adults 75–105 cm (30–41 in) TL but reaching up to 143 cm (56 in). Body scales are smooth and the anal plate is undivided. Dorsal ground color is tan to grayish brown to yellowish brown with a series of dark-edged reddish brown to brown to greenish brown spots or blotches and two alter-nating rows of similarly colored, smaller spots running along the sides that might occasionally fuse. Some older adults darken, thus obscuring the dorsal patterning. Some adults that darken often develop four longitudinal dusky stripes. The head is relatively small with a distinctive broken V-shaped, arrowhead-like marking on the crown. The venter is white to yellowish to pale brown with a series of alternating, darker brown, squarish spots that form a checker-board pattern. Hatchlings are 17–31 cm (7–12

Left: Yellow-Bellied Kingsnake adult, Limestone County, Alabama. (Photo by Ritchie King) *Right:* Mole Kingsnake juvenile, Pender County, North Carolina. (Photo by Andrew Durso)

in) TL. Both hatchlings and juveniles are more boldly patterned than adults. Three subspecies are recognized and two occur in Tennessee: the Prairie Kingsnake (*L. c. calligaster*) and Mole Kingsnake (*L. c. rhombomaculata*). Prairie Kingsnakes typically have 50–65 dorsal blotches or spots and 25–27 dorsal scale rows, whereas Mole Kingsnakes have up to 71 dorsal blotches or spots and 21–23 dorsal scale rows. These two subspecies are known to intergrade in Tennessee.

Etymology: The specific and subspecific epithet, *calligaster* [kal-leh-GAS-ter], is from the Greek *kallos* for "beauty" and *gaster* for "belly." The subspecific epithet, *rhombomaculata* [rom-boh-mak-you-LAH-tah], is from the Greek *rhombos* (square shapes) and *maculata* (spotted), which refers to the greater number of dorsal blotches in this subspecies.

Similar Species: Yellow-Bellied Kingsnakes can easily be confused (especially when young) with several other snake species in Tennessee, including Gray Ratsnakes (*Pantherophis spiloides*), Red Cornsnakes (*P. guttatus*), and juvenile Racers (*Coluber constrictor*). All three of these species have a divided anal plate, while Yellow-Bellied Kingsnakes have an undivided anal plate. Likewise, both Gray Ratsnakes and Red Cornsnakes from some regions have keeled body scales. Racers do have smooth scales but lack the conspicuous checkerboard patterning on the venter.

Distribution: Yellow-Bellied Kingsnakes are found throughout much of the southeast-

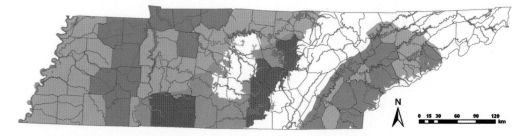

Two subspecies of Yellow-Bellied Kingsnakes are recognized in Tennessee: the Mole Kingsnake (*Lampropeltis calligaster rhombomaculata*), whose range appears in orange, and the Prairie Kingsnake (*L. c. calligaster*), whose range appears in blue. A contact zone between the subspecies occurs in Middle Tennessee, shown in purple.

ern United States, except the higher elevations of the Appalachian Mountains, from Nebraska, Oklahoma, and eastern Texas, eastward into Mississippi, Alabama, Kentucky, Tennessee, Illinois, Indiana, and Georgia then northward along the Piedmont and Atlantic Coastal Plain into Maryland. In Tennessee, Yellow-Bellied Kingsnakes are likely found throughout much of the state, except for the Ridge and Valley of northeastern Tennessee and the higher elevations of the Blue Ridge Mountains, Cumberland Plateau, and Cumberland Mountains. However, records for this species are scattered and the distribution of Yellow-Bellied Kingsnakes is not well known in the state. Mole Kingsnakes occur east of the Cumberland Plateau, whereas Prairie Kingsnakes occur to the west of the Cumberland Plateau. However, intergrades are known from the southern tier counties in Middle and West Tennessee.

Habitat: Yellow-Bellied Kingsnakes are inhabitants of open fields, cultivated farmland, barnyards, pastures, prairies, rocky hillsides and open woodlands. Adults are occasionally found crossing roads, both during the day and at night, particularly after rainfall or on warm spring or summer nights. Yellow-Bellied Kingsnakes are confirmed burrowers and occasionally are found while plowing fields or excavating land. Yellow-Bellied Kingsnakes are very secretive and spend much of the day in animal burrows or under rocks, logs or several inches of loose soil.

Natural History: Not much is known about the natural history of Yellow-Bellied King-snakes because of their fossorial and secretive nature. This species is active from March into October. Mating occurs during spring after which females lay 5–20 eggs in an earthen cavity below the surface usually during June into July. Eggs hatch in 7–11 weeks, usually in August or September. Sexual maturity is reached in three years. Yellow-Bellied Kingsnakes will constrict struggling prey, and they feed on skinks, other lizards, small snakes, bird eggs, and rodents. Hatchlings and young juveniles particularly prey on skinks. Predators include hawks, predatory mammals, and other snakes, including Eastern Black Kingsnakes (*L. nigra*) and Racers (*Coluber constrictor*). When threatened, Yellow-Bellied Kingsnakes will vibrate their tails, release a foul-smelling musk, and attempt to bite. However, this species calms down quickly and is quite docile when handled.

Conservation Status: Although secretive and rarely seen, Yellow-Bellied Kingsnakes are thought to be secure and are not listed in Tennessee. However, the status of this species is difficult to assess because of its reclusive nature. Habitat destruction, fragmentation and disturbance by farming equipment are thought to be significant threats to this species in Tennessee. Also, by far the most frequently encountered specimens are found dead on roads, so road mortality might have a significant effect on local populations.

Comments: Yellow-Bellied Kingsnakes are called "mole snakes" and "brown kingsnakes."

Lisa Powers and
Matthew L. Niemiller

Scarlet Kingsnake adult, Coffee County. (Photo by G. Richie Wyckoff)

Scarlet Kingsnake

Lampropeltis elapsoides

Description: Scarlet Kingsnakes are colorful, small kingsnakes, with most adults 36–50 cm (14–20 in) TL but reaching at least 68.6 cm (27 in). Body scales are smooth and the anal plate is undivided. Dorsal coloration consists of rings or bands of alternating colors that continue onto the venter and completely encircle the body. Narrow yellow or off-white rings are bordered on both sides by usually wider black rings, and these are interspersed with much wider scarlet red rings. The snout is narrow, pointed, and red, and the eyes are small and black. Hatchlings are 12–20 cm (4.7–7.9 in) TL and resemble miniature adults, although the pale bands might be white rather than yellow. This species was recently elevated from a subspecies of Milksnake (*L. triangulum;* see **Comments**), and no subspecies are recognized. However, identification is complicated because some individuals from populations of this species interbreed with Milksnakes and, consequently, hybrid individuals have been found in Tennessee.

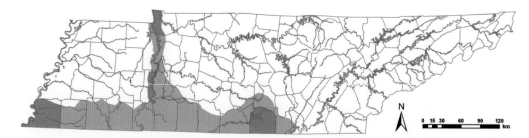

Left: Scarlet Kingsnake adult, Trigg County, Kentucky (Photo by David F. Frymire) *Right:* Head of a Scarlet Kingsnake adult, Coffee County. (Photo by G. Richie Wyckoff)

Etymology: The specific epithet, *elapsoides* [ee-lap-SOY-dees], is derived from the now-defunct genus name *Elaps,* from the Greek *ellops* for "coral," which used to refer to the coral snakes (now genus *Micrurus*), which Scarlet Kingsnakes superficially resemble in coloration. *Elaps* is now used for the family Elapidae, which contains the coral snakes, kraits, cobras, and relatives.

Similar Species: Scarlet Kingsnakes are sometimes confused with several other similarly colored and patterned snakes in Tennessee. Two subspecies of Milksnakes occur in Tennessee, the Eastern Milksnake (*L. t. triangulum*) and Red Milksnake (*L. t. syspila*), both of which have similar dorsal coloration to Scarlet Kingsnakes. Eastern Milksnakes from East Tennessee have brownish blotches and a gray to tan background, whereas individuals from Middle Tennessee often have dull to bright red blotches and a series of scattered flank spots of either black or miniature versions of the dorsal spots interspersed with the large dorsal spots. Juvenile Eastern Milksnakes usually have a much brighter red than adults. Red Milksnakes have bright red saddles bordered in black across the dorsum that extend well onto the sides and generally lack flank spots. A generally reliable

characteristic to distinguish Milksnakes is that the venter is usually of a checkerboard pattern and the dorsal coloration does not extend onto the venter, although MLN found an unusual Red Milksnake in Obion County with an immaculate creamcolored venter. Neither Milksnake subspecies have a narrow, pointed snout. Although Northern Scarletsnakes (*Cemophora coccinea copei*) are similarly colored, their venter is immaculate and their snout is more narrow and pointed compared to milksnakes. Red Cornsnakes (*Pantherophis guttatus*) and juvenile Gray Ratsnakes (*P. spiloides*) have blotches, but both of these species have keeled scales in some populations, a divided anal plate, and a patterned venter. Some people mistake Scarlet Kingsnakes for the venomous Coral Snake (*Micrurus fulvius*), although this species does not presently occur in Tennessee and will have red saddles bordered by yellow instead of black.

Distribution: Scarlet Kingsnakes are a species of the southeastern Coastal Plain of the United States, ranging from southern Kentucky to the Gulf Coast, skirting the Southern Appalachians, and ranging up the Atlantic seaboard to the New Jersey Pine Barrens. In Tennessee this species is narrowly distributed around the north-flowing Tennessee River, from Land

Between the Lakes south, roughly dividing the ranges of the Eastern Milksnake to the east and the Red Milksnake to the west. County records exist in this region for Stewart, Houston, and Humphreys counties. Additionally, a population is known from Arnold Air Force Base in Coffee County, and populations might extend south from this area. Characterizing the actual distribution of this species in Tennessee is difficult, as specimens are infrequently encountered and a documented intergrade with a Red Milksnake has been found in Decatur County. Interestingly, a single preserved specimen exists for the Great Smoky Mountains National Park, which was found in 1932 underneath a large boulder on a xeric, south-facing slope in Sevier County.

Habitat: Scarlet Kingsnakes are found primarily, although certainly not exclusively, in sandy soil and pine forest habitat. They prefer wooded areas with mature forest and ample upright snags with peeling bark. The bark provides a good refuge, and it frequently houses their preferred prey of skinks. Scarlet Kingsnakes are also found underneath other cover, such as logs and rocks, and will also use artificial cover. This species is rarely seen in the open, although individuals have been found crossing roads at night.

Natural History: Scarlet Kingsnakes are arguably the most attractive and one of the most infrequently encountered snakes in Tennessee. This species is nocturnal and is usually most active after heavy rains in the warmer months. Scarlet Kingsnakes likely thermoregulate by adjusting their position beneath peeling bark, either remaining on the sunny side of the snag or shifting to the shaded side. This behavior allows them to remain hidden but still achieve an appropriate body temperature. Almost nothing is known about Scarlet Kingsnakes in Tennessee, but populations have been well studied elsewhere in their range. Mating occurs in April and May, followed by females laying 2–12 eggs in rotten logs or under rocks about a month later. Eggs hatch in 8–10 weeks during late summer. Age at sexual maturity and longevity is unknown, although a captive individual lived for 22 years and four months. Scarlet Kingsnakes feed on most small vertebrates, including (like other kingsnakes) other snakes and young mice, although they seem to prefer feeding on lizards, especially skinks and anoles. Scarlet Kingsnakes are preyed on by birds, predatory mammals, and other snakes. Although they have tiny mouths and are not aggressive, this species will smear their captor with foul-smelling musk when threatened and captured.

Conservation Status: The status of Tennessee populations of Scarlet Kingsnakes has not been assessed and would, indeed, be a difficult undertaking, as individuals are secretive and infrequently encountered. This species is thought to be common in other parts of its range. Threats to Scarlet Kingsnakes include severe habitat loss and degradation, collection for the pet trade, persecution, and road mortality.

Comments: Scarlet Kingsnakes were previously considered to be a subspecies of Milksnake (*L. triangulum elapsoides*), as the two species were known to hybridize and morphological intergrades were not uncommon. However, several studies have demonstrated that Scarlet Kingsnakes are best placed in their own species. Behavioral studies along intergrade zones indicate that the two species generally maintain a species boundary, with only occasional intergrades. In addition, recent molecular studies have indicated that Scarlet Kingsnakes represent a unique evolutionary lineage and should, therefore, be considered as a distinct species. Finally, many herpetologists have long recognized that this species is both ecologically and morphologically distinct from Milksnakes. We therefore include them here as the unique species *L. elapsoides.*

R. Graham Reynolds and
Matthew L. Niemiller

Eastern Black Kingsnake adult, Knox County. (Photo by Matthew L. Niemiller)

Eastern Black Kingsnake

Lampropeltis nigra

Description: Eastern Black Kingsnakes are medium to large snakes with most adults 90–120 cm (35–47 in) TL, with some individuals reaching at least 150 cm (59 in). This species is slightly smaller than the related Eastern King-snake (*L. getula*), which can reach 208 cm (82 in) TL. Body scales are smooth and the anal plate is undivided. Coloration of the dorsum is variable, related in large part to how much of the juvenile color pattern is retained. Most of the dorsum of hatchlings is a rich blue-black, but a series of yellow or white dotted crossbands are present on the back from the neck to the tip of the tail. Similar markings are present on the sides, but typically these side bands are located between two adjacent dorsal bands; thus the dorsal and sidebands do not meet. As individuals age, the lighter coloration fades, and the pigments migrate to the lower edges of the sides. Thus, the dorsal colorations of adults range from completely black, to predominantly black with faint

Left: Eastern Black Kingsnake juvenile, Wilson County. (Photo by Matthew L. Niemiller) *Right:* Eastern Black Kingsnake x Speckled Kingsnake adult, Lake County. (Photo by Brad M. Glorioso)

yellow to whitish speckles or crossbands on the dorsum, to that described for juveniles. Regardless of dorsal coloration, the venter is checkered with alternating squares of yellow and black. The upper and lower labial scales are yellow or white, except at their borders with other labial scales. The anterior and posterior margins of the labial scales are black. The labial scale pattern produces a series of bands on the lips, reminiscent of teeth. Males grow to larger sizes than females. Hatchlings are 23–30 cm (9–12 in) TL. Eastern Kingsnakes (*L. getula*) and Speckled Kingsnakes (*L. holbrooki*) are extremely similar to Eastern Black Kingsnakes (*L. nigra*), with the former usually possessing distinct white to yellowish crossbands on the dorsum and little speckling, whereas Speckled Kingsnakes are heavily speckled with yellow across the dorsum and usually have quite a bit of yellow across the flanks. Although these species were all recently considered subspecies of *L. getula,* the subspecies were found to be genetically distinct enough to warrant separation to individual species. Thus no current subspecies are recognized for this group (see **Comments** below). Eastern Black Kingsnakes likely intergrade with Speckled Kingsnakes in the western part of the state.

Etymology: The specific epithet, *nigra* [NYE-grah], is from the Latin *niger,* meaning "black," which refers to the dorsal coloration of this species.

Similar Species: Eastern Black Kingsnakes are easily and often confused with other large black snakes, including Gray Ratsnakes (*Pantherophis spiloides*) and Racers (*Coluber constrictor*). Gray Ratsnakes attain a much larger size, have keeled scales, are more robust, and have a divided anal plate. Racers are more

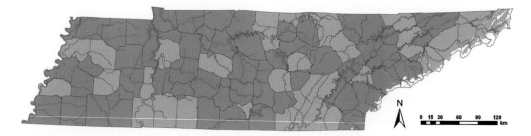

slender, have a white chin, a divided anal plate, and an immaculate venter.

Distribution: Eastern Black Kingsnakes occur from southern Indiana and Illinois to central Alabama and are found statewide in Tennessee. Speckled Kingsnakes (*L. holbrooki*) do not occur east of the Mississippi River; however, individuals of *L. nigra* in West Tennessee closely resemble *L. holbrooki* and possibly represent intergrades between the two species. Eastern Kingsnakes (*L. getula*) occur along the Atlantic Coast, including North Carolina's border with Tennessee in Swain, Graham, and Cherokee counties, and the border of Tennessee and Georgia in Polk County. Some individuals from extreme eastern Polk County and other eastern corners of the state resemble *L. getula*. Molecular confirmation is required to determine if this species occurs in Tennessee. Presumably, neither *L. holbrooki* nor *L. getula* occurs as a unique non-interbreeding species in Tennessee; hence we treat them as peripheral species to *L. nigra* in Tennessee (see **Comments** below).

Habitat: Eastern Black Kingsnakes occur in most habitats in Tennessee below 800 m (2,600 ft) in elevation. They are more common in lower elevations found outside of the Cumberland Mountains and Blue Ridge Mountains. Preferred habitats include areas with access to loose soil for burrowing, large rocks, and other cover, such as downed trees or root stumps. They are often found in both open and forested areas near water, such as in piles of retaining rock along ponds, lakes, and streams, and they are also frequently found in urban areas, for example, around the downtown Knoxville Greenway. They are common in cedar glades of the Inner Nashville Basin of Middle Tennessee.

Kingsnakes are diurnally active during warmer months of the year. Most individuals are encountered during the spring and early summer, and individuals retreat into underground burrows during the heat of summer and the cold of winter.

Natural History: Eastern Black Kingsnakes mate in the spring from March to June, and females lay 5–30 eggs in moist substrates six weeks after mating. Hatchlings emerge after 8–10 weeks of incubation, usually in late summer, and sexual maturity is reached in about three years. Kingsnakes are generalist predators and will consume small rabbits, rodents, lizards, amphibians, birds, reptile and bird eggs, and other snakes. They are perhaps most famous for the latter and are known to feed on venomous species, such as Copperheads (*Agkistrodon contortrix*), and they are immune to the venom of pit vipers. Kingsnakes subdue other snakes with a quick bite behind the head and then by encircling and constricting them. Kingsnakes are capable of consuming snakes longer than themselves by folding the prey over itself before swallowing. Predators include other kingsnakes, Racers (*Coluber constrictor*), birds, and predatory mammals. Eastern Black Kingsnakes have a variety of dispositions. Some are very aggressive and will coil and strike; others are calm. Most will bite when first handled, and almost all individuals will smear their captor with a pungent musk released from anal glands. This species will frequently vibrate the tail when threatened.

Conservation Status: The status of Tennessee populations of Eastern Black Kingsnakes has not been assessed. Populations appear stable and this species remains one of the more commonly encountered snakes in the state. However, populations of Eastern Kingsnakes (*L. getula*) are declining precipitously across the Southeast, largely for unknown reasons. Threats to Eastern Black Kingsnakes include habitat loss and degradation, persecution, collection for the pet trade, loss of cover due to decorative rock collection, and road mortality. This is one of the most commonly encountered road-killed snakes in Middle Tennessee.

Comments: The taxonomy of kingsnakes has been in a state of flux. Two papers published during 2009 (Pyron and Burbrink 2009a, 2009b) indicate that subspecies of *L. getula* should be elevated to full species. Thus the species *holbrooki* and *nigra*, formerly considered subspecies, are now treated as unique taxa from *L. getula* (Crother 2011). Furthermore, these authors found that pure *L. holbrooki* do not occur east of the Mississippi River, a find-ing which contrasts with previous range maps (Conant and Collins 1998) and morphological studies, which considered populations from West Tennessee to be *holbrooki* due to their morphological resemblance to Speckled Kingsnakes. Although molecular evidence is minimal to our knowledge, we treat kingsnakes in Tennessee as *L. nigra*, and we acknowledge that *L. getula* might be present in pockets in extreme East Tennessee. King (1939) indicated that a *getula* x *nigra* intergrade was found just across the North Carolina border in Swain County, and Palmer and Braswell (1995) treat this record as an intergrade. We also consider that kingsnakes from southwestern Tennessee are morphological intergrades with *holbrooki;* hence we consider them as *L. nigra* and not *L. holbrooki,* as has been previously suggested.

R. Graham Reynolds and
Matthew L. Niemiller

Eastern Milksnake adult, Wilson County. (Photo by Matthew L. Niemiller)

Milksnake

Lampropeltis triangulum

Description: Milksnakes are medium-sized snakes. Most adults are 50–90 cm (20–35 in) TL, but individuals can be much larger, reaching at least 132 cm (52 in). Body scales are smooth and the anal plate is undivided. Dorsal coloration is highly variable, although most individuals have a series of large, irregularly shaped dorsal blotches that are various shades of brown, grayish brown, reddish brown, or red. The dorsal blotches are bordered by black. The most anterior dorsal blotch located on the back of the head is usually Y- to V-shaped and, therefore, differs in shape from the more posterior blotches. Small lateral blotches, similar in color to the dorsal blotches, extend from the neck to the tail tip. The dorsal

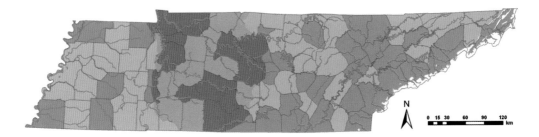

Two subspecies of milksnakes are recognized in Tennessee: the Eastern Milksnake (*Lampropeltis triangulum triangulum*), whose range appears in orange, and the Red Milksnake (*L. t. syspila*), whose range appears in blue. A contact zone between the subspecies occurs in middle Tennessee, shown in purple.

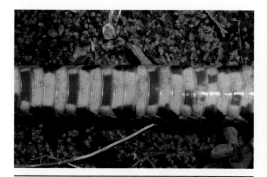

Top left: Venter of an Eastern Milksnake, Wilson County. (Photo by Matthew L. Niemiller) *Top right:* Red Milksnake adult, Lake County. (Photo by Brad M. Glorioso) *Bottom right:* Eastern Milksnake adult, Trousdale County. (Photo by Brad M. Glorioso)

and lateral blotches alternate in position such that a lateral blotch is found between two dorsal blotches. However, in some populations the dorsal and lateral spots coalesce into a single large blotch. The body or background color between the blotches is gray to tan. The contrast between blotches and ground coloration can be dramatic, particularly if the ground color is white and the blotches are bright red. The venter is irregularly checkered with black and white. The iris of the eye is red to reddish brown. Hatchlings are 20–26 cm (8–10 in) TL and generally more brightly colored than adults. Eight subspecies are recognized, but only two occur in Tennessee: the Eastern Milksnake (*L. t. triangulum*) and Red Milksnake (*L. t. syspila*). Eastern Milksnakes from East Tennessee are dull colored as adults, usually having brownish blotches and a gray to tan background, whereas individuals from Middle Tennessee often have dull to bright red blotches and a series of scattered flank spots of either black or miniature versions of the dorsal spots interspersed with the large dorsal spots. There appears to be a distinct cline

of increasing brightness of dorsal spots from east to West Tennessee. Red Milksnakes have bright red saddles bordered in black across the dorsum that extend well onto the sides and generally lack flank spots. The venter is usually of the checkerboard pattern, although an unusual individual was found by MLN in Obion County with an immaculate creamcolored venter.

Etymology: The specific and subspecific epithet, *triangulum* [try-AYNG-you-lum], is from the Latin words *tri,* meaning "three," and *angulum,* meaning "corner," which refers to the three-angled pattern on the head. The subspecific epithet, *syspila* [sis-PYE-lah], is from the Greek words *sys,* meaning "together," and *spilos,* meaning "spots," which refers to the enlarged dorsal spots of many individuals extending onto their flanks, as opposed to smaller dorsal spots and even smaller flank spots.

Similar Species: Milksnakes are sometimes confused with several other similarly patterned snakes in Tennessee. Scarlet Kingsnakes (*L. elapsoides*) are similarly patterned but have a red, noticeably pointed snout and red, black, and yellow coloration that form bands that continue onto the venter. Likewise, Northern Scarletsnakes (*Cemophora coccinea copei*) are similarly colored, although the venter is immaculate and the snout is noticeably pointed. Red Cornsnakes (*Pantherophis guttatus*) and juvenile Gray Ratsnakes (*P. spiloides*) have blotches, but both of these species have a body shape distinct from that of Milksnakes. In cross section, the body of a milksnake is circular, whereas the cross sectional shape of rat and cornsnakes resembles a slice of store-bought bread. Furthermore, ratsnakes and some populations of cornsnakes have keeled scales and a divided anal plate. Some people mistake Eastern Milksnakes for the venomous Coral Snake (*Micrurus fulvius*), although this species does not occur in Tennessee and will have red saddles bordered by yellow instead of black.

Distribution: Milksnakes occur from southern New England west to the Great Lakes region south to the Southern Appalachians into northern Alabama and Georgia and west into the Great Plains. Milksnakes are replaced to the south by Scarlet Kingsnakes (*L. elapsoides*), which occur throughout the Southeast but potentially overlap with Milksnakes in central Tennessee. Milksnakes occur statewide in Tennessee, although they are infrequently encountered in some areas. Eastern Milksnakes occur primarily east of the north-flowing Tennessee River, whereas Red Milksnakes are found primarily across the Coastal Plain ecoregions of West Tennessee. Intergrades between these two subspecies have been reported from Stewart and Houston counties. Red Milksnakes may intergrade with Scarlet Kingsnakes along the southern stretch of the north-flowing Tennessee River in West Tennessee. Milksnakes can be locally common in some areas of the Inner Nashville Basin and the Blue Ridge Mountains.

Habitat: Milksnakes occur in a wide variety of habitats, including forests, open old fields, cedar glades, near human habitations, and in upland areas near cypress swamps. They are most frequently encountered in the spring underneath cover objects, such as rocks, logs, and artificial cover such as roofing tin and boards. They are also found (alas, usually dead) on roads on warm rainy spring nights. Milksnakes are largely fossorial and spend most of the year in burrows underground and hence are infrequently seen in most regions.

Natural History: Milksnakes are one of the more attractive and infrequently seen species in our state. In the warmer months, Milksnakes are primarily nocturnal. Rather than basking in direct sunlight, this species will often warm on rocks already heated by the sun. This behavior often occurs at night, which explains the frequency that individuals are found as roadkill. Mating occurs in April and May, and females lay 10–25 eggs in burrows, rotten logs, or under rocks about a month later. Eggs hatch in 7–10 weeks, during late summer. Sexual maturity probably is reached in three years. Longevity is unknown in the wild, but captive individuals are known to live 20 years or more. Milksnakes are opportunistic and generalist predators consuming almost any small vertebrate or invertebrate (such as worms) that they can find. Large or struggling prey is usually subdued by constriction. Although they prefer small mammals, Milksnakes will also eat reptiles, including other snakes, birds, amphibians, fish, reptile and bird eggs, and even carrion. Young milksnakes in particular seem to have a fondness for skinks. Like the kingsnakes, this species is immune to pit viper venom and will consume venomous species, such as Copperheads (*Agkistrodon contortrix*). Milksnakes are preyed on by birds, predatory mammals, American Bullfrogs, and other snakes. They are not generally aggressive, but this species will occasionally bite when first captured, and will smear their captor with foul-smelling musk released from anal glands.

Conservation Status: The status of Tennessee populations of Milksnakes has not been assessed. Individuals are secretive and infrequently encountered, except in some areas such as the Inner Nashville Basin. Threats to Eastern Milksnakes include severe habitat loss and degradation, collection for the pet trade, persecution, and road mortality.

Comments: Milksnakes are named for an unusual rural belief, which presumed that they drank milk from dairy cows. This myth may be attributed to their proclivity for hunting rodents in barns and old fields. Recent analyses of nuclear and mitochondrial DNA suggests that the eight subspecies of *L. triangulum* do not properly reflect evolutionary relationships; hence the taxonomy of this group probably will be reorganized in the near future. However, we use the current accepted taxonomy and consider both Eastern and Red Milksnakes to be subspecies of *L. triangulum*. A further complication is the propensity for members of this group to interbreed, as intergrades have been found in several areas of Tennessee.

R. Graham Reynolds and
Matthew L. Niemiller

Eastern Coachwhip adult, Osceola County, Florida. (Photo by Ken Wray)

Coachwhip

Masticophis flagellum

Description: Coachwhips are long and slender terrestrial snakes. Adults are 100–150 cm (39–59 in) TL but can reach at least 260 cm (102 in). The body scales are smooth and the anal plate is divided. The body coloration of adults is highly variable. The dorsum is reddish to dark brown or black anteriorly, and transitions to tan toward the tail. The sides of the tail often have a reddish tint. The ventral coloration also varies down the length of an individual snake, generally in correspondence with the change in the dorsal coloration. The dorsal scales are dark but bordered by a distinct thin light margin, at least in scales anterior to the tail. The light border of the scales breaks up the dark coloration and makes the snakes look dusty or dirty. On the tail, the anterior half of a scale is often white and the posterior half dark. This scale color pattern, coupled with the organization of scales, creates

Left: Eastern Coachwhip juvenile, Liberty County, Florida. (Photo by Ken Wray) *Right:* Eastern Coachwhip adult, Benton County, Missouri. (Photo by Brad M. Glorioso)

a distinct color pattern such that the tail resembles a braided whip (hence the common name). The eyes are large, as is the head. Hatchlings are 30–44 cm (12–17 in) TL. Hatchling and juvenile Coachwhips have 50 or more narrow, irregular, dark transverse lines on the dorsum that fade or become less distinct toward the tail. The background dorsal coloration is typically pale brown or tan. Two conspicuous rows of dark spots, which also fade toward the tail, are present on the pale brown venter. The juvenile patterning fades each time the snake sheds; therefore, the juvenile phenotype gradually transitions into the adult phenotype. Six subspecies are recognized (but see **Comments**), and only the Eastern Coachwhip (*M. f. flagellum*) occurs in Tennessee.

Etymology: The specific and subspecific epithet, *flagellum* [flah-JEL-lum], is Latin for "whip" in reference to the long, slender body, dark head (the whip handle) and the braided appearance of the tail.

Similar Species: Coachwhips are most similar in appearance (and behavior) to Racers (*Coluber constrictor*), as both are dark, slender snakes with large eyes and are fast moving. However, adult Racers in Tennessee have a blue-black dorsum from head to tail; in comparison,

adult Coachwhips have a dorsum that is not monochromatic. Even melanistic Coachwhips usually have a reddish tint along the sides of the tail that is lacking in Racers. The number of dorsal scale rows immediately in front of the cloaca can also be used to distinguish these two species, as Coachwhips have 13 dorsal scale rows and Racers have 15 dorsal scale rows. Juveniles of both species also are similar, but Coachwhips have a braided rope-like appearance with many transverse, irregular lines, whereas Racers are blotched in pattern.

Distribution: Coachwhips occur from southern North Carolina south to Florida then west into southern Missouri, eastern Oklahoma, Arkansas, and eastern Texas. However, Coachwhips are absent from much of the lower Mississippi River Valley. In Tennessee, Eastern Coachwhips range into the extreme southwestern part of the state, with individuals documented from Hardeman, McNairy, and Shelby counties. Most records are from the Southeastern Plains and Hills of McNairy County.

Habitat: Coachwhips occur in a variety of habitats, including old fields, pine forests, upland forests, scrub, agricultural lands, and even power-line cuts. These habitats typically are dry and open with sandy soil. Coachwhips are active during the day and are known to climb into scrub and trees. Adults and juveniles can be seen actively basking, foraging, and moving in and around such habitats or under cover before retreating underground in animal burrows, old root cavities, or beneath rocks and logs at night and during colder weather. They, along with Racers, will often "periscope" to examine their surroundings, poking their heads out of tall grass that conceals the body.

Natural History: Little is known about the natural history of Coachwhips in Tennessee. In adjacent states, Coachwhips are seasonally active from late March into November. Individuals often are observed basking on the ground or on rocks and logs in direct sunlight. Mating probably occurs from April into May. Females lay 4–24 eggs in nests that they construct in loose soil, mammal burrows, or in rotting logs and stumps. Eggs have distinctive nodules on the outer surface that resemble salt grains. Eggs typically hatch in 6–10 weeks. Coachwhips are active predators, consuming a variety of prey, including grasshoppers, cicada nymphs, lizards, other snakes, birds, small turtles, and small rodents. Lizards consumed include Six-Lined Racerunners (*Aspidoscelis sexlineata*), Broad-Headed Skinks (*Plestiodon laticeps*), Eastern Fence Lizards (*Sceloporus undulatus*), and Little Brown Skinks (*Scincella lateralis*). Coachwhips do not use constriction to subdue their prey. Predators include birds of prey and predatory mammals, such as coyotes. Coachwhips are

extremely fast and agile snakes, often making hand capture difficult. When threatened, an individual will raise the anterior third of its body off the ground in a defensive posture, and is then quick to strike or bite. Coachwhips are prone to bite when captured and will whip their bodies around forcefully in an attempt to escape.

Conservation Status: Little is known about the status of Tennessee populations, and individuals are extremely difficult to find. In other states, Coachwhips are frequently encountered and are not thought to be of conservation concern. Loss of habitat likely is the leading threat facing Tennessee populations. Additionally, because Coachwhips are long and active snakes, many snakes are killed crossing roads by motor vehicles.

Comments: Coachwhips are sometimes referred to as "whipsnakes." The taxonomy of Coachwhips has been the subject of debate in recent years. Coachwhips and Racers (genus *Coluber*) are closely related, and some evidence suggests that the genus *Masticophis* is paraphyletic with respect to, and should therefore be synonymized with, *Coluber*. Accordingly, some authors refer to Coachwhips as *Coluber flagellum*. However, other molecular evidence suggests this might not be the case. Until a more thorough analysis of the relationships between *Coluber* and *Masticophis* is conducted, we refrain from implementing the new taxonomy.

Matthew L. Niemiller and
R. Graham Reynolds

Mississippi Green Watersnake adult, Lake County. (Photo by Greg Sievert)

Mississippi Green Watersnake

Nerodia cyclopion

Description: Mississippi Green Watersnakes are large, robust, aquatic snakes. Most adults are 70–110 cm (27.5–43 in) TL, but individuals can grow to at least 130 cm (51 in). Body scales are keeled and the anal plate is divided. The dorsum is olive brown to olive to greenish gray with around 50 narrow, darker transverse bars, which are often are indistinct. The venter is dark brown to gray with numerous white to yellow irregular, semicircular (half-moon) spots. Darker pigmentation becomes more prominent toward the tail. Two rows of scales are present on the undersurface of the tail. The head is relatively large and the eyes are noticeably high on the head. Unlike other species of *Nerodia* in Tennessee, Mississippi Green Watersnakes have a pair of subocular scales between the eye and upper labial scales. Females typically are larger than males. Young are 23–28 cm (9–11 in) TL at birth. Young and juveniles resemble adults, but the darker transverse bars are more prominent. No subspecies are recognized.

Etymology: The specific epithet *cyclopion* [sigh-KLOH-pee-on] is from Greek *kycklos,* meaning "circle," *ops,* meaning "eye," and *ion,* meaning

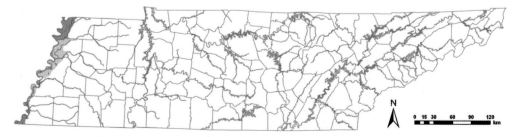

Left: Head of a Mississippi Green Watersnake, Lake County. (Photo by Collin Jaeger) *Right:* Venter of a Mississippi Green Watersnake, Lake County. (Photo by Collin Jaeger)

to "come or go." This is in reference to the sub-ocular scales, which surround the lower part of the eye and are absent in other species of *Nerodia.*

Similar Species: Mississippi Green Watersnakes most closely resemble other watersnakes and Cottonmouths (*Agkistrodon piscivorus*). Distinguishing between the watersnake species typically relies on dorsal patterning and ventral coloration. Mississippi Green Watersnakes generally lack the distinctive dorsal patterns (stripes, spots, or blotches) observed in other species of *Nerodia.* Likewise, Mississippi Green Watersnakes possess two subocular scales between the eye and labial scales that are absent in the other species. Plain-Bellied Watersnakes (*N. erythrogaster*) also lack distinctive dorsal patterning as adults but have immaculate (un-patterned) venters. As juveniles, Plain-Bellied Watersnakes have distinct dark crossbands that are complete nearest the head, but these crossbands break up into alternating dorsal and lateral blotches as individuals age. Both Broad-Banded Watersnakes (*N. fasciata confluens*) and Northern Watersnakes (*N. sipedon*) also have a banded appearance but have distinct spots on the ventral scales. Diamond-Backed Watersnakes (*N. rhombifer*) have a chainlike dorsal pattern that forms lighter diamond-shaped blotches, and the venter has irregularly spaced, dark half moons or spots. Cottonmouths inhabit similar habitats and resemble several species of *Nerodia,* including *N. cyclopion;* however, Cottonmouths have a single row of scales on the underside of the tail, vertically elliptical pupils, and loreal (heat-sensory) pits on the head.

Distribution: Mississippi Green Watersnakes range throughout the Mississippi River Valley from southern Illinois southward into Louisiana and along the Gulf Coast from the western panhandle of Florida to eastern Texas. In Tennessee, this species is confined to the Northern Mississippi Alluvial Plain, where it is known from just three counties: Lake and Obion counties around Reelfoot Lake and along the Mississippi River in Shelby County.

Habitat: Mississippi Green Watersnakes have been observed in a number of habitats throughout their range, including cypress swamps, oxbows, sloughs, ditches, bayous, lakes, ponds, and sluggish streams. They are even occasionally found in brackish water along the Gulf Coast. Individuals are often observed basking on logs and in vegetation during the day and seen active in the water at night, especially during summer. During colder weather or when inactive, Mississippi Green Watersnakes will retreat underground near water under logs and other debris, beneath root masses, and in animal burrows. It has been noted that this species is particularly active during and after rainfall.

Natural History: Little is known about the natural history of Tennessee populations. Mississippi Green Watersnakes likely are active on the surface from March into October. Mating occurs in April and May on land or in water. Females give birth to 6–18 (but up to 37) live young in late summer. Litter size is directly related to female body size. Sexual maturity likely is reached in 3–4 years. Mississippi Green Wa-

tersnakes are known to eat crayfish, aquatic salamanders, tadpoles, frogs, and several species of fishes, such as mosquitofish, minnows, gizzard shad, and sunfishes. Unlike Diamond-Backed Watersnakes, catfishes do not appear to be an important part of the diet. Prey is seized and swallowed alive. Predators likely include wading birds, American Alligators, otters, raccoons, and other snakes, such as Cottonmouths and Kingsnakes. Like other watersnakes, this species will bite aggressively when captured and release a pungent musk from anal glands.

Conservation Status: Because of its limited distribution and few occurrences, Mississippi Green Watersnakes are considered very rare and imperiled in Tennessee. Accordingly, TWRA lists this species as "Deemed in Need of Management." However, the status of Tennessee populations is largely unknown and warrants study. Presumed threats include loss of habitat and road mortality. This species is listed as "Endangered" in Missouri.

Matthew L. Niemiller and
R. Graham Reynolds

Yellow-Bellied Watersnake adult, Hardeman County. (Photo by Matthew L. Niemiller)

Plain-Bellied Watersnake

Nerodia erythrogaster

Description: Plain-Bellied Watersnakes are medium- to large-sized aquatic snakes with adults 75–120 cm (30–47 in) TL but reaching 150 cm (59 in). Body scales are heavily keeled and the anal plate is divided. Dorsal coloration of adults is typically a uniform greenish gray to brownish gray to black, although there may be evidence of the juvenile patterning in some in-

dividuals. The venter is unpatterned and ranges from yellow to red in coloration but the edges of the ventral scales may exhibit an incursion of dorsal pigmentation. The head is distinctly larger than the neck. The labial scales are colored like the venter but have darker pigment at the edges that form dark vertical bars. Two rows of scales are present on the undersurface of the tail. Females are generally much larger and more robust than males. Young are 20–30 cm (8–12 in) TL at birth. Young and juveniles

Top left: Yellow-Bellied Watersnake juvenile, Lake County. (Photo by Matthew L. Niemiller) *Top right:* Venter of a Yellow-Bellied Watersnake adult, Hardeman County. (Photo by Matthew L. Niemiller) *Bottom right:* Copper-Bellied Watersnake adult, Cheatham County. (Photo by Lisa Powers)

are boldly patterned with dark brown to black dorsal blotches or bands that begin as complete crossbands near the head but transition to alternating blotches on the dorsum and sides toward the tail. In the past, four subspecies were recognized, including two in Tennessee: the Yellow-Bellied Watersnake (*N. e. flavigaster*) and Copper-Bellied Watersnake (*N. e. neglecta*). However, current research indicates that previously recognized subspecies are not genetically distinct (see **Comments**).

Etymology: The specific epithet, *erythrogaster* [eh-WRITH-roh-gas-ter], is from the Greek *erythros* (red) and *gaster* (belly).

Similar Species: Plain-Bellied Watersnakes are confused with the venomous Western Cottonmouth (*Agkistrodon piscivorus leucostoma*) and hence are often killed on sight. Cottonmouths are generally stockier, have paired loreal pits between the nostrils and the eyes, have vertically elliptical pupils, and have a single row of scales on the underside of the tail. Cottonmouths will gape when threatened,

a behavior Plain-Bellied Watersnakes usually never exhibit. Plain-Bellied Watersnakes also resemble other watersnakes in the genus *Nerodia*. Distinguishing between these species typically relies on dorsal patterning and ventral coloration. Diamond-Backed Watersnakes (*N. rhombifer*) have a chainlike dorsal pattern that forms lighter diamond-shaped blotches and the venter has irregularly spaced, dark half moons or spots. Mississippi Green Watersnakes (*N. cyclopion*) lack a distinctive dorsal pattern and possess two subocular scales between the eye and labial scales that are absent in other watersnakes. Both Southern Watersnakes (*N. fasciata*) and Northern Watersnakes (*N. sipedon*) have incomplete crossbands down the entire dorsum and have markings on the venter. Young and juvenile Plain-Bellied Watersnakes may resemble these two species but have dorsal blotches that are mostly offset from the lateral blotches and have immaculate (unpatterned) venters.

Distribution: Plain-Bellied Watersnakes range throughout much of the southeastern United States with the core of the distribution from southern Kansas, Oklahoma, and Texas eastward throughout much of the Mississippi River Valley as far north as Iowa, the Gulf Coast Plain, and the Atlantic Coastal Plain into Virginia. In Tennessee, this species occurs throughout the Coastal Plain ecoregions of West Tennessee but also along the lower Cumberland River in Middle Tennessee. An isolated population exists in the Barrens region of the Eastern Highland Rim in Coffee County. Populations in West Tennessee have been identified in the past as Yellow-Bellied Watersnakes, while those along the Cumberland River have been identified as Copper-Bellied Watersnakes.

Habitat: Plain-Bellied Watersnakes are found in a variety of habitats, including rivers, streams, swamps, sloughs, ponds, ditches, lakes, and reservoirs. Individuals are most common along the edges of such habitats with heavy vegetation. This species also can found in more ephemeral wetlands in mesic forests. Plain-Bellied Watersnakes are more terrestrial than most other watersnake species and can be found crossing roads, particularly at night during rainy weather in summer. Adults are often observed basking in the early morning during the warmer months in low vegetation or on logs in or overhanging water.

Natural History: Plain-Bellied Watersnakes are active from March through October and begin breeding soon after emergence from hibernation. Many males may court a single female at the same time. Females give birth to 8–25 (up to 55) live young from late June into early September. Litter size is positively correlated with female body size. Sexual maturity probably is reached in 2–3 years. Plain-Bellied Watersnakes are active hunters and feed mainly on amphibians; however, they will also eat several species of fish, crayfish, hatchling turtles, small mammals, and invertebrates, including earthworms, leeches, and aquatic insects. These snakes do not constrict but simply swallow prey alive. Predators include hawks, wading birds, raccoons, river otters, large fish (such as largemouth bass), and other snakes, such as Eastern Black Kingsnakes (*Lampropeltis nigra*) and Cottonmouths (*Agkistrodon piscivorus*). Plain-Bellied Watersnakes are wary when basking and will quickly drop or slide into water when disturbed. However, this species also is known to try and flee on land. When threatened, they will often flatten their jaws and bodies, strike viciously, and release a foul-smelling musk. Although harmless and generally painless, their bite can bleed profusely, particularly one inflicted by a larger individual.

Conservation Status: Populations of Plain-Bellied Watersnakes in Tennessee appear stable, as this species can be quite common in some habitats in West Tennessee. However, Copper-Bellied Watersnakes (*N. e. neglecta*) are tracked by TWRA and protected in the state (but see **Comments**). Habitat destruction and degradation, agricultural runoff, pollution, road mortality, and persecution by humans because of similarity in appearance to venomous Cottonmouths are significant threats to this species.

Comments: A recent molecular study (Makowsky et al. 2010) based on mitochondrial DNA found little support for currently recognized subspecies of Plain-Bellied Watersnakes across the entire range of the species, including both subspecies reported from Tennessee. Consequently, the most recent checklist of North American reptiles no longer recognizes any subspecies (Crother et al. 2011), and we follow this taxonomy. However, it is worth noting that a morphological examination of Plain-Bellied Watersnakes in Tennessee generally supported recognition of two subspecies in our state (Bufalino and Scott 2008).

Lisa Powers and
Matthew L. Niemiller

Southern Watersnake adult, Lake County. (Photo by Brad M. Gloroso)

Southern Watersnake

Nerodia fasciata

Description: Southern Watersnakes are medium-sized aquatic snakes. Most adults are 56–90 cm (22–35 in) TL but can reach lengths of at least 114 cm (45 in). Body scales are heavily keeled, especially along the middle scale rows of the dorsum, and the anal plate is divided. Dorsal coloration is variable. Most adults in Tennessee have 11–17 irregularly shaped dark dorsal blotches or crossbands. These blotches are dark, usually brown, reddish brown or even black, and they are separated from each other by a ground color of yellow to brown to gray. A dark stripe extends from the eye to the corner of the mouth. The venter is cream to yellow and has prominent squarish, dark reddish brown to black markings that often are clustered. Two rows of scales are present on the undersurface of the tail. Females are generally much larger and more robust than males. Young, 16–19 cm (6–8 in) TL at birth, are typically more brightly colored and patterned than adults. *Nerodia fasciata* is a complex of three subspecies known collectively as Southern Watersnakes, but only the Broad-Banded Watersnake (*N. f. confluens*) occurs in Tennessee.

Etymology: The specific epithet, *fasciata* [fas-see-AH-tah], is Latin for "banded" and refers to the dorsal pattern of this species. The subspecific epithet, *confluens* [kon-FLUE-ens], is from the Latin *con* (together) and *fluens* (flowing), referring to the description of this subspecies from an area where the St. Francis and the Black rivers flow parallel to each other in Butler County, Missouri.

Similar Species: Tragically, Southern Watersnakes are easily confused with the venomous Western Cottonmouth (*Agkistrodon piscivorus leucostoma*) and hence are often killed on sight. Cottonmouths are generally stockier, and have paired loreal pits between the nostrils and the eyes, vertically elliptical pupils, and a single row of scales on the underside of the tail. Cottonmouths will also gape when threatened, a behavior watersnakes typically never exhibit. Southern Watersnakes also resemble other watersnakes in the genus *Nerodia*. Distinguishing between these species typically relies on differences in dorsal patterning and ventral coloration. Diamond-Backed Watersnakes (*N. rhombifer*) have a chain-like dorsal pattern that forms lighter diamond-shaped blotches, and their venter has irregularly spaced, dark half moons or spots. If complete dark bands are present in Northern Watersnakes (*N. sipedon*), they do not extend down the entire length of the dorsum, and the venter of Northern Watersnakes often has half-moon-shaped markings. Juvenile Plain-Bellied Watersnakes (*N. erythrogaster*) have dorsal blotches that are mostly offset from the lateral blotches, and they have immaculate (unpatterned) venters. Mississippi Green Watersnakes (*N. cyclopion*) lack a distinctive dorsal pattern and are unique in possessing two subocular scales between the eye and labial scales.

Distribution: Southern Watersnakes occur throughout much of the Atlantic and Gulf Coastal Plains in the southeastern United States. Broad-Banded Watersnakes range from extreme southeastern Illinois south along the Mississippi River to the Gulf of Mexico and into central Texas. In Tennessee this species is largely restricted to the counties immediately bordering the Mississippi River. Additionally a record exists along the headwaters of the Wolf River in Fayette County. This species is most common in and around Reelfoot Lake in Lake and Obion counties.

Habitat: Southern Watersnakes are frequently found basking on exposed vegetation or logs, either in or above the water, along the edges of cypress swamps, sloughs, ditches, bayous, and lakes. They are generally active during most of the warmer months, from March to October, and are particularly easy to see in the morning when basking. Indeed, we have sat

Left: Head of a Southern Watersnake, Lake County. (Photo by Brad. M. Glorioso) *Right:* Venter of a Southern Watersnake, St. Martin Parish, Louisiana. (Photo by Matthew L. Niemiller)

and observed dozens of individuals crawl from the water onto exposed vegetation in a small area of cypress swamp early in the morning on Reelfoot Lake. This species is active both at night and day during summer. A warm rain will bring snakes out searching for food, and they are often seen crossing roads near water. This species retreats underneath logs or in crayfish burrows during cold weather.

Natural History: Southern Watersnakes mate in April and May, and females produce a litter of 7–40 (but up to 80) live young in late summer. Mating consists of a mating ball of several males surrounding a female. Litter size is positively correlated with female body size. Southern Watersnakes are generalist predators and prey on fish, crayfish, and larval and adult amphibians. Predators include large fish, wading birds, other snakes, and predatory mammals. Southern Watersnakes generally have an irascible disposition if captured and will bite and smear a very foul musk. They also exhibit a threat display that includes flattening the head to resemble a venomous Cottonmouth. A bite will draw blood but is usually painless.

Conservation Status: The status of Tennessee populations of Broad-Banded Watersnakes has not been assessed. Although this species is locally common, it has a limited distribution in our state and should probably be monitored. Threats to Broad-Banded Watersnakes include habitat loss and degradation, heavy persecution, and road mortality.

Comments: Southern Watersnakes are sometimes called "yellow moccasins" or "pink flamingo snakes" in reference to the yellow or red coloration on the dorsum.

R. Graham Reynolds and
Matthew L. Niemiller

Diamond-Backed Watersnake adult, Lake County. (Photo by Brad M. Glorioso)

Diamond-Backed Watersnake

Nerodia rhombifer

Description: Diamond-Backed Watersnakes are large, robust, aquatic snakes. Most adults are 70–120 cm (27.5–47 in) TL but are known to reach lengths of at least 160 cm (63 in). Body scales are keeled and the anal plate is divided. The dorsum has a yellowish brown, greenish brown, pale brown, or gray ground color overlain with a darker brown chainlike pattern, which forms lighter diamond-shaped blotches down the length of the body. The darker dorsal blotches are connected by thin, oblique lines, which also connect with the dark dorsal blotches to lateral blotches. The venter is cream to yellow, with dark brown to black half moons or spots on each of the ventral scales. Two rows of scales are present on the undersurface of the tail. The head is distinctly larger than the neck. Unlike other snakes in Tennessee, male Diamond-Backed Watersnakes have numerous protuberances under the chin. Young are 20–33 cm (8–13 in) TL at birth. Young and juveniles

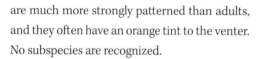

Top left: Diamond-Backed Watersnake basking, Lake County. (Photo by Collin Jaeger) *Top right:* A mating ball of Diamond-Backed Watersnakes in Reelfoot Lake, Lake County. (Photo by Brad M. Glorioso) *Bottom right:* Protuberances on the chin of a male Diamond-Backed Watersnake, Johnson County, Illinois. (Photo by Andrew Durso)

are much more strongly patterned than adults, and they often have an orange tint to the venter. No subspecies are recognized.

Etymology: The specific epithet, *rhombifer* [rom-BEH-fur], is from the Greek *rhombos* (rectangle) and *ifera* (bearing), referring to the blotches on the dorsum.

Similar Species: Diamond-Backed Watersnakes most closely resemble other watersnakes and Cottonmouths (*Agkistrodon piscivorus*). Distinguishing between these species typically relies on differences in dorsal pattern and in ventral coloration. Broad-Banded Watersnakes (*N. fasciata confluens*) and Northern Watersnakes (*N. sipedon*) have distinct blotches on the dorsum, but these typically are much broader and not connected by thin oblique lines to lateral blotches. Additionally, both of these species have markings on the ventral scales, but these half moons or spots typically contain some reddish pigment. Juvenile Plain-Bellied

Watersnakes (*N. erythrogaster*) also have a distinct dorsal pattern but lack conspicuous spots or other markings on the ventral scales. Mississippi Green Watersnakes (*N. cyclopion*) lack a distinctive dorsal pattern but are unique in possessing two subocular scales. Cottonmouths inhabit similar habitats and resemble several species of *Nerodia,* but they have a single row of scales on the underside of the tail, vertically elliptical pupils, and loreal (heat-sensory) pits on the head.

Distribution: Diamond-Backed Watersnakes occur throughout the Mississippi River Valley, the Gulf Coast from Alabama into Texas and Mexico, and from Missouri and eastern Kansas south into Oklahoma and eastern Texas. In Tennessee, this species occurs largely in the western part of the state, west of the north-flowing Tennessee River in the Coastal Plain ecoregions. However, populations exist also along the lower Tennessee and Cumberland rivers in the Western Highland Rim and Western Pennyroyal Karst. This species is particularly common around Reelfoot Lake in Lake and Obion counties.

Habitat: Diamond-Backed Watersnakes can found in any body water throughout their range, including rivers, streams, ditches, sloughs, ponds, lakes, oxbows, and reservoirs. This species is often seen around spillway areas of lakes. Individuals are often observed basking on logs and in vegetation during the day and active in the water at night, especially during summer. During colder weather or when inactive, Diamond-Backed Watersnakes retreat to underground hiding places near water, such as under logs and other debris, beneath root masses, and in animal burrows.

Natural History: Diamond-Backed Watersnakes become active in late March but sometimes are forced from their hibernation sites during flood conditions. This species is largely nocturnal during warm weather. Mating occurs during April and May, either on land or in water. Males will follow scent trails produced by females, and many times a single female will be surrounded by several males at one time in a mating ball. Depending on when breeding occurs, females give birth as early as August and as late as September. Litter size is directly related to female body size. A litter typically consists of 15–30 live young, but litters of up to 50 young have been reported. Diamond-Backed Watersnakes consume a diversity of aquatic prey, including frogs, toads, tadpoles, and several species of fishes, such as catfish, sunfish, eels, and mosquitofish. Adults predominantly feed on catfishes and bullheads. When feeding underwater, Diamond-Backed Watersnakes often will move their heads from side to side with mouth agape, snapping at anything they happen to touch. Individuals also extend their bodies down from tree limbs overhanging water to catch fish. Predators include American Alligators, Cottonmouths, large catfishes, and wading birds. Individuals bite aggressively when captured and release a pungent musk from anal glands.

Conservation Status: Diamond-Backed Watersnakes are common and occur in large numbers where they are found in Tennessee.

Matthew L. Niemiller and
R. Graham Reynolds

Northern Watersnake adult, Polk County. (Photo by Matthew L. Niemiller)

Northern Watersnake

Nerodia sipedon

Description: Northern Watersnakes are medium-sized, thick-bodied, semiaquatic snakes. Adults are 55–110 cm (22–43 in) TL but can reach lengths of at least 150 cm (59 in). The body scales are heavily keeled, especially along the middle of the back, and the anal plate is divided. Dorsal coloration is highly variable. Most adults have irregularly shaped dark blotches on the dorsum. The dorsal blotches often resemble transverse bands near the head because they are narrow and connected to lateral blotches on each side of the body. The connection between dorsal and lateral blotches does not occur posteriorly much behind the neck; rather, more posteriorly located dorsal blotches are

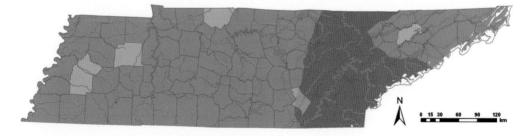

Two subspecies of Northern Watersnakes are recognized in Tennessee: the Northern Watersnake (*Nerodia sipedon sipedon*), whose range appears in blue, and the Midland Watersnake (*N. s. pleuralis*), whose range appears in orange. A contact zone between the subspecies occurs in East Tennessee, shown in purple.

offset from the lateral blotches. Regardless of shape or location, the blotches are various shades of brown, reddish brown, gray, or black, and they usually have a thin black border. The ground color separating blotches is variable shades of tan, brown, or gray. In some individuals the blotched pattern becomes obscured with age, either because the snakes become melanistic or for some other genetic reason. Note that the pattern often is obscured because the snake is covered in a thin coat of mud, which can be washed off easily. The venter is usually cream or yellow, and prominent red to brown half-moon markings are present on each ventral scale. These markings may be arranged in a row centered along the midline of the venter, scattered at random, appear as undefined dusky smudges, or be completely absent. Two rows of scales are present on the undersurface of the tail. Females grow to longer lengths and are more robust than males. Young are blotched and otherwise patterned as the adults, but their colors are brighter and more vivid, particularly in the neonates. Northern Watersnakes are ovoviparous, and the neonates are 19–27 cm (7.5–11 in) TL at birth. Four subspecies are recognized, and two subspecies occur in Tennessee: the Northern Watersnake (*N. s. sipedon*) and the Midland Watersnake (*N. s. pleuralis*). Northern Watersnakes usually have 30 or more dorsal bands or blotches from the neck to the anus that are separated by one scale row or less, whereas Midland Watersnakes usually have 30 or fewer dorsal bands or blotches from the neck to the anus that are separated by more than one scale row. The ventral markings typically are arranged in pairs in Midland Watersnakes and do not break up toward the tail, as they typically do in Northern Watersnakes. However, these two subspecies are known to hybridize in Tennessee; therefore, a combination, or intermediates, of the scalation and color patterns described for each subspecies is possible.

Etymology: The specific and subspecific epithet, *sipedon* [SIP-peh-don], is from the Greek *sepedon* (rottenness or decay) and probably refers to the mistaken belief that this snake has a venomous bite. The subspecific epithet, *pleuralis* [pleh-ur-RAL-es], is from the Greek *pleura* (side) and *allis* (of the), in reference to the often-distinct blotches present on the sides of individuals of this subspecies.

Similar Species: Northern Watersnakes tragically are easily confused with the venomous Western Cottonmouth (*Agkistrodon piscivorus leucostoma*) and hence often killed on sight. Cottonmouths are generally stockier, have paired loreal pits between the nostrils and the eyes, have vertically elliptical pupils, and have a single row of scales on the underside of the tail. Cottonmouths will also gape when threatened, a behavior watersnakes usually never exhibit. In addition, Cottonmouths occur only in western and parts of Middle Tennessee. Northern Watersnakes also resemble other watersnakes in the genus *Nerodia*. Distinguishing between these species typically relies on examination of dorsal patterning and ventral coloration. Diamond-Backed Watersnakes (*N. rhombifer*) have a chainlike dorsal pattern that forms lighter diamond-shaped blotches, and their venter has irregularly spaced, dark half moons or spots. Southern Watersnakes (*N. fasciata*) have complete crossbands down the entire dorsum, and they tend to have more square-shaped markings on their venter. Additionally, Southern Watersnakes have a dark stripe from the eye to the corner of the jaw. Juvenile Plain-Bellied

Top left: Venter of a Northern Watersnake adult, Polk County. (Photo by Matthew L. Niemiller) Top right: Northern Watersnake juvenile, Sevier County. (Photo by Matthew L. Niemiller) *Bottom right:* Midland Watersnake in defensive posture, Rutherford County. (Photo by Brad M. Glorioso)

Watersnakes (*N. erythrogaster*) have dorsal blotches that are mostly offset from the lateral blotches, and they have immaculate (unpatterned) venters. Mississippi Green Watersnakes (*N. cyclopion*) lack a distinctive dorsal pattern, and they possess two subocular scales between the eye and labial scales that are absent in other watersnakes.

Distribution: Northern Watersnakes occur throughout much of the eastern United States and Canada from southern Ontario and Quebec in Canada, the Great Lakes, and southern New England south into Mississippi, Alabama, and Georgia. In Tennessee, Northern Watersnakes are found nearly statewide, absent in appropriate habitat only from the highest elevations of the Blue Ridge Mountains. Northern Watersnakes occur in the Ridge and Valley and Blue Ridge Mountains, whereas Midland Watersnakes occur throughout the rest of the state. However, an intergrade zone of these two subspecies exists where their ranges contact in eastern Tennessee.

Habitat: Northern Watersnakes inhabit nearly any aquatic habitat type, including lakes, reservoirs, ponds, streams, rivers, bayous, swamps, ephemeral wetlands, and bogs. This species also is commonly found in urban areas around people. In fact, they can be abundant in streams passing through parks and greenways, and have been known to establish residence in garden ponds in suburban yards. They bask on exposed vegetation, logs, or rocks either in or above the water as high as 3 m (9.8 ft) above the water. They are generally active during most of the warmer months, from March to October, and are particularly easy to observe in the morning when basking. Juveniles and neonates can be quite common in smaller creeks and streams. This species is active both at night and during the day during summer. A warm rain will bring snakes out searching for food and they are often seen crossing roads near water. Individuals seek refuge during cold weather beneath logs, in root masses, or underground in rock rubble, crayfish, or mammal burrows. Often these refuges are at or near the water level.

Natural History: The natural history of Northern Watersnakes has been well studied, although Tennessee populations have not received as much attention as populations in neighboring states. Northern Watersnakes mate from late spring into early summer (April–June). Females may mate with several males before giving birth to 5–60 (but up to 99) live young in late summer or early autumn. Most females breed every year. Litter size is positively correlated with female body size. Sexual maturity is reached in 2–3 years. Northern Watersnakes are generalist predators and will consume nearly any prey they can swallow. Fishes and amphibians constitute the bulk of the diet, with more than 80 species of fish and 30 species of amphibians having been documented as prey. Northern Watersnakes also will eat small mammals, crayfish, earthworms, mollusks, snails, slugs, and aquatic insects. Predators include Eastern Black Kingsnakes (*Lampropeltis nigra*), Cottonmouths (*Agkistrodon piscivorus*), Snapping Turtles (*Chelydra serpentina*), largemouth bass, raccoons, and various predatory birds and mammals. Northern Watersnakes are typically ill-tempered when captured; they will repeatedly attempt to bite, and they exude a foul-smelling musk from their anal glands onto the captor. Although larger individuals can draw blood with their recurved teeth adapted to holding onto slippery prey, the bite is harmless and usually painless.

Conservation Status: Northern Watersnakes are one of the most abundant and thus one of the most commonly encountered snakes in Tennessee. However, people who mistake this species for venomous Cottonmouths unjustly kill many individuals each year. Many people in eastern Tennessee mistakenly believe that Cottonmouths occur in this region because they confuse them with Northern Watersnakes, and they will often kill or run from any watersnake they encounter. Likewise, Northern Watersnakes will travel overland between aquatic habitats and often are killed crossing roads.

Comments: Northern Watersnakes are sometimes called "water adders" or, more frequently, "water moccasins" in Tennessee because they are mistaken for venomous Cottonmouths, particularly in eastern Tennessee.

R. Graham Reynolds and
Matthew L. Niemiller

Rough Greensnake adult, Lake County. (Photo by Matthew L. Niemiller)

Rough Greensnake

Opheodrys aestivus

Description: Rough Greensnakes are medium-sized, very slender snakes that appear vine-like, with adults 55–80 cm (22–32 in) TL but occasionally reaching as long as 116 cm (46 in). Body scales are weakly keeled and the anal plate is divided. The body is a darker to bright lime green on the dorsum and uniformly white, yellow, or greenish yellow on the venter. The ventral coloration also extends to the labials and the lower sides of the neck. The bright green coloration quickly transitions to a dull blue after death. The eyes are large and the head is fairly distinct from the slender neck, while the tail tapers to a very thin tip. Females are larger than males, but males have longer tails at maturity. Hatchlings are 17–23 cm (7–9 in) TL and resemble adults but may be more grayish green. Two subspecies are recognized, but only the Northern Rough Greensnake (*O. a. aestivus*) occurs in Tennessee.

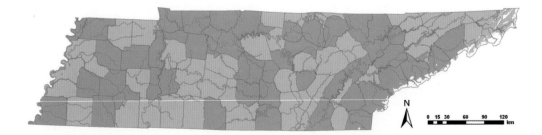

Left: Rough Greensnake adult, Coffee County. (Photo by Brad M. Glorioso) *Right:* Rough Greensnake juvenile, Lake County. (Photo by Matthew L. Niemiller)

Etymology: The specific and subspecific epithet, *aestivus* [ES-teh-vus], is Latin for "pertaining to summer."

Similar Species: There are no other snakes in Tennessee that are bright green in coloration like Rough Greensnakes. Dead Rough Greensnakes, often found on roads, turn a dull blue and might be confused with small Racers (*Coluber constrictor*). However, Racers have smooth dorsal scales, whereas the dorsal scales of Rough Greensnakes are keeled.

Distribution: Rough Greensnakes are found throughout the southeastern United States except the higher elevations of the Appalachian Mountains. They occur along the Piedmont and Atlantic Coastal Plain as far north as Pennsylvania, Maryland, Delaware, and New Jersey and as far west as Kansas, Oklahoma, Arkansas, Texas, and northern Mexico. In Tennessee, Rough Greensnakes are likely found throughout most of the state, except for the higher elevations of the Blue Ridge Mountains. However, in northeast Tennessee, records for this species are sparse and the exact distribution is not well known.

Habitat: Rough Greensnakes are mostly found in association with riparian vegetation along streams, lakes, and other bodies of water. Their long slender bodies are adapted to an arboreal lifestyle and they sometimes imitate vegetation, such as vines and branches. Indeed, their well-camouflaged bodies can be quite difficult to see. It is also not uncommon to see them in or near vegetation along disturbed areas, such as in backyards and road edges, and they even occur in urban areas, such as the campus of the University of Tennessee. Rough Greensnakes are excellent climbers and can be found several meters above the ground or water, although they are more easily spotted when they move between trees on the ground or across a road. This species is also often located at night while resting in vegetation by shining a flashlight up into trees and vegetation around bodies of water and looking for the pale bellies of this snake.

Natural History: Rough Greensnakes are active during the day and do not emerge from hibernation until late March into April. This species is one of the first to go into hibernation underground or in stumps, usually in October. At night, Rough Greensnakes remain inactive usually well hidden within vegetation up off the ground. Mating occurs primarily during spring but will sometimes occur in autumn. During summer, females lay 3–12 eggs on the ground in concealed locations, such as old decaying logs or under rocks. More than one female might lay her eggs in the same location and females are also known to use the same nesting location in multiple years. Eggs hatch in 6–7 weeks. Sexually maturity is reached in 2–3 years. Rough Greensnakes feed mostly on insects, such as hairless caterpillars, longhorn grasshoppers, crickets, and dragonflies, but they will also take spiders, millipedes, land snails, and even the occasional young treefrog or lizard. Predators include Racers (*C. constrictor*), Eastern Black Kingsnakes (*Lampropeltis nigra*), domestic cats, blue jays, hawks, and other predatory birds. Broad-Headed Skinks (*Plestiodon laticeps*) and Racers are known to eat the eggs. Rough Greensnakes rarely bite when handled but may gape or expel a musky secretion from the vent when threatened.

Conservation Status: Rough Greensnakes are thought to be secure and are not listed as a species of conservation concern in Tennessee. However, pesticides, habitat destruction, habitat fragmentation, and road mortality are thought to be significant threats to this species.

Comments: Rough Greensnakes are sometimes called "keeled greensnakes" in reference to the keeled body scales and in contrast to the Smooth Greensnake (*O. vernalis*), which has smooth scales.

Lisa Powers and
Matthew L. Niemiller

Red Cornsnake adult, Knox County. (Photo by Matthew L. Niemiller)

Red Cornsnake

Pantherophis guttatus

Description: Red Cornsnakes are relatively large, slender snakes. Most adults are 75–140 cm (29.5–55.0 in) TL, but older individuals can reach lengths of at least 183 cm (6 ft). Most of the dorsal scales are smooth but may be weakly keeled near the dorsal midline. However, even the dorsal midline scales are smooth in many individuals, at least in Middle Tennessee. The anal plate is divided. In cross section, the bodies of Cornsnakes and other species of this genus and closely related genera, are shaped like a loaf of bread—the venter is flat, the sides are straight, and the top is rounded. The overall coloration of the dorsum is orange to red, but less colorful (e.g., orangish brown to reddish brown to brown) individuals are not uncommon. The dorsum (back) of all life stages (hatchlings, juveniles, and adults) is characterized by a series of rectangular spots or blotches that extends from

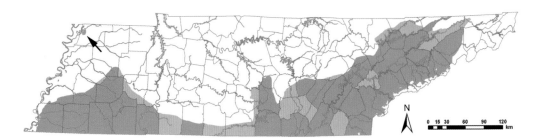

Top left: Head of a Red Cornsnake, Knox County. (Photo by Matthew L. Niemiller) *Top right:* Venter of a Red Cornsnake, Knox County. (Photo by Matthew L. Niemiller) *Bottom right:* Red Cornsnake adult, Cannon County. (Photo by Brian T. Miller)

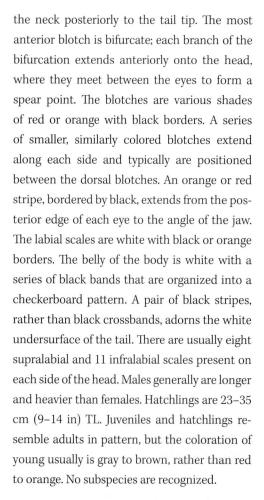

the neck posteriorly to the tail tip. The most anterior blotch is bifurcate; each branch of the bifurcation extends anteriorly onto the head, where they meet between the eyes to form a spear point. The blotches are various shades of red or orange with black borders. A series of smaller, similarly colored blotches extend along each side and typically are positioned between the dorsal blotches. An orange or red stripe, bordered by black, extends from the posterior edge of each eye to the angle of the jaw. The labial scales are white with black or orange borders. The belly of the body is white with a series of black bands that are organized into a checkerboard pattern. A pair of black stripes, rather than black crossbands, adorns the white undersurface of the tail. There are usually eight supralabial and 11 infralabial scales present on each side of the head. Males generally are longer and heavier than females. Hatchlings are 23–35 cm (9–14 in) TL. Juveniles and hatchlings resemble adults in pattern, but the coloration of young usually is gray to brown, rather than red to orange. No subspecies are recognized.

Etymology: The specific epithet *guttatus* [gew-TAY-tus] is from the Latin *gutta* for "spotted" and refers to the conspicuous blotched pattern characteristic of this species, particularly adult individuals (the blotched pattern becomes obscured with age in some closely related species).

Similar Species: Red Cornsnakes most closely resemble other large, blotched snakes, including Milksnakes (*Lampropeltis triangulum*), Scarletsnakes (*Cemophora coccinea*), Pinesnakes (*Pituophis melanoleucus*), Yellow-Bellied Kingsnakes (*L. calligaster*), several species of watersnakes in the genus *Nerodia,* and Gray Ratsnakes (*Pantherophis spiloides*). However, no other blotched species has the following combination of characters: loaf of bread cross-sectional shape of the body, spear-point marking on the head, and longitudinal stripe on the venter of the tail. Watersnakes have heavily keeled dorsal scales, Milksnakes, Yellow-Bellied Kingsnakes, and Scarletsnakes, which most closely resemble Cornsnakes in coloration, have round bodies in cross section and undivided anal plates. Gray Ratsnakes and Pinesnakes have neither spear-point markings on their head nor longitudinal stripes on the venter of their tail. Furthermore, Pinesnakes have an undivided anal plate.

Distribution: Red Cornsnakes are found throughout much of the southeastern United States, but populations occur west of the Mississippi River in eastern Texas, western Louisiana, and southern Arkansas. Small, isolated populations exist in Kentucky, and disjunct populations occur in Virginia, Maryland, and New Jersey. Records for this species in Tennessee are scattered and the exact distribution is poorly understood. Red Cornsnakes have been found through much of the Ridge and Valley and lower elevations of the Blue Ridge Mountains in East Tennessee, southern sections of the Eastern Highland Rim and Cumberland Plateau in Middle Tennessee, and southern sections of the Coastal Plain ecoregion in West Tennessee. This species can be locally common, but populations appear fragmented.

Habitat: Red Cornsnakes are associated with mesic forests and old fields. They are found along the edges of forests, old fields, power-line cuts, and agricultural fields. However, Red Cornsnakes also can be common in residential areas. They are often particularly fond of old buildings, which usually harbor healthy rodent populations. They are largely terrestrial during the spring mating season and often found actively searching for mates. During spring, Cornsnakes are also commonly found under old boards, sheets of tin, and other surface cover objects. They are arboreal, but less so than their congener the Gray Ratsnake, and spend a large amount of time underground in rodent burrows, rotting root tunnels, stump holes, and behind bark of dead trees.

Natural History: Red Cornsnakes emerge from hibernation during late winter or early spring and mate soon thereafter. Males actively seek females with use of pheromone trails. Red Cornsnakes have a ritualized courtship that involves the male quivering his body when on top of a female. He eventually wraps his tail around the females so that the vents are in close approximation to each other to allow for copulation. Females deposit white, elliptical and leathery eggs a month or two after breeding, typically during June or early July. Clutch size is variable, with reports ranging from as few as 3 to as many as 40 eggs; however, most clutches have 10–20 eggs. Larger females generally lay more eggs than smaller females, and some individuals lay two clutches during a breeding season. Eggs have been found in a variety of locations, including within rotting logs and tree stumps, in mulch and saw dust piles, beneath surface objects such as rocks, logs, and hay bales, and in rodent burrows. Hatching occurs in mid- to late summer (August through September), after an incubation period of about 1–3.5 months. Hatchlings and juveniles grow rapidly if well fed, and can breed during their second year of life. Captive individuals have lived for more than 20 years.

The diet consists largely of mammals and birds, but snakes, lizards, amphibians, and even insects are occasionally eaten. Small prey, such as young hairless rodents and featherless birds, are typically eaten alive; larger prey, prone to struggling, is subdued by constriction. Constriction typically involves the Red Cornsnake rapidly wrapping its body around the prey and slowly tightening the coils. This squeezing action disrupts heart function (extreme hypertension), which rapidly incapacitates the struggling prey. As with many species of large snakes, Red Cornsnakes often vibrate their tail rapidly when frightened, which may startle predators, including humans. When striking dried leaves, the vibrating tail sounds similar to the rattling tail of a rattlesnake! Although some individuals are feisty, strike when first encountered, and bite when handled, most individuals, at least in Middle Tennessee, are docile and rarely attempt to bite. Predators include various birds of prey, carnivorous mammals, and snakes (including other Red Cornsnakes).

Conservation Status: Red Cornsnakes are locally abundant. The greatest threats to this species include habitat destruction, fragmentation, and overharvest for the pet trade. I often find them crossing roads in forested regions interspersed with pastures in the Eastern Highland Rim and Outer Nashville Basin regions of Middle Tennessee; unfortunately, far too often they are killed by automobiles.

Comments: It is not legal to keep as pets most species of snakes that are native to Tennessee. However, Red Cornsnakes can be kept as pets if they are obtained in a legal manner (i.e., through pet dealers). The owner must keep receipts or other forms of documentation to verify that their snakes were not caught in the wild in Tennessee and should always strive to obtain responsibly captive-bred individuals.

Brian T. Miller

Gray Ratsnake adult, Wilson County. (Photo by Brad M. Glorioso)

Gray Ratsnake

Pantherophis spiloides

Description: Gray Ratsnakes are generally very large snakes. Most adults are 100–180 cm (39–71 in) TL, but individuals exceeding 210 cm (82.5 in) have been reported. However, encounters with larger individuals are becoming increasingly infrequent. Dorsal and lateral body scales are slightly keeled and the anal plate is divided. Dorsal coloration is variable, with adults either being mostly black to grayish black to grayish brown, or with some faint patterning of cream to white or yellowish coloration. The skin between the scales is cream colored and sometimes gives the illusion of a dorsal pattern, particularly in a snake that has recently eaten a large meal. The venter is checkered with either gray or black on an otherwise white background. The throat and labial scales are white, and the eyes are relatively large with a dark pupil. The body is shaped like a slice of commercial bread in cross-section: the top is rounded, the sides extend straight down, and the venter

Top left: Gray Ratsnake juvenile, Gibson County. (Photo by Matthew L. Niemiller) *Top right:* Venter of a juvenile Gray Ratsnake, Gibson County. (Photo by Matthew L. Niemiller) *Bottom right:* Gray Ratsnake adult, Blount County. (Photo by Matthew L. Niemiller)

is flat. A distinct angle is formed where the belly meets the sides, forming an edge that extends down the length of the body. Juveniles have a much more distinct pattern than adults, with a series of gray, brown, or black blotches that begin on the neck and continue to the tail tip. Small lateral blotches, similar in color to the dorsal blotches, also extend down the length of the body. The dorsal and lateral blotches alternate in position such that a lateral blotch is found between two dorsal blotches. Hatchlings are 25–40 cm (10–16 in) TL. This species was recently elevated from a subspecies of Ratsnakes (*P. obsoletus;* see **Comments**), and no subspecies are recognized.

Etymology: The specific epithet, *spiloides* [speh-LOYE-dees], is from the Greek *spilos* (spot) and *eidos* (similar to), which refer to the blotched pattern on the dorsum of some individuals.

Similar Species: Juvenile Gray Ratsnakes are sometimes confused with Red Cornsnakes (*P. guttatus*), as both have a blotched pattern. Red Cornsnakes in Tennessee generally have some reddish coloration and a backward-pointing V- or U-blotch on the head and neck. Additionally, Cornsnakes have a postocular stripe that continues onto the neck. This stripe stops at the mouth in Gray Ratsnakes. Older adults might be confused with Racers (*Coluber constrictor*) and Eastern Black Kingsnakes (*Lampropeltis nigra*), as all three species are large and darkly colored as adults. Racers are more slender, have smooth scales, do not have a "slice of bread shaped" cross-section, and have an immaculate venter. Eastern Black Kingsnakes are less robust, have smooth scales, and have an undivided anal plate.

Distribution: Gray Ratsnakes range generally between the Mississippi River to the west and the Appalachian Mountains to the east, from the Great Lakes and southern Ontario to the Gulf Coast of the Florida Panhandle. In Tennessee, this species is found nearly statewide, presumably absent only from the highest elevations of the Blue Ridge Mountains; however, confirmed records are lacking for some counties.

Habitat: Gray Ratsnakes occupy most habitats in Tennessee, including urban settings where they can be quite common. They are diurnal and active during the warmer months, especially the spring, and are one of the most frequently encountered snakes in our state. They prefer areas of deciduous or mixed forest, although not necessarily with a closed canopy, and often patrol forest edges, which are preferred nesting sites for many bird species, a major component of their diet. They are also frequently found in barns and outbuildings, attics, and other manmade structures that attract rodents, and they will readily seek cover under artificial objects, which also attract nesting rodents. Gray Ratsnakes are frequently seen crossing roads in late morning, and juveniles are common on roadways at dusk, especially after a rain.

Natural History: Gray Ratsnakes are common to abundant in most areas and are active from March to November, although they are more frequently encountered in the warmer months. They forage widely during the day and are excellent climbers and are frequently seen in trees searching for birds' nests. Indeed, they can sometimes be located by the distress calls of birds, as Blue Jays and American Crows will sometimes audibly harass an arboreal snake. Mating occurs in late spring, and males are known to engage in combat wrestling, where they attempt to pin the other male's head to the ground. Five to 40 eggs are laid with an average clutch size of 12–15, and the eggs are generally laid in moist areas, such as rotten logs, shallow holes, piles of sawdust or loose dirt, and underneath rocks or artificial cover such as boards. Hatching occurs in late summer, and sexual maturity is reached in four years. Females often return to the same nest site in subsequent years, and territories that include a hibernation site are adhered to, although not defended, year after year. Gray Ratsnakes are generalist predators, consuming most small- to medium-sized vertebrates, although they particularly favor rodents, birds, and bird eggs. Juveniles feed on amphibians, lizards, and reptile eggs. Predators include predatory mammals, birds such as hawks and owls, and other snakes such as Eastern Black Kingsnakes. Gray Ratsnakes are generally of an irascible disposition when cornered, as they will readily bite and vibrate their tail in leaf litter. If captured they emit a foul-smelling musk from their anal glands and continue their attempts to bite. Larger individuals can draw blood, but bites are painless and not of any concern if washed with soap or otherwise disinfected.

Conservation Status: Gray Ratsnakes are common and widespread in Tennessee. Larger individuals are harder to come by now compared to previous decades, perhaps because the longer they get, the more susceptible they are to vehicle strikes, and the older they get the more likely they are to have a fatal encounter with a misguided human. Threats to Gray Ratsnakes include frequent persecution and road mortality.

Comments: Gray Ratsnakes are most frequently called by the colloquial name "black snakes" or "oak snakes" in Tennessee and throughout the South, although some refer to them as "chicken snakes" for their proclivity for raiding chicken coops. The genus *Pantherophis* was recently erected to distinguish the Old World Ratsnakes (genus *Elaphe*) from the

New World Ratsnakes, whereas previously all ratsnakes were all included in *Elaphe*. Furthermore, a number of previous subspecies of *P. obsoletus,* including *P. o. spiloides,* were elevated to full species based on molecular and morpho-logical work, which more accurately reflects evolutionary relationships than did historical morphological work alone.

R. Graham Reynolds and
Matthew L. Niemiller

Northern Pinesnake adult, Coffee County. (Photo courtesy of Arnold Air Force Base)

Pinesnake

Pituophis melanoleucus

Description: Pinesnakes are the largest snakes in Tennessee, with most adults 110–180 cm (43–71 in) TL but reaching 229 cm (7.5 ft) TL. Body scales are strongly keeled and the anal plate is single. The dorsum is pale gray to dull white to tan to yellow with dark black to brown to reddish brown blotches. These blotches are typically black toward the head but may become brown toward the tail. The blotches might form saddles around the midbody but usually form bands on the tail. A series of smaller blotches run along the sides. The venter is cream to yellow with light to moderate darker spotting. Four prefrontal scales are present. The snout is somewhat pointed, and the enlarged rostral scale extends between the internasal scales. Hatchlings are 32–50 cm (12.5–20 in) TL. Hatchlings and juveniles are similar in pattern to adults, but the ground coloration typically is paler and often with an orangish or pinkish tint. Three subspecies are recognized, but only the Northern Pinesnake (*P. m. melanoleucus*) occurs in Tennessee.

Etymology: The specific and subspecific epithet, *melanoleucus* [meh-lan-no-LEW-kus], is from the Greek *melaina* (black or dark) and *leukos* (white), which refers to the two main colors of this subspecies.

Similar Species: Pinesnakes are most easily confused with Gray Ratsnakes (*Pantherophis spiloides*), as Gray Ratsnakes also are large-bodied and have a similar pattern. However, Gray Ratsnakes have weakly keeled body scales, do not have an enlarged rostral scale, have a divided anal plate, and only have two prefrontal scales.

Northern Pinesnake adult, Coffee County. (Photo by Gary W. Gerald)

Distribution: Pinesnakes occur in several disjunct units throughout the mid-Atlantic and southeastern United States. In Tennessee, Northern Pinesnakes presumably range throughout much of the state but are apparently absent from northwest Tennessee, the Nashville Basin, northeast Tennessee, and higher elevations of the Cumberland and Blue Ridge Mountains. This species is infrequently encountered and might have a larger distribution in Tennessee than currently thought.

Habitat: Pinesnakes are associated with well-drained, xeric, sandy habitats, including sandy pine barrens, scrub, sand hills, agricultural lands, and xeric, rocky mountain ridges with stands of pine. In a study at Arnold Air Force Base in Middle Tennessee, Northern Pinesnakes primarily were found in disturbed habitats with little to no forest canopy. Although Pinesnakes have been documented to climb trees, most spend the majority of their time underground in mammal burrows, stump holes, and rotted root channels and, consequently, are seldom encountered on the surface. Pinesnakes will even excavate their own burrows in sandy soil.

Natural History: Pinesnakes become surface active in early March but do not begin wider movements until late march or early April. Individuals enter hibernacula from late October through early November and do not hibernate communally. Most individuals are encountered on the surface in May and June during the day, although some individuals may move at night. Pinesnakes have large home ranges and often move long distances over land. Mating occurs

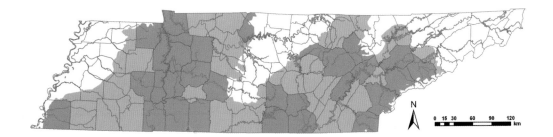

in April and May, and females typically lay 6–12 eggs (up to 27) in June or July in underground cavities, often in open sandy areas. Eggs hatch in 8–10 weeks in September or October. Pinesnakes feed on small mammals, such as rats, mice, squirrels, and rabbits, ground-nesting birds and their eggs. Documented predators include coyotes, foxes, skunks, large snakes, and birds of prey. Scarletsnakes (*Cemophora coccinea*) are known to eat Pinesnake eggs. When threatened, Northern Pinesnakes have an impressive defensive display that includes inflating the body and raising the body up to appear larger, vibrating the tail to make noise, hissing, and striking with the mouth open or closed. Pinesnakes will often wrap themselves around a person's arm and occasionally bite when handled.

Conservation Status: Because of the presumed limited distribution and few documented occurrences, populations of Northern Pinesnakes are considered vulnerable in Tennessee. Accordingly, TWRA lists this species as "Threatened." However, the status of most Tennessee populations is largely unknown and warrants study. Like most other snakes, habitat loss and fragmentation, as well as road mortality, are the primary threats faced by Northern Pinesnakes in Tennessee. Northern Pinesnakes also are occasionally killed by humans who mistake them for Timber Rattlesnakes because of their large size and impressive defensive display, which includes vibrating the tail.

Matthew L. Niemiller and
R. Graham Reynolds

Queensnake adult, Coffee County. (Photo by Matthew L. Niemiller)

Queensnake

Regina septemvittata

Description: Queensnakes are medium-sized, slender, primarily aquatic snakes 36–60 cm (14–24 in) TL but capable of reaching lengths of at least 102 cm (40 in). Body scales are keeled and the anal plate is divided. Adults are gray to brown to olive green, with a cream to yellow, longitudinal stripe running along each side of the body on dorsal scale rows 1 (upper half) and 2. Three narrow and darker longitudinal stripes run along the dorsum but are often difficult to see except in individuals that have recently shed. The venter is grayish white to cream to yellow with four dark brown longitudinal stripes that can be boldly marked or become obscured with age. The two outer stripes are

Top left: Queensnake juvenile, Polk County. (Photo by Matthew L. Niemiller) *Top right:* Venter of a juvenile Queensnake, Polk County. (Photo by Matthew L. Niemiller) *Bottom right:* Venter of an adult Queensnake, Rutherford County. (Photo by Brad M. Glorioso)

situated along the edges of the ventral scales, whereas the two inner stripes are situated along the midline of the venter. These stripes are most prominent toward the head. The head is only slightly wider than the neck. Nineteen dorsal scale rows are present at midbody. Young are 17–26 cm (6.5–10 in) TL and resemble adults but have bold ventral stripes that are clearly defined all the way to the tail. No subspecies are recognized.

Etymology: The specific epithet, *septemvittata* [sep-tem-veh-TAH-tah], is from the Latin *septem* (seven) and *vittata* (banded or ribbon striped), in reference to the dorsal and ventral patterning, especially on younger individuals.

Similar Species: Queensnakes most closely resemble Common Gartersnakes (*T. sirtalis*), Eastern Ribbonsnakes (*T. sauritus*), and Western Ribbonsnakes (*T. proximus*), as each of these species has longitudinal stripes running along the body and often occur in similar habitats. However, each of these species has an undivided anal plate and a light-colored mid-dorsal stripe. Likewise, both species of Ribbonsnakes have immaculate venters.

Distribution: Queensnakes are found primarily east of the Mississippi River from southwestern Ontario in Canada, southern Michigan, and southwestern New York southward to the panhandle of Florida. There also is a large disjunct population west of the Mississippi River in Arkansas and southwestern Missouri. Queensnakes are found throughout the eastern two-thirds of Tennessee, east of the north-flowing Tennessee River to the lower elevations of the Blue Ridge Mountains. However, records are lacking throughout much of the Cumberland Mountains.

Habitat: Queensnakes are most common in and around cool, rocky streams and rivers but are occasionally found in heavily vegetated lakes and other aquatic habitats, such as river backwaters, ponds, and reservoirs. Areas where they are most common have cold and clear flowing water, with rocky to sandy bottoms and an abundance of crayfish, their primary food source. Queensnakes are most often found underneath rocks and logs along the water's edge or swimming in shallow water where they might seek refuge underneath a submerged rock. Queensnakes will also bask in the limbs of shrubs and trees adjacent to or overhanging water.

Natural History: Queensnakes are active during the warmer months in Tennessee and will retreat into animal burrows, root masses, or under vegetation along the water's edge during cold weather. Queensnakes are primarily active during the day and may occasionally be found crossing roads near aquatic habitats at dusk. Mating primarily occurs during spring and females give birth to 6–20 young in late summer into early autumn. Sexual maturity is reached in 2–3 years. Queensnakes are highly aquatic and very adept swimmers. Queensnakes feed almost exclusively on recently molted crayfish that have soft exoskeletons, avoiding those with hard exoskeletons that can defend themselves.

However, Queensnakes will also eat small fish, tadpoles, and aquatic invertebrates. Predators include foxes, raccoons, otters, wading birds, Hellbenders, large fish, and other snakes, such as Eastern Black Kingsnakes (*Lampropeltis nigra*), and Northern Watersnakes (*Nerodia sipedon sipedon*). Queensnakes will quickly drop into water from overhanging vegetation when disturbed while basking and hide underneath submerged rocks. They will often writhe and release a foul-smelling musk when captured but rarely attempt to bite.

Conservation Status: Queensnakes are thought to be secure in Tennessee in habitats where crayfish are abundant. However, stream degradation, pollution, and siltation may pose serious threats for some Queensnake populations and the crayfish they feed upon.

Comments: Queensnakes are sometimes called "willow snakes," "moon snakes," or "leather snakes."

Lisa Powers and
Matthew L. Niemiller

DeKay's Brownsnake adult, Monroe County. (Photo by Matthew L. Niemiller)

DeKay's Brownsnake

Storeria dekayi

Description: DeKay's Brownsnakes are relatively small, slender snakes. Most adults are 22–39 cm (9–15 in) TL, but older adults have reached lengths of at least 49 cm (19 in). Body scales are keeled and the anal plate is divided. Loreal scales are absent. Coloration of the dorsum and tail is various shades of brown or gray, but nearly black individuals also have been reported. Except in melanistic individuals, the top of the head is darker than the body. A thin tan stripe, bordered by a series of paired brown spots, extends from the back of the head to the base of the tail. The first pair of spots is darker than more posterior pairs of spots, giving the neck a dark banded appearance. The dorsum

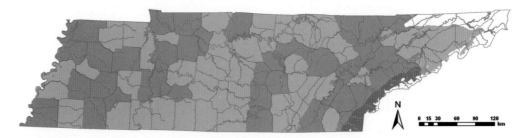

One subspecies of DeKay's Brownsnake is recognized in Tennessee: the Midland Brownsnake (*Storeria dekayi wrightorum*), whose range appears in orange. Possible intergrades with the Northern Brownsnake (*S. d. dekayi*) exist in extreme eastern Tennessee, shown in purple.

may appear banded because a thin line often, but not always, interconnects the paired spots. The venter is typically off white, tan-yellow, or pale pink, and it is always paler than the dorsum and sides. Sexual dimorphism is not obvious in this species, but females have slightly larger heads and bodies than males. Young are 7–11 cm (2.8–3.5 in) TL, and because they lack the middorsal stripe, they are uniformly brown, gray, or black above with a pale yellow, gray, or cream collar around the neck and pale tan below. Juveniles attain the coloration characteristic of adults during their first year of life. Four subspecies are recognized, with the Midland Brownsnake (*S. d. wrightorum*) occurring throughout most of Tennessee. Populations of DeKay's Brownsnakes in eastern Tennessee potentially represent intergrades between Midland Brownsnakes and Northern Brownsnakes (*S. d. dekayi*).

Etymology: The specific and subspecific epithet, *dekayi* [deh-KAY-i], honors James E. Dekay, a prominent 19th-century natural historian who authored volume 3 (*Reptiles and Amphibians*) of the *Zoology of New York*. The subspecific epithet, *wrightorum* [rye-TORE-um], honors the husband and wife team of Albert Hazen Wright and Anna Allen Wright, prominent early mid-20th-century herpetologists who coauthored the *Handbook of Snakes*.

Similar Species: DeKay's Brownsnakes most closely resemble Northern Red-Bellied Snakes (*Storeria occipitomaculata occipitomaculata*), but might also be confused with other small-bodied snakes with keeled dorsal scales. The ventral and neckband colorations (light in Northern Red-Bellied Snakes, dark in Brownsnakes) readily distinguish DeKay's Brownsnakes from Northern Red-Bellied Snakes.

Ring-Necked Snakes (*Diadophis punctatus*) have a slate gray dorsum, a pale neckband, and a yellow venter. Rough Earthsnakes (*Virginia striatula*) have weakly keeled dorsal scales, a more uniformly colored dorsum, and a more pointed snout. The divided anal plate distinguishes DeKay's Brownsnakes from young Common Gartersnakes (*Thamnophis sirtalis*), Eastern Ribbonsnakes (*T. sauritus*), and Western Ribbonsnakes (*T. proximus*).

Distribution: DeKay's Brownsnakes are found throughout much of the eastern half of the United States; however, records are lacking for the higher elevations of the Appalachian Mountains in southwestern Ohio, Pennsylvania, West Virginia, North Carolina, South Carolina, and extreme eastern Tennessee. Furthermore, records are lacking for areas in upper New England, and an isolated area that includes southeastern Alabama, southwestern Georgia, and the Florida Panhandle. Midland Brownsnakes are likely found throughout most of Tennessee. They are likely absent from the northern Ridge and Valley and Blue Ridge Mountains in northeast Tennessee, and from higher elevations of the Blue Ridge Mountains elsewhere. However, records for this species are scattered, and few records exist for much of central Tennessee. The absence of records from this region is puzzling, and it is not known if the apparent gap in distribution is real or an artifact of limited collection. Midland Brownsnakes may intergrade with Northern Brownsnakes in extreme southeastern Tennessee.

Habitat: DeKay's Brownsnakes are associated with humus-rich soils and can be found in nearly every type of terrestrial and wetland habit present in Tennessee. They are particularly common along the edges of forests

Top left: Venter of a DeKay's Brownsnake, Polk County. (Photo by Matthew L. Niemiller) *Top right:* DeKay's Brownsnake adult, Polk County. Possible intergrade with Northern Brownsnake (*Storeria dekayi dekayi*). (Photo by Matthew L. Niemiller) *Bottom right:* DeKay's Brownsnake adult, Stewart County. (Photo by Matthew L. Niemiller)

and grasslands, including parks and yards. Although considered to be a terrestrial or fossorial species, they are also commonly found in and around water and are at least occasionally arboreal. They are apparently commonly found in bushes in wetlands of western Kentucky and similar behavior has been reported in other areas. I have twice encountered DeKay's Brownsnakes several feet off the ground in ornamental evergreens used in landscaping my yard.

Natural History: DeKay's Brownsnakes are primarily encountered from late March through mid-October, but individuals can be found during every month of the year. They hibernate in Tennessee during cold winter months but become surface active during warm winter weather. Hibernacula are varied, including rotting logs and stumps, anthills, stone walls, limestone-lined cisterns, and other decaying or rocky structures. Mating occurs during early spring (March through April) and involves males tracking pheromone trails to locate females. When females are encountered, the male rubs his chin against the female's body and pursues her as she slithers away. Eventually, the male wraps his tail around the female's, which helps align his vent to hers, and coitus occurs. DeKay's Brownsnakes are ovoviviparous and typically give birth from June to late September, after about a four-month gestation period. Females have reportedly given birth to as few as three and to as many as 41 young. The small neonates grow rapidly. Individuals become sexually mature by their third year, when females are $\geq$17 cm (6.4 in) SVL and males $\geq$15 cm (6.0 in) SVL. These snakes can live for more than seven years. DeKay's Brownsnakes are active predators that rely heavily on chemoreception to locate prey. They use flicks of their tongue to scan the ground for prey trails, which they then track. Their diet consists primarily of earthworms and slugs, but spiders, small woodland salamanders, small frogs and toads, and even small fish are occasionally eaten. Predators include a variety of mammals (e.g., shrews, raccoons, weasels, skunks, opossums, foxes, and domestic cats), birds (e.g., crows, shrikes, and screech owls), other snakes, large frogs, and some invertebrates (e.g., spiders and centipedes). DeKay's Brownsnakes often release a foul-smelling musk mixed with feces when captured, and they have been reported to feign death when alarmed. They are inoffensive snakes and rarely attempt to bite when handled; however, I have unintentionally prompted several individuals to take defensive postures and strike at a camera lens.

Perhaps this defensive behavior is a typical response to small predators.

Conservation Status: DeKay's Brownsnakes have a wide distribution and are abundant in suitable habitat. Consequently, this species is thought to be secure and is not listed as a species of conservation concern in Tennessee. Because they tolerate disturbed habitats associated with subdivisions and parks, DeKay's Brownsnakes are too often victims of lawnmowers and "weed whackers."

Brian T. Miller

Red-Bellied Snake adult, Cannon County. (Photo by Matthew L. Niemiller)

Red-Bellied Snake

Storeria occipitomaculata

Description: Red-Bellied Snakes are relatively small, slender snakes. Most adults are about 21–29 cm (8–11 in) TL but can reach lengths of at least 40.6 cm (16 in). Body scales are distinctly keeled and the anal plate is divided. Lo-real scales are absent. Dorsal coloration of the body and tail is various shades of brown, gray, or, occasionally, black. Three small pale spots are located at the base of the head and often form a conspicuous pale neck ring. A pale mid-dorsal stripe also is often present. The venter is typically salmon pink but occasionally is bright red, orange, or yellow. Sexual dimorphism is

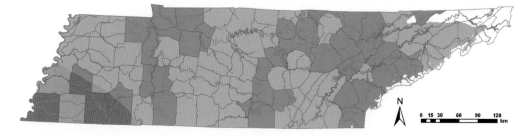

One subspecies of Red-Bellied Snake is recognized in Tennessee: the Northern Red-Bellied Snake (*Storeria occipitomaculata occipitomaculata*), whose range appears in orange. Possible intergrades with the Florida Red-Bellied Snake (*S. o. obscura*) exist in southwestern Tennessee, shown in purple.

Top left: Red-Bellied Snake adult, Hardeman County. (Photo by Matthew L. Niemiller) *Top right:* Red-Bellied Snake adult, Jackson County. (Photo by Matthew L. Niemiller)

not obvious in this species, although the tail is usually larger in males than in females, and pregnant females are plumper than males. Neonates are 7–10 cm (2.8–4.0 in) TL and typically have darker, more intense coloration than adults. Three subspecies are recognized, but only the Northern Red-Bellied Snake (*S. o. occipitomaculata*) occurs in Tennessee. However, populations in southwestern Tennessee might intergrade with Florida Red-Bellied Snakes (*S. o. obscura*), which typically have the three pale spots at the base head fused together and have black pigmentation at the top and sides of the head. Individuals from southwest Tennessee certainly resemble Florida Red-Bellied Snakes (see photo).

Etymology: The specific and subspecific epithet, *occipitomaculata* [ock-sip-pe-toe-mack-you-LAH-tah], is from the Latin *occiput,* meaning "back of head," and *maculata,* meaning "spot" and refers to the three conspicuous pale spots located at the base of the head that often form a pale neck ring.

Similar Species: Red-Bellied Snakes most closely resemble DeKay's Brownsnakes (*Storeria dekayi*) but might also be confused with other small-bodied snakes, including Rough Earthsnakes (*Virginia striatula*), Smooth Earthsnakes (*Virginia valeriae*), Eastern Wormsnakes (*Carphophis amoenus*), Kirtland's Snake (*Clonophis kirtlandi*) and young Common Gartersnakes (*Thamnophis sirtalis*), Eastern Ribbonsnakes (*T. sauritus*), and Western Ribbonsnakes (*T. proximus*). However, the ventral coloration quickly distinguishes Northern Red-Bellied Snakes from most species, except for Kirtland's Snake and some populations of Eastern Wormsnakes. However, the three conspicuous pale spots at the base of the head, and the absence of spots at the edges of the ventral scales distinguish Northern Red-Bellied snakes from Kirtland's Snakes. The large eyes, keeled scales, distinct neck, and absence of a spine at the tip of the tail readily distinguish Northern Red-Bellied Snakes from Eastern Wormsnakes with a similarly colored venter.

Distribution: Red-Bellied Snakes are found throughout much of the eastern half of the United States and southeastern Canada; however, records are lacking for the higher elevations of the Appalachian Mountains and portions of the Ohio River Valley. Furthermore, records are lacking for much of Mississippi and for northwestern Alabama. In Tennessee, Northern Red-Bellied Snakes likely inhabit most regions of the state except for the Ridge and Valley of northeastern Tennessee and the highest elevations of the Blue Ridge Mountains. However, records for this species in Tennessee are scattered, and the exact distribution is not well known. Northern Red-Bellied Snakes may intergrade with Florida Red-Bellied Snakes in southwestern Tennessee.

Habitat: Red-Bellied Snakes are primarily fossorial and burrow into leaf litter and humus-laden soil in mesic hardwood forests, but they are also found in fields and wetlands. Although they are typically found under rocks, logs, and other surface objects, they occasionally are found climbing or basking in weeds or bushes or moving on the forest floor during the day. They are also found inside rotting logs and root stumps and under sheets of tin or old boards in pastures, old fields, and lawns, and they are occasionally found on roads on rainy nights.

Natural History: Red-Bellied Snakes are primarily encountered during spring and autumn but can be found surface active during summer following rainstorms. Furthermore, they become surface active during warm winter weather (I have found them under logs during late January in Middle Tennessee). However, they do hibernate during the winter months, using a variety of subsurface habitats, including anthills, rodent burrows, rotting roots, stumps, and logs, rocky crevices, stone walls, and other decaying or rocky structures. Red-Bellied Snakes become active during late winter or early spring and mate shortly thereafter (late March through early May). Males locate females by tracking pheromone trails. Red-Bellied Snakes are ovoviviparous, and females give birth usually during summer (July through August), but early autumn births are known. Litter size is variable, with reports of females giving birth to as few as

one and to as many as 23 young; however, litters typically consist of fewer than 10 young. Neonates are small but grow relatively quickly. Individuals become sexually active during their third year (second full growing season), when males are $\geq$ 13 cm (5 in) SVL and females are $\geq$ 15 cm (6 in) SVL. These snakes are known to live about five years but possibly live twice that long. Feeding behavior of Northern Red-Bellied Snakes is similar to that described for Dekay's Brownsnakes: they use flicks of their tongue to pick up the scent of prey, which they then track. Their diet consists primarily of soft-bodied invertebrates, such as slugs and earthworms, but also includes insects, snails, centipedes, newly transformed frogs, and small salamanders. Predators include other snakes, large frogs, birds, and mammals. Red-Bellied Snakes are inoffensive snakes and rarely attempt to bite when handled. However, when disturbed they often release a foul-smelling musk mixed with feces, and they sometimes curl their lips. Furthermore, they occasionally feign death when alarmed; this behavior involves rolling onto their back and exposing their brightly colored belly.

Conservation Status: As with other small woodland snakes, the status of Tennessee populations has not been assessed but is assumed to be secure. Northern Red-Bellied Snakes are secretive and seldom seen, which confounds attempts to determine population size and density, and to detect shifts in these parameters. Threats to Red-Bellied Snakes include loss and fragmentation of habitat, such as conversion of forests and fields into subdivisions and shopping malls.

Brian T. Miller

Southeastern Crowned Snake adult, Wilson County. (Photo by Matthew L. Niemiller)

Southeastern Crowned Snake

Tantilla coronata

Description: Southeastern Crowned Snakes are relatively small, shiny, and slender snakes with small eyes. Most adults are about 18 to 25 cm (7–10 in) TL, but older adults have reportedly reached a TL of 33 cm (13 in). Body scales are smooth and the anal plate in divided. Loreal scales are absent. The dorsum is beige, tan, or pinkish brown, whereas the venter is pale beige, pale pink, cream, or white. The top of the head is dark brown or black, with a pale collar separating the dark head cap from a dark brown to black neckband that is 3–5 scales wide. The head is relatively flat and slightly pointed. Sexual dimorphism is not obvious in this species, but females are heavier and grow to greater lengths than males; however, males have longer tails, and consequently more subcaudal scales,

Top left: Southeastern Crowned Snake adult, Blount County. (Photo by Matthew L. Niemiller) *Top right:* Venter of a Southeastern Crowned Snake, Wilson County. (Photo by Matthew L. Niemiller) *Bottom right:* Southeastern Crowned Snake juvenile, Wilson County. (Photo by Brad M. Glorioso)

than females. Young are 7–8 cm (2.7–3.1 in) TL at hatching, and they resemble adults in general coloration. No subspecies are recognized.

Etymology: The specific epithet, *coronata* [koh-roh-NAH-tah], is Latin for "crown" and refers to the conspicuous head coloration characteristic of this species.

Similar Species: Southeastern Crowned Snakes are difficult to confuse with any other snakes in Tennessee, but most closely resemble other small fossorial snakes, such as Eastern Wormsnakes (*Carphophis amoenus*), Ring-Necked Snakes (*Diadophis punctatus*), Smooth Earthsnakes (*Virginia valeriae*), and Rough Earthsnakes (*V. striatula*). However, none of these species has the prominent black neckband, which is a conspicuous feature of Southeastern Crowned Snakes. Ring-Necked Snakes have smooth scales, but their neck ring is colored yellow to cream, their dorsum is blue-gray, and they grow larger than Southeastern Crowned Snakes.

Distribution: As the common name implies, this species is endemic to the southeastern United States from Tennessee and Kentucky southward to the Florida Panhandle and along the Atlantic Coast to North Carolina and parts of Virginia. The distribution of Southeastern Crowned Snakes in Tennessee is poorly understood. Although possibly occurring statewide in suitable habitat, the secretive nature of this fossorial species hampers collection efforts and records are limited to a few scattered locations across the state. Southeastern Crowned Snakes have been reported from only a few localities in the Ridge and Valley and the Blue Ridge Mountain regions of eastern Tennessee, the Loess Plains and Southeastern Plains and Hills of western Tennessee, and the Eastern Highland Rim and Nashville Basin of central Tennessee.

Habitat: Southeastern Crowned Snakes are found in a variety of habits, including mesic hardwood forests, xeric pinewoods, and xeric cedar glades. In the Blue Ridge Mountains, they are found on exposed south-facing slopes in power-line cuts and other more open situations. These snakes are nocturnal but can be found under rocks, logs, and other surface cover during daylight hours. Although fossorial, they can be found on roads and the surface at night.

Natural History: Southeastern Crowned Snakes are primarily encountered during early spring and early autumn, although they are occasionally encountered during the summer following heavy rains. Presumably they hibernate during winter, but data on hibernacula are scarce. Nonetheless, Southeastern Crowned Snakes become active during warmer weather in late March or early April. Courtship has not been reported, but mating occurs shortly after individuals become active and extends into late spring. Females lay 1–3 small (18–22 mm [0.7–0.9 in] long) cylindrical eggs during May and June under rocks or logs, or in rotting logs, stumps, mulch, and sawdust piles. The eggs hatch after about 60 days of incubation. These snakes grow fairly rapidly, more than doubling their size during their first two years of life. They become sexually mature during their third year, and live for at least five years. Southeastern Crowned Snakes have very small home ranges. Adults are encountered more frequently than the young, perhaps because adults are more often surface active at night than are neonates and juveniles. Southeastern Crowned Snakes are rear-fanged; the enlarged posterior teeth of the upper jaws are grooved, which facilitates injection of venom into prey. The venom potentially immobilizes prey as it is being ingested, but few data exist on venom effect on invertebrates. The diet consists largely of centipedes and spiders, although small insects, earthworms, and slugs are also eaten. These inoffensive snakes seldom if ever attempt to bite humans and their teeth are too small to puncture the skin. Consequently, Southeastern Crowned Snakes pose no threat to people or their pets. Predators include snakes, frogs, birds, spiders, centipedes, and mammals.

Conservation Status: This secretive and fossorial snake is possibly more common in Tennessee than scattered records of occurrence indicate. Few records exist for Middle Tennessee, but my students and I routinely find this species at several localities in the Nashville Basin. Indeed, this is one of the more commonly encountered species of small snakes in the cedar glades of Rutherford and Wilson counties in the spring and autumn. Although locally abundant, we know too little about the population size or density of this species. Major threats include habitat fragmentation and destruction. Currently, this species is not listed as a species of conservation concern in Tennessee.

Brian T. Miller

Western Ribbonsnake adult, Lake County. (Photo by Brad M. Glorioso)

Western Ribbonsnake

Thamnophis proximus

Description: Western Ribbonsnakes are slender, medium-sized snakes. Adults are usually 50–75 cm (20–30 in) TL, but individuals can reach at least 99 cm (39 in). Body scales are keeled and the anal plate is undivided. The tail is long, typically accounting for slightly less than one-third the total length. Three prominent longitudinal stripes extend the length of the body, including a yellow to orange middorsal stripe and two pale yellow to bluish white lateral stripes. One lateral stripe is found on dorsal scale rows 3 and 4 on each side of the body. The dorsal ground color is black, but sometimes white, bluish white, or mint green flecks are present between scales. The venter and chin are a uniform color, usually cream, bluish white, or mint green. The narrow head is only slightly wider than the neck. A pair of white to yellow parietal spots is located on top of the head, but

these two spots sometimes fuse into a single larger spot. A short, conspicuous, white vertical bar is found on the preocular scales immediately in front of each eye. The upper lips are an immaculate white or yellow and formed by usually eight upper labial scales. Females typically are longer and more robust than males. Young are 20–30 cm (8–12 in) TL at birth and resemble adults in body coloration. Four subspecies are recognized, but only the Orange-Striped Ribbonsnake (*T. p. proximus*) occurs in Tennessee.

Etymology: The specific and subspecific epithet, *proximus* [prahx-EH-mus], is Latin for "next" or "nearest" and might refer to the range of this species abutting that of the Eastern Ribbonsnake (*T. sauritus*) to the east.

Similar Species: Western Ribbonsnakes are most similar to the two congeners that occur in Tennessee: Eastern Ribbonsnakes (*T. sauritus*) and Common Gartersnakes (*T. sirtalis*). Eastern Ribbonsnakes typically either lack the parietal spots on top of the head or, if present, the parietal spots are never fused together. Furthermore, Eastern Ribbonsnakes typically have seven upper labial scales, whereas Western Ribbonsnakes typically have eight upper labial scales. Western Ribbonsnakes are much more slender and have a longer tail than Common Gartersnakes. For example, in Western Ribbon snakes the tail accounts for about 30% of TL, whereas in Common Gartersnakes the tail is usually 19–25% of TL, with only males being toward the upper end of this range. Western Ribbonsnakes also more commonly appear to have three stripes, one on the dorsum and two lateral stripes. These lateral stripes might either be true stripes, in that they are bordered by darker background coloration, or they might mark the transition between the darker dorsum

Venter of a Western Ribbonsnake, Obion County. (Photo by Matthew L. Niemiller)

and lighter venter, although the transition point is usually lighter than the venter color and thus appears to be a stripe. Additionally, the lateral stripes occur on the third and fourth scale rows, as opposed to the second and third scale rows in Common Gartersnakes. Finally, Common Gartersnakes have darker, vertical bars on the upper labial scales that are lacking in Western Ribbonsnakes.

Distribution: Western Ribbonsnakes are found from southwest Wisconsin and southern Iowa to central Louisiana and Texas, and from western Tennessee into eastern New Mexico. Also, there are several disjunct populations in Illinois, Indiana, and southeast Wisconsin. Tennessee populations are limited to just three counties in the Mississippi Alluvial Plain around Reelfoot Lake (Lake and Obion counties) and the Memphis area (Shelby County). A literature record exists from Hardeman County. Records for this species are sparse and the exact distribution in Tennessee is not well known.

Habitat: Western Ribbonsnakes are at home in a variety of habitats but are almost always associated with water. Western Ribbonsnakes can be found in and along the margins of ponds, marshes, swamps, sloughs, streams, rivers, lakes, and reservoirs. Individuals are often found climbing in low vegetation around permanent bodies of water and along bases of nearby rock outcrops and animal burrows where some individuals may hibernate. When not active, Western Ribbonsnakes can be found underneath logs, trash, and other debris near aquatic habitats. They are often seen foraging during the day and commonly found crossing roads at night, especially after a rain, when they might be pursuing amphibians to eat.

Natural History: Western Ribbonsnakes are diurnal predators and can be found active during the warmer months as well as on warm, sunny days during winter around their hibernacula. Mating occurs in early to mid-spring, and females give birth to 10–15 young (up to 36) from late June into September. Western Ribbonsnakes feed primarily on amphibians, including larval salamanders and several species of frogs and toads; however, they also eat small fish, lizards, and invertebrates. Western Ribbonsnakes capture aquatic prey by swimming with their mouth open and use both vision and smell to track potential meals. Once grabbed, prey is either shaken vigorously until subdued or simply grasped until the prey tires from struggling to escape. Predators include Eastern Black Kingsnakes (*Lampropeltis nigra*), Cottonmouths (*Agkistrodon piscivorus*), semi-aquatic mammals, herons, bullfrogs, and predatory fish. Western Ribbonsnakes are wary and will head toward water and submerge when threatened. When captured, they will release a foul-smelling musk from their anal glands, and they occasionally bite.

Conservation Status: Although only found within a limited area in Tennessee, Western Ribbonsnakes are quite common where they occur and populations are thought to be secure. Consequently, this species is not listed in Tennessee. However, habitat destruction, habitat fragmentation, and road mortality are significant threats to populations.

Lisa Powers and
Matthew L. Niemiller

Eastern Ribbonsnake adult, Cheatham County. (Photo by Ritchie King)

Eastern Ribbonsnake

Thamnophis sauritus

Description: Eastern Ribbonsnakes are slender, medium-sized snakes. Most adults are 45–70 cm (18–27.5 in) TL but can reach lengths of at least 97 cm (38 in). Body scales are keeled and the anal plate is undivided. The tail is long, typically about one-third the TL. Three prominent longitudinal stripes extend the length of the body, a yellow to orange middorsal stripe and two pale yellow lateral stripes. One lateral stripe is found on dorsal scale rows 3 and 4 on each side of the body. The dorsal ground color is dark gray or black. The venter and chin are a uniform color, usually cream, pale yellow, or pale green. The narrow head is only slightly wider than the neck. Occasionally, a pair of white or yellow parietal spots is located on top of the head. A conspicuous white or otherwise light-colored vertical bar is found on the preocular scales immediately in front of each eye. There are usually seven upper labial scales that are an immaculate white or yellow white, forming distinctly

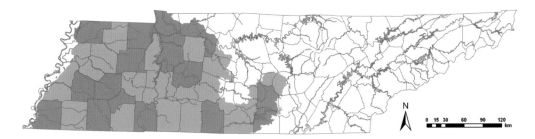

light-colored upper lips. Females typically are longer and more robust than males. Young are 18–23 cm (7–9 in) TL at birth and resemble adults. Four subspecies are recognized, but only the Common Ribbonsnake (*T. s. sauritus*) occurs in Tennessee.

Etymology: The specific and subspecific epithet, *sauritus* [saw-RYE-tus], is a Latinized word for "lizard-like," although the reason for this name is unclear.

Similar Species: Eastern Ribbonsnakes are most similar to the two congeners that occur in Tennessee: Western Ribbonsnakes (*T. proximus*) and Common Gartersnakes (*T. sirtalis*). Eastern Ribbonsnakes typically lack the parietal spots on top of the head, but if present they are always paired—they are never fused into a single spot. Furthermore, Eastern Ribbonsnakes typically have seven upper labial scales, whereas Western Ribbonsnakes typically have eight upper labial scales. Eastern Ribbonsnakes are much more slender and have a longer tail than Common Gartersnakes. In Eastern Ribbonsnakes the tail accounts for about 33% of the TL, whereas in Common Gartersnakes the tail is usually 19–25% of TL, with only males being toward the upper end of this range. Eastern Ribbonsnakes also more commonly appear to have three stripes, one on the dorsum and two lateral stripes. These lateral stripes might either be true stripes, in that they are bordered by darker background coloration, or they might mark the transition between the darker dorsum and lighter venter, although the transition point is usually paler than the venter color and thus appears to be a stripe. Additionally, the lateral stripes occur on the third and fourth scale rows, as opposed to the second and third scale rows

Head of an Eastern Ribbonsnake, Cheatham County. (Photo by Ritchie King)

in Common Gartersnakes. Finally, Common Gartersnakes have darker, vertical bars on the upper labial scales that are lacking in Eastern Ribbonsnakes.

Distribution: Eastern Ribbonsnakes are found throughout much of eastern North America east of the Mississippi River from Nova Scotia and Ontario in Canada southward to the Gulf Coast and Florida. However, this species appears to be absent from much of the Interior Plateau, Southwestern Appalachians, Central Appalachians, Ridge and Valley, and Blue Ridge Mountains. In Tennessee, Eastern Ribbonsnakes are known from the Mississippi Valley Loess Plains and Southeastern Plains, and Western Highland Rim in western and west-central Tennessee. Eastern Ribbonsnakes also have been infrequently documented from the Outer Nashville Basin and Eastern Highland Rim of Coffee, Franklin, and Giles counties in Middle Tennessee. The exact distribution of Eastern Ribbonsnakes in Tennessee is unknown and warrants further study.

Habitat: Eastern Ribbonsnakes are found in a variety of habitats, but they never venture too far from water. This semiaquatic snake occurs around streams, rivers, ponds, bogs, swamps, ditches, and lakes, where individuals can be found climbing in and around low bushes and grasses along the water's edge. Eastern Ribbonsnakes are adept swimmers but avoid open and deeper water. In the Outer Nashville Basin and Eastern Highland Rim, Eastern Ribbonsnakes are often found around rocky streams in mesic forests, presumably on the hunt for salamanders, which are common in such habitat. They are primarily diurnal and are most often observed on the move, but they also can be seen basking in open areas on rocks and logs, particularly during the early morning. When not active, Eastern Ribbonsnakes can be found underneath logs, trash, and other debris near aquatic habitats. They hibernate in abandoned rodent or crayfish burrows or in anthills.

Natural History: Eastern Ribbonsnakes are active from late March into October in Tennessee but can also be found during warm, sunny days during winter basking around hibernacula. Mating occurs soon after emergence, and females give birth to 3–12 (but up to 36) live young from July into early October. Sexual maturity is typically reached in two years for males and three years for females. Eastern Ribbon-snakes feed primarily on amphibians, such as tadpoles, frogs, and salamanders, but they will also eat fish and invertebrates, such as spiders and earthworms. Eastern Ribbonsnakes are active predators and catch their prey by stalking or chasing them. Documented predators include wading birds, birds of prey, predatory mammals, large fish, and other snakes, such as Cottonmouths (*Agkistrodon piscivorus*) and Eastern Black Kingsnakes (*Lampropeltis nigra*). Eastern Ribbonsnakes rely on their speed and agility to escape predation. When threatened, individuals will flatten their heads and bite at the attacker while also thrashing their bodies violently and smearing the attacker with a foul-smelling musk secreted from anal glands.

Conservation Status: Although infrequently encountered in some parts of the state (i.e., Eastern Highland Rim), Eastern Ribbonsnakes are thought to be secure and are not listed in Tennessee. Habitat destruction and fragmentation, agricultural runoff, pollution, and road mortality are thought to be significant threats to this species throughout its range.

Lisa Powers and
Matthew L. Niemiller

Eastern Gartersnake adult, Lake County. (Photo by Brad M. Glorioso)

Common Gartersnake

Thamnophis sirtalis

Description: Common Gartersnakes are small- to medium-sized snakes. Most adults are 45–75 cm (18–30 in) TL but are capable of reaching close to 130 cm (51 in). Dorsal and body scales are keeled and the anal plate is undivided. Coloration and patterning is extremely variable in this species. Dorsal background coloration usually is greenish yellow to black, and most Common Gartersnakes in Tennessee have a distinct white, yellow, yellow-green, or bluish middorsal stripe that starts at the neck and extends to the tail tip. This stripe is usually lighter than the background color. The stripe is bordered by the background color, which is occasionally solid but more often contains checkers of lighter scales if the background color is dark, or darker scales if the background color is light. Occasionally, these checkers line up and resemble faint lateral stripes that are usually

similar in color to the middorsal stripe but confined to the second and third body scale rows. Individuals with light lateral coloration usually have darker-colored scales interspersed among the light scales, forming a checkered pattern that often continues onto the venter. Some individuals appear checkered because the lateral stripes are indistinct. The top of the head is green to black, usually matching the dorsal background color, and the sides of the back of the neck are usually dark brown or black. The labials, sides of the neck, and chin are light, usually matching the lateral and ventral color, and the eyes are large relative to the head and have a light-colored iris surrounding the dark pupil. In general, females grow to larger size than males. Furthermore, females have larger heads but, relative to TL, slightly shorter tails than males. Neonates are 12–22 cm (4.7–8.6 in) TL and resemble adults in color and pattern. Twelve subspecies of Common Gartersnakes are recognized; however, molecular evidence suggests that these subspecies may not accurately reflect true evolutionary relationships. Only the Eastern Gartersnake, (*T. s. sirtalis*) is found in Tennessee.

Etymology: The specific epithet, *sirtalis* [ser-TAH-lis], refers to a 19th-century garment known as a garter, which usually was striped and reflects the common name of this species.

Similar Species: Common Gartersnakes are most likely to be confused with Eastern and Western Ribbonsnakes (*T. sauritus* and *T. proximus*), both of which also have three light-colored stripes on a darker background. Ribbonsnakes are much more slender and have a longer tail constituting about 30% of TL, whereas in Common Gartersnakes the tail is usually 19–25% of TL, with only males being toward the upper

end of this range. Ribbonsnakes also more commonly appear to have three stripes, one on the dorsum and two lateral stripes. These lateral stripes might either be true stripes, in that they are bordered by darker background coloration, or they might mark the transition between the darker dorsum and lighter venter, although the transition point is usually lighter than the venter color and thus appears to be a stripe. Additionally, the lateral stripes occur on the third and fourth scale rows, as opposed to the second and third scale rows in Common Gartersnakes.

Distribution: Common Gartersnakes occupy a huge range, through most of the United States outside of the Southwest, high deserts of the American West, and much of Canada. Eastern Gartersnakes occur across much of the eastern United States, although exact boundaries are complicated by incomplete species definitions and possible hybridization with other species of *Thamnophis*. In Tennessee this species is found statewide, including the high peaks of the Blue Ridge Mountains, although some counties still lack confirmed records. Indeed, we have found a specimen in the parking lot of Clingman's Dome at Great Smoky Mountains National Park in October, the highest peak in Tennessee.

Habitat: Common Gartersnakes occupy most habitats in Tennessee, including urban and suburban situations, where they can be extremely common; consequently, they are one of the most frequently encountered snakes in our state. They are diurnal and active during the warmer months, especially the spring. Common Gartersnakes prefer areas near water, either permanent or ephemeral, but they will wander far from water sources. They are capable of tolerating very cold temperatures and are

Top left: Eastern Gartersnake adult, Sevier County. (Photo by Matthew L. Niemiller) *Top right:* Venter of an Eastern Gartersnake, Sevier County. (Photo by Matthew L. Niemiller) *Bottom right:* Head of an Eastern Gartersnake, Campbell County. (Photo by Matthew L. Niemiller)

one of the few snake species occasionally encountered on warmer winter days. They readily seek cover under artificial objects, indeed Common Gartersnakes frequent stacks of lumber or tin roofing in yards or urban areas. They are often seen foraging during the day on the forest floor and in fields and are also commonly found crossing roads at night, especially after a rain, when they might be pursuing frogs or other amphibians.

Natural History: Common Gartersnakes are common to abundant in most areas and are potentially active year around in most of Tennessee; however, winter activity is usually restricted to basking near a hibernaculum. They forage widely during the day and are often seen around the margins of ponds, lakes, or streams. Common Gartersnakes often become more nocturnal in extremely warm weather. Mating usually occurs in early spring, just after emergence from an overwintering site, and multiple males might simultaneously court a female. Females can store sperm for up to a year, and

a related species (*T. marcianus*) has been reported to engage in facultative parthenogenesis, or virgin birth. Live young are born in July and August, with large females capable of producing 50–100 neonates, although the average litter size is likely less than 50. Sexual maturity is achieved in about 2 years, and the life span is likely around 6–8 years. Common Gartersnakes are supreme generalists, consuming most any vertebrate or invertebrate they encounter. They have a particular preference for amphibians but will also feed on carrion. We have encountered Common Gartersnakes on roads at night consuming dead frogs that were killed by passing cars. Unfortunately, in some areas on warm rainy nights, amphibian carrion represents a cornucopia of alimentary options. Juveniles feed on small and larval amphibians, as well as on small invertebrates. Predators are well documented and include most predatory birds and mammals as well as American Bullfrogs, other snakes, Eastern Box Turtles, large salamanders, fish, and large spiders that have been found eating neonates. Common Gartersnakes engage in a defensive display that includes flattening of the head, presumably to resemble a pit viper. Captured or restrained individuals are quick to bite, although bites rarely break the skin, and they emit a very foul musk from their anal glands. Common Gartersnakes, while harmless, do possess various enzymes in the saliva that aide in either digestion or prey immobilization. Technically, this saliva is venom, albeit mild. Very rarely, saliva from a Common Gartersnake causes minor local swelling if the person bitten is allergic to the enzymes.

Conservation Status: Common Gartersnakes are abundant and widespread in Tennes-see and can occupy most habitats, even in close proximity to people. They are useful inhabitants of gardens, as they will consume pests, such as snails and slugs. However, this does not often protect them from the sharp end of a hoe from those who fail to recognize their beneficial presence. Threats to Common Gartersnakes include frequent persecution, road mortality, and being killed, inadvertently or otherwise, by lawn and garden equipment.

Comments: Common Gartersnakes are most frequently called by the incorrect names of "gardener snakes" or "garden snakes" in Tennessee, probably due to both linguistic error and the areas where most people are likely to find this species—in their yards and gardens.

R. Graham Reynolds and
Matthew L. Niemiller

Rough Earthsnake adult, San Jacinto County, Texas. (Photo by Ken Wray)

Rough Earthsnake

Virginia striatula

Description: Rough Earthsnakes are one of Tennessee's smallest snakes. Most adults are 16–25 cm (6–10 in) TL but can reach lengths of at least 32 cm (12.5 in). Body scales are keeled and the anal plate is usually divided. The dorsum is uniformly colored beige, tan, brown, or gray and often appears glossy. However, a faint middorsal stripe is present in some individuals, and in others the head is darker than the body. Other individuals might have an indistinct pale neckband at the base of the head. The venter is pinkish tan or cream colored and lacks patterning. The head is small with a pointed snout. The sexes appear similar to each other, but females do grow to larger sizes than males. Young are about 812 cm (3–5 in) TL at birth. Both neonates and juveniles resemble adults, but generally they are darker in overall body coloration and have a pale collar or neckband that is often lost as they age. No subspecies are recognized.

Etymology: The specific epithet, *striatula* [stry-AT-you-lah], is from the Latin *striatus* (stripe) and *ula* (little) and refers to striated appearance of the dorsum, which is caused by the alignment of the keeled scales.

Similar Species: Rough Earthsnakes most closely resemble Smooth Earthsnakes (*V. valeriae*) but might also be confused with several other small, fossorial snakes, including Eastern Wormsnakes (*Carphophis amoenus*), DeKay's Brownsnakes (*Storeria dekayi*), and Southeastern Crowned Snakes (*Tantilla coronata*). Rough Earthsnakes differ from Smooth Earthsnakes in labial scale counts and degree of dorsal scale keeling. Rough Earthsnakes have keeled dorsal body scales, and five upper labial scales on each side of their head, whereas Smooth Earthsnakes have mostly smooth or weakly keeled dorsal body scales and six upper labials. The keeled dorsal scales also help distinguish Rough Earthsnakes from Southeastern Crowned Snakes, Ring-Necked Snakes, and Eastern Wormsnakes. Their overall glossy appearance, pointed snout, and uniformly colored dorsum help distinguish them from DeKay's Brownsnakes and Red-Bellied Snakes (*S. occipitomaculata*).

Distribution: Rough Earthsnakes are found throughout the eastern half of Texas and Oklahoma eastward through southern Missouri, most of Arkansas, and Louisiana. East of the Mississippi River, this species is found throughout much of the Gulf Coastal Plain and southern Atlantic Coastal Plain in southwestern Tennessee, most of Mississippi, southern Alabama, southern Georgia, eastern South Carolina and eastern North Carolina. Most Rough Earthsnake records are limited to four counties in southwest Tennessee: Fayette, Hardeman, Perry, and Shelby counties. This species might eventually be found in other western counties.

Several Rough Earthsnakes, Wake County, North Carolina. (Photo by Andrew Durso)

Habitat: Rough Earthsnakes are fossorial and associated primarily with mesic to xeric open habitats, including open grasslands, pastures, woodlots, glades, rocky hillsides, pine flatwoods, gardens, along rail lines, and in vacant lots in urban environments. Most records from Tennessee are from residential and urban areas around the greater Memphis area in Shelby County. They are often found underneath a variety of cover objects during daylight hours, including rocks, logs, boards, rotting carpets, cardboard, sheets of tin, and leaf litter, but they also can be found on the surface at night during warmer months.

Natural History: Rough Earthsnakes are fossorial snakes that are primarily encountered during spring and autumn but are also found during summer following rainstorms. They hibernate during the winter months, using ant nests, rotting roots, stumps, and logs, rocky crevices, compost piles, and other decaying or rocky structures as hibernacula. Rough Earthsnakes mate shortly after exiting hibernacula during late winter or early spring (late March through early May). Very little is known about mating or courtship behavior in this species, but as in Smooth Earthsnakes, males presumably track females by use of pheromone trails. Rough Earthsnakes are ovoviviparous. Females give birth to as many as 12 (typically 3–7) young during summer (July through September). Individuals presumably become sexually mature during the start of their second (male) or third (female) year, when they are ≥ 140 (5.5 in, male) or 180 (7.0 in, female) mm TL. Little is known about the lifespan of this species; however, an individual lived more than seven years in captivity. Their diet consists primarily of earthworms, but they also eat ant eggs, slugs, and other soft-bodied invertebrates. As with other small fossorial snakes, they use flicks of their tongue to pick up the scent of prey, which they then follow by tracking chemical cues. Predators include other snakes, large frogs, birds, mammals, insects, spiders, and centipedes. Rough Earth-snakes are inoffensive and rarely attempt to bite when handled. Defense mechanisms include feigning death (rolling on back with tongue protruding from a gaping mouth) and release of a foul-smelling musk, often mixed with feces.

Conservation Status: These secretive snakes are rarely seen, but populations are thought to be secure and Rough Earthsnakes are not listed as a species of conservation concern in Tennessee. However, this species has a limited distribution in Tennessee, with most records from the extreme southwestern counties, mostly from Shelby County. Although this species is known to inhabit appropriate habitat in urban environments, threats to its survival include habitat destruction and fragmentation, pesticides, and road mortality.

————————Brian T. Miller

Smooth Earthsnake adult, Wilson County. (Photo by Matthew L. Niemiller)

Smooth Earthsnake

Virginia valeriae

Description: Smooth Earthsnakes are one of Tennessee's smallest snakes. Most adults are 16–22 cm (6.3–8.7 in) TL but can grow to at least 39 cm (15 in). Body scales are usually smooth and the anal plate is divided. The short body is relatively plump, and the small head has a distinctly pointed snout. The superficially uniformly colored dorsum is beige to tan and has pairs of tiny black spots adjacent to the midline that extend from the neck to the base of the tail. The coloration of the dorsum extends to the sides, where the color gradually lightens to a pinkish tan, white, or yellowish venter.

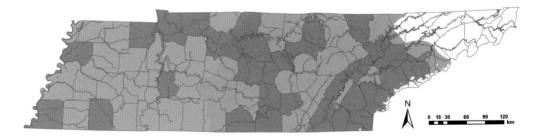

Two subspecies of Smooth Earthsnakes are recognized in Tennessee: the Eastern Smooth Earthsnake (*Virginia valeriae valeriae*), whose range appears in orange, and the Western Smooth Earthsnake (*V. v. elegans*), whose range appears in blue.

In most populations, the dorsal scales lack a keel (hence Smooth Earthsnake); however, in some populations the dorsal scales are weakly keeled. Males have proportionally longer tails than females. Six upper labial scales are present. Young are 6–12 cm (2.4–4.7 in) TL at birth and resemble adults. Three subspecies are recognized, two of which occur in Tennessee: the Eastern Smooth Earthsnake (*V. v. valeriae*) and the Western Smooth Earthsnake (*V. v. elegans*). The body scales are weakly keeled in Western Smooth Earthsnakes, whereas most of the body scales are smooth in Eastern Smooth Earthsnakes. In some Eastern Smooth Earthsnakes, dorsal scales are weakly keeled near the tail. Additionally, Eastern Smooth Earthsnakes have 15 dorsal scale rows, whereas Western Smooth Earthsnakes have 17 dorsal scale rows.

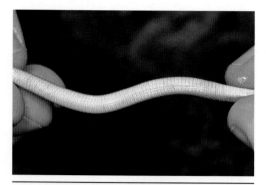

Venter of a Smooth Earthsnake, Wilson County. (Photo by Matthew L. Niemiller)

Etymology: The specific and subspecific epithet, *valeriae* [vah-leh-REE], honors Valery Blany, who collected the specimen used by her first cousin, Spencer Baird, and his colleague, Charles Girard, to describe the species. The subspecific epithet, *elegans,* is Latin for "elegant."

Similar Species: Smooth Earthsnakes most closely resemble Rough Earthsnakes (*V. striatula*) but might also be confused with other small-bodied snakes with tan to beige backs, including Eastern Wormsnakes (*Carphophis amoenus*), DeKay's Brownsnakes (*Storeria dekayi*), and Southeastern Crowned Snakes (*Tantilla coronata*). However, the tiny eyes and spine on the tail tip distinguish Eastern Wormsnakes from Smooth Earthsnakes, the dark neckband and black head spot distinguish Southeastern Crowned Snakes from Smooth Earthsnakes, and the rounded snout, dark neck spots, strongly keeled scales, and prominent middorsal stripe distinguish Brownsnakes from Smooth Earthsnakes. Smooth Earthsnakes differ from Rough Earthsnakes in labial scale counts and degree of dorsal scale keeling. Rough Earthsnakes have keeled dorsal body scales, and five upper labial scales on each side of their head, whereas Smooth Earthsnakes have mostly smooth or weakly keeled dorsal body scales and six upper labials. Rough Earthsnakes are known only from southwest Tennessee.

Distribution: Smooth Earthsnakes are found throughout much of the eastern half of the United States from eastern Texas and eastern Oklahoma eastward to the Atlantic Coast and northward into parts of Illinois, Indiana, Ohio, West Virginia, Pennsylvania, and New Jersey. Smooth Earthsnakes are absent from much of peninsular Florida. They are likely found statewide in Tennessee, except for the higher elevations of the Blue Ridge Mountains; however, the exact distribution is not well established because records are sparse for West and Middle Tennessee as well as northeastern Tennessee.

Habitat: Although commonly associated with humus-rich soils in mesic woodlands, Smooth Earthsnakes are also frequently encountered in old fields, pastures, and xeric cedar glade habitats. They are a fossorial species and, consequently, often are found underneath rocks, logs, boards and other surface material (including old carpets and rotting cardboard) and in rotting root stumps. Smooth Earthsnakes are surface active at night, particularly during or after rainstorms. I often find them when tending my gardens or mowing my lawn. Furthermore, they are found in association with DeKay's Brownsnakes (*Storeria dekayi*), Red-bellied Snakes (*S. occipitomaculata*), and Ring-Necked Snakes (*Diadophis punctatus*) in the Eastern Highland Rim and with Southeastern Crowned Snakes (*Tantilla coronata*), Eastern Worm-snakes (*Carphophis amoenus*), and Ring-necked Snakes in the Inner Nashville Basin.

Natural History: Smooth Earthsnakes are primarily encountered during spring and autumn but can be found surface active during summer following rainstorms. They hibernate during the winter months, using rodent burrows, rotting roots, stumps, and logs, rocky crevices, compost piles, and other decaying or rocky structures as hibernacula. They exit hibernacula during late winter or early spring and mate shortly thereafter (late March through early May). Very little is known about mating or courtship behavior in this species, but males presumably track females by use of pheromone trails. Regardless, Smooth Earthsnakes are ovoviviparous, and females give birth during summer (July through September), after carrying the young for about 2.5–3 months. Litter size is variable, with reports of females giving birth to as few as two and to as many as 14 young; however, litters typically consist of fewer than 10 young. Individuals presumably become sexually mature during their second or third year, when they are $\geq$ 18 cm (7 in) TL. Little is known about the lifespan of this species; however, a captive individual reportedly lived more than six years. Their feeding behavior is poorly documented, but they undoubtedly use flicks of their tongue to pick up the scent of prey, which they then track. Their diet consists primarily of earthworms but also includes slugs, snails, and small soft-bodied insects. Predators include other snakes, large frogs, birds, mammals, insects, and spiders. Smooth Earthsnakes are inoffensive and rarely attempt to bite when handled; however, they occasionally release a foul-smelling musk mixed with feces. Furthermore, they occasionally feign death when alarmed.

Conservation Status: This secretive and fossorial snake is possibly more common in Tennessee than scattered records of occurrence indicate. Few records exist for Middle Tennessee, but my students and I routinely find this species at several localities in the Nashville Basin, and I often find this snake in my yard and gardens in the Eastern Highland Rim. Although locally abundant, little is known about the population size or density of this species. Major threats include habitat fragmentation and destruction. Currently, this species is not listed as a species of conservation concern in Tennessee.

Brian T. Miller

15
Family Viperidae (Vipers)

Vipers are a family of potentially dangerous, venomous snakes found in primarily tropical and temperate regions of the world. Notably, they are not found (presently or historically) in Antarctica or Australia. Around 270 species are recognized in 37 genera, although these numbers differ depending on the author. Several distinct lineages of vipers are recognized. One of these lineages is commonly known as the "pit vipers," so named because the snakes have a pair of infrared organs associated with a distinct pit located on each side of the head between their eye and nostril. The infrared organs essentially serve as heat-sensing devices that produce a thermal image of the snakes' environment and allow them to detect prey or potential threats. Pit vipers occur in North America, South America, and parts of Eurasia. The three genera (*Agkistrodon, Crotalus,* and *Sistrurus*) and four species of vipers found in Tennessee are all pit vipers. Fossil vipers are known from the Miocene of Eurasia, and molecular estimates place the origin of this family to at least 50 million years ago. In comparison, pit viper fossils are known from the Miocene of North America. Some species of vipers are relatively small, obtaining lengths of only about 30 cm (12 in) as adults; other species are relatively long and can reach lengths of about 3.6 m (12 ft). Most species give birth to live young (ovoviviparity). All vipers are venomous and possess a pair of long, hollow (solenoglyphous) fangs used to inject venom produced in glands located along the posterior end of the upper jaw. These fangs fold up against the roof of the mouth when not in use. In general, vipers are stout-bodied snakes with heads that are conspicuously wider than the neck; also, their eyes have vertically elliptical pupils. Most vipers are ambush predators that rely on their cryptic patterning and coloration to blend into their surrounding environment. This family includes the rattlesnakes (genera *Crotalus* and *Sistrurus*), which have evolved specialized scales at the end of the tail that produce sound when vibrated and are used as a warning device to alert intruders of the snake's presence. Other species vigorously vibrate their tail (which lacks a rattle), rub their keeled scales together, or even hiss in defense. Viper venom varies in potency among species. Venom is a solution of many different enzymes, each of which can affect a different aspect of a bite victim's physiology. Most species possess enzymes that act on the vascular system, causing destruction of red blood cells, coagulation of the blood, clotting of blood vessels and resultant hemorrhaging, and internal organ breakdown and failure. Intense pain, swelling, and even necrosis occur around the bite location. Although many species are capable of killing a human with a single bite, death from viper bites are extremely rare in Tennessee. Regardless, one should seek medical attention if bitten by any venomous snake.

Several species are abundant and commonly encountered, such as Copperheads (*Agkistrodon contortrix*), whereas other species have experienced dramatic declines and extirpations of local populations throughout their respective ranges. Many species of Rattlesnakes are persecuted and killed on sight. Thousands of individuals are captured, sold, and killed for food or the production of snakeskin wallets, hats, boots, and so forth, each year at rattlesnake roundups. Rattlesnake roundups, highly controversial local events held at various localities in the Midwest and southern United States, result in millions of dollars of revenue for local communities but result in the destruction of local wildlife.

The family name, Viperidae [vye-PEAR-rih-dee], is from the Latin *vipera* (snake). The etymology of the three genera that occur in Tennessee is as follows. The genus name *Agkistrodon* [ag-KISS-troh-don] is from the Greek *ankistron* (fishhook) and *odontos* (teeth) and describes the shape of the enlarged fangs. The genus name *Crotalus* [KROW-tal-iss] is from the Greek *krotalon* (rattle) and describes the unique sound-producing scales on the tip of the tail. The genus name *Sistrurus* [sis-TRUE-russ] is from the Latin *sistrum* (rattle) and the Greek *oura* (tail), another description of the rattle.

Northern Copperhead adult, Wilson County. (Photo by Matthew L. Niemiller)

Copperhead

Agkistrodon contortrix

Description: Copperheads are handsomely patterned, medium-sized pit vipers, with most adults 50–120 cm (20–47 in) TL, but older adults approach lengths of 137 cm (54 in). Females are generally smaller than males and reach maturity at a smaller size. They are moderately stout snakes, especially when threatened, as they will flatten out their heads and bodies. Body scales are distinctively keeled and the anal plate is undivided. Dorsal coloration consists of a background of brown or orangish brown to grayish brown. A distinct pattern of dark brown hourglass or saddle shapes occurs down the dorsum, with the pattern being wide on the sides and constricted in the middle. Each band is lighter brown in the middle with darker brown borders, and often contains a dark brown spot in the center of each wider lateral element of the band. Lateral band elements are occasionally off center relative to the opposite side when viewed from above. There are occasionally additional small dark spots interspersed with the bands. The head is distinctly wider than the neck and appears arrowhead shaped. The top of the head is the same color as the dorsum, although a pair of dark spots is usually present on top of the head. Underneath the eyes and nares typically is lighter brown than the top of the head, and the venter is a similar lighter brown to gray color with dark brown blotches. The eyes have vertically elliptical pupils, although the pupils will appear more rounded when the snake is first encountered at night; the eyes will then become more elliptical after light has been shined on them. Loreal (heat-sensory) pits are present on each side of the head between the eye and nostril. A single row of scales is present under the tail. Young are 20–25 cm (8–10 in) TL at birth and have a distinct yellow tail tip, which persists into the juvenile stage. Five subspecies

are recognized, and two occur in Tennessee: the Southern Copperhead (*A. c. contortrix*) in extreme West Tennessee and the Northern Copperhead (*A. c. mokasen*) across the rest of the state. A large contact zone occurs between these two subspecies in West Tennessee, and it is not clear where the range of the Northern Copperhead ends and that of the Southern Copperhead begins, although most individuals from the southwestern counties resemble Southern Copperheads. Southern Copperheads generally have narrower crossbands connecting lateral pattern elements, whereas Northern Copperheads have crossbands generally at least three scales wide, although this is a highly variable character. Southern Copperheads also usually lack the dark spots between and within the larger dorsal saddles.

Etymology: The specific and subspecific epithet, *contortrix* [kon-TORE-tricks], is Latin for "a female contortionist," although this reference is unclear. The subspecific epithet, *mokasen* [MOCK-ah-sin], is an Anglicized translation of the Native American (dialect unknown) *moccasin*, a type of footwear made from animal skins, which might refer to the coloration of this subspecies.

Similar Species: Copperheads might be mistaken for Milksnakes (*Lampropeltis*

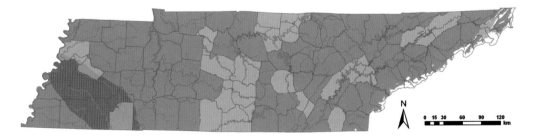

Two subspecies of copperhead are recognized in Tennessee: the Northern Copperhead (*Agkistrodon contortrix mokasen*), whose range appears in orange, and the Southern Copperhead (*A. c. contortrix*), whose range appears in blue. A contact zone between the subspecies occurs in southwestern Tennessee, shown in purple.

Top left: Dorsal patterning of a Southern Copperhead, Hardeman County. (Photo by Matthew L. Niemiller) *Top right:* Venter of a Southern Copperhead, Hardeman County. (Photo by Matthew L. Niemiller) *Bottom right:* Head of a Northern Copperhead adult, Blount County. (Photo by Matthew L. Niemiller)

triangulum), although the pattern on Milksnakes is rounded blotches and not distinct hourglass or saddle markings. The patterning of Copperheads also resembles that of juvenile Gray Ratsnakes (*Pantherophis spiloides*), which have dark blotches but not saddles. Both Milksnakes and Gray Ratsnakes have round pupils and lack loreal pits. Juvenile Cottonmouths (*A. piscivorus*) closely resemble Copperheads, including the presence of a yellow tail tip, but they have a broad, dark stripe running from the snout through the eyes to the back of the head, as opposed to a narrow, dark stripe in Copperheads. Juvenile Copperheads also tend to be more reddish than juvenile Cottonmouths.

Distribution: Copperheads are widely distributed across the eastern United States, from southwest Texas north to southeastern Nebraska and Missouri east to eastern Massachusetts, although they are absent from peninsular Florida. In Tennessee, this species occurs statewide and is one of the most abundant snakes in some areas, such as the Great Smoky Mountains National Park. However, some areas in Middle and West Tennessee lack records, potentially indicating a low abundance or lack of populations in these areas, as Copperheads are usually well documented due to the high volume of road mortality and human encounters.

Habitat: Copperheads occur in a wide variety of habitats, from primary forest around Joyce Kilmer Memorial Forest (border with North Carolina) to disturbed areas, such as back yards and, occasionally, urban areas. They prefer some forest cover and often are found near, although rarely in, water. Interestingly, individuals in Middle Tennessee were found to obtain more food when they occupied open old field areas in contrast to forested areas. Juveniles are particularly fond of more mesic areas where they hunt amphibians. Populations, or at least individuals, can occupy small forest patches surrounded by agriculture or urban development, hence they are frequently found by homeowners or farmers in close proximity to buildings or areas with lots of cover, such as barns or among farm implements or wood-piles. However, this species can most easily be observed when crossing roads at night after or during a warm rain. Indeed, in some areas, such as the Smoky Mountains, driving slowly along a road on a rainy summer night is virtually guaranteed to produce a Copperhead, or sadly, at least one that has been run over.

Natural History: Copperheads are nocturnal during the warmer months, usually from late April to October. They appear to be most active during or after warm rains, and might be encountered on trails or roads after dark. Mating generally occurs in the spring, although mating will also take place in autumn, and male combat sometimes occurs when multiple males are tracking a receptive female. Like most snake combat, males wrestle with their upper bodies, attempting to pin the other male, although they never bite each other. One to 20 young are born live in early autumn. Sexual maturity is reached in three years. Copperheads are generalist feeders and will consume most small vertebrate and invertebrate prey. Copperheads, especially juveniles, feed frequently on amphibians and will either ambush or pursue prey. Adults in Middle Tennessee consume mostly small mammals, such as voles (*Microtus*) and mice (*Pero-*

myscus), as well as the caterpillars of Io moths (*Automeris io*). Interestingly, females in Middle Tennessee consume mostly lizards and shrews. Juveniles have a particularly interesting method of attracting prey. They wave their yellow tail tip to attract amphibians, as it resembles a caterpillar or other soft-bodied invertebrate. When the prey approaches close enough, the Copperhead strikes and envenomates, then releases the prey and waits for it to cease struggling before consuming it. Copperheads are eaten by small predatory mammals and birds that are fast enough to avoid being bitten, as well as by other snakes, including Eastern Black Kingsnakes (*Lampropeltis nigra*) and Racers (*Coluber constrictor*). Copperheads vary in their disposition; many are quite shy and docile, although they will occasionally strike when harassed or cornered. Bites should be considered dangerous, although they are rarely fatal unless the person has an allergic reaction to the venom and goes into anaphylactic shock. Nevertheless, any bite should receive prompt medical attention. Copperheads, like most snakes, often exude a potent musk when handled.

Conservation Status: Copperheads are generally common where they occur and may be abundant in relatively undisturbed areas such as national parks or national forests. However, this species suffers tremendous persecution when discovered by humans, and it frequently is the most common snake killed on roads.

R. Graham Reynolds and
Matthew L. Niemiller

Western Cottonmouth adult, Lake County. (Photo by Brad M. Glorioso)

Cottonmouth

Agkistrodon piscivorus

Description: Cottonmouths are medium-sized, stout-bodied pit vipers, with most adults 75–120 cm (29.5–47 in) TL but reaching 157 cm (62 in). Body scales are heavily keeled and the anal plate is undivided. Dorsal coloration is variable, although most adults are dark and have an olive brown to black dorsal coloration with some crossbands or blotches of lighter brown to black, especially on the sides. Many adults are melanistic and have no pattern vis-ible. The venter is brownish cream and has a large amount of dark brown or black mottling, such that it sometimes appears black with a few white blotches. The head is distinctly wider than the neck and appears arrowhead shaped. The pupils are vertically elliptical, although the iris is dark and not easily distinguished from the pupil, and a dark mask extends from the nares through the eyes to the back of the head. The upper labials are usually white, and the interior of the mouth, which is frequently on display, is pure white. Loreal (heat-sensory) pits are present on each side of the head between the eye

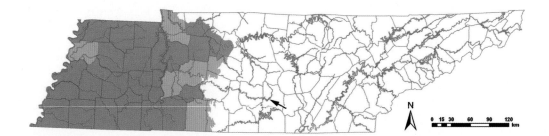

and nostril. A single row of scales is present under the tail. Males are typically longer and heavier than females. Young are 15–28 cm (6–11 in) TL, and both young and juveniles are much more strongly patterned than adults. They are generally pale brown with dark brown or black saddles, a dark brown mask, and a bright yellow tail tip. There are often small black spots in and/or between the saddles. The juvenile patterning becomes obscured with age, although a few individuals may retain the juvenile patterning into adulthood. Three subspecies are recognized, but only the Western Cottonmouth (*A. p. leucostoma*) occurs in Tennessee.

Etymology: The specific epithet, *piscivorus* [piss-SIV-or-rus], is from the Latin *piscis* (fish) and *voro,* (eat), in reference to the diet of some populations of this species.

Similar Species: Cottonmouths are universally confused with watersnakes in the genus *Nerodia*. In fact, very few untrained people can accurately differentiate Cottonmouths from nonvenomous watersnakes. The default reaction, even in eastern Tennessee, is to assume all watersnakes are venomous Cottonmouths. In spite of this, Cottonmouths can be easily distinguished from watersnakes by a number of characters. Cottonmouths are generally stockier, have paired loreal pits between the nostrils and the eyes, have vertically elliptical pupils, and have a single row of scales on the underside of the tail. Cottonmouths will also gape when threatened, a behavior watersnakes usually never exhibit. Young Cottonmouths can be easily confused with Copperheads (*A. contortrix*), as the dorsal patterns are similar. Copperheads and Cottonmouths also have the same profile when crossing roads—straight body with the head elevated. However, Cottonmouths have a broad dark stripe that appears like a mask from the snout through the eyes and white or cream-colored upper labials, whereas Copperheads have a narrow dark stripe extending posteriorly from the snout through the eyes. Additionally, juvenile Copperheads tend to be more reddish than juvenile Cottonmouths.

Distribution: Cottonmouths occur throughout much of the southeastern United States from central Texas east through the Florida peninsula and north to southern coastal Virginia. In Tennessee, Cottonmouths are found mainly in the western part of the state, west of the north-flowing Tennessee River in the Coastal Plain ecoregions. However, populations do occur to the east, along the Cumberland River in Montgomery, Cheatham, and Dickson counties, as well as Lewis County to the south. Individuals also have been reported along the Harpeth River in Davidson County. An isolated population occurred along the Duck River in Coffee County and was widely believed to be an introduction. However, it's unknown whether this population is still extant. Contrary to popular belief, Cottonmouths do not occur east of the Cumberland Plateau. Isolated reports are either misidentifications, or perhaps very rarely represent individuals that have been transported unintentionally or otherwise. However, Cottonmouths would likely not be able to become established in the cold mountain streams and rivers in eastern Tennessee.

Habitat: Cottonmouths inhabit nearly any aquatic habitat, including lakes, reservoirs, ponds, streams, rivers, bayous, swamps, ephemeral wetlands, and bogs. This species also is commonly found in urban waterways around people. They bask on exposed vegetation, logs, rocks, and on grassy and muddy banks. Cottonmouths

Top left: Western Cottonmouth juvenile, Cheatham County. (Photo by Lisa Powers) *Top right:* A Western Cottonmouth swimming, Lake County. (Photo by Collin Jaeger) *Bottom right:* Western Cottonmouth in defensive posture, Obion County. (Photo by Brad M. Glorioso)

are particularly easy to observe in the morning when basking or at night along the margins of waterways. Indeed the editors (RGR and MLN) have routinely encountered this species in swamps while searching for amphibians. Cottonmouths characteristically swim with their head above the water surface. This species can be locally common in appropriate habitat. Cottonmouths are active both at night and during the day during summer. A warm rain will bring snakes out searching for food, and they are often seen crossing roads near water. This species retreats underneath logs, in root masses, or in crayfish or mammal burrows during cold weather.

Natural History: The natural history of Cottonmouths has been well studied, although Tennessee populations have not. They are generally active during most of the warmer months, from March to October. Cottonmouths are diurnal in cooler months but are mainly a nocturnal species throughout most of the active season. Mating occurs most frequently in the spring once they emerge from hibernation sites. Males engage in combat over a female, which involves a wrestling match in which males attempt to pin rivals with their upper bodies. Two to 15 live young are born at the end of summer, and sexual maturity is reached in three years. Females might reproduce either every year (annual) or every other year (biennial). Cottonmouths are truly generalist predators and will consume small mammals, small amphibians, fish, insects, birds, and other snakes. They will also consume carrion and can frequently be found attempting to consume road-killed frogs on rainy summer nights. Adult Cottonmouths have few predators, although they are eaten by American Alligators (*Alligator mississippiensis*). Juveniles are preyed upon by other Cottonmouths and wading birds. Cottonmouths are quick to engage a defensive display that includes throwing the head back and gaping the mouth to show the immaculate white interior (hence the common name). They will usually not strike unless actually picked up, although if the threat display fails and they continue to be harassed, they will certainly attempt to strike. Cottonmouths might also strike if heavily tread upon. If captured they will smear a foul-smelling musk from their vent. Cottonmouths should be considered dangerously venomous, and any bite should receive immediate emergency medical attention. Although rarely fatal, bites are very painful often becoming septic if not properly treated and can result in the loss of fingers or toes.

Conservation Status: Cottonmouths tend to be common where they occur and thus are not tracked by state agencies. However, this species is heavily persecuted and is almost always killed on sight. Cottonmouths are also killed on roadways when attempting to cross.

Comments: Cottonmouths are frequently referred to as "water moccasins," "water adders," "gaping vipers," "trapjaws," or "moccasins." In spite of local lore, this species does not actually pursue boats; instances of Cottonmouths climbing into boats are likely due to confusion or being weary from a long swim. Additionally, reports of people being chased by this species are likely attributable to instances where people blocked a Cottonmouth from retreating to the water. They are indeed bold and will head straight towards a person to get to the water.

R. Graham Reynolds and
Matthew L. Niemiller

Timber Rattlesnake adult, Wilson County. (Photo by Brad M. Glorioso)

Timber Rattlesnake

Crotalus horridus

Description: Timber Rattlesnakes are large, robust vipers, usually 100–150 cm (39–59 in) TL, but specimens over 180 cm (70 in) TL have been recorded. Body scales are keeled and the anal plate is undivided. Their pattern and coloration varies dramatically with location (or latitude) and elevation, but they consistently exhibit the following characteristics: dark, wavy or otherwise angular crossbands on a pale gray, yellow, tan, pinkish, or otherwise pale background; a brown, rusty, reddish, or orange dorsal stripe extending most of the length of the body; a single, dark, post-orbital band extending from the eye to the corner of the mouth on a relatively unpatterned, pale-colored head; and a relatively pale

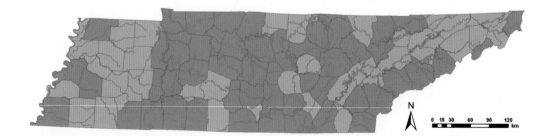

colored rattle when compared to the tail region that gradually transitions to a darker color than the rest of the body (hence the vernacular name "velvet-tail" in some regions). Older individuals are often highly melanistic, even entirely black. The ventral scales are white or cream colored but may be infused with melanin in old individuals. The head is distinctly wider than the neck and appears arrowhead shaped. The pupils are vertically elliptical. Loreal (heat-sensory) pits are present on each side of the head between the eye and nostril. The tail ends in a segmented rattle. Sexual dimorphism is not dramatic; males are slightly larger than females of the same age, have a slightly more swollen tail base, and have a slightly longer tail, but this would be very difficult to distinguish without a hemipenal probe or without having both sexes in hand (or preferably, in a snake tube!). Neonates are 20–40 cm (8–16 in) TL and always have a series of dark bands across their back. No subspecies are currently recognized, despite the huge amount of variation in color and pattern across their range, molecular genetics studies indicate that this species possesses minimal "geographic genetic structure" (variation) across their range.

Etymology: The specific epithet, *horridus* [HORE-reh-dus], is Latin for "horrible" or, more literally, "dreadful." The vernacular name refers to its habitat; other vernaculars include "canebrake," which refers loosely to a valley or ravine, "velvet-tail," which refers to the darker tail, and "rattler."

Similar Species: Juvenile and young adult Timber Rattlesnakes are fairly colorful and could easily be mistaken for similarly sized Western Pygmy Rattlesnakes (*Sistrurus miliarius streckeri*), as the latter also have a broad reddish or orange stripe along the dorsal midline; however, the former has complete dorso-lateral bands on the body and a single eye stripe, whereas the latter has unconnected dorsal and lateral spots and usually has two dark, wavy stripes on the dorsal surface of the head. Melanistic Timber Rattlesnakes might be confused with Cottonmouths (*Agkistrodon piscivorus*), but the light-colored rattle, if not heard first, will be easy to see, and Cottonmouths, upon being seen, often quickly open their mouths in a defensive display. From a distance, some watersnakes (genus *Nerodia*) may look similar if they are found flattened out on land (basking or defensive display), but watersnakes are much narrower in profile and have a much smaller, less well defined head (with a wider neck) and horizontal or round pupils, and they lack the rattle.

Distribution: Timber Rattlesnakes occur throughout the eastern half of the United States, from New England to northern Florida in the east and from Minnesota to eastern Texas in the west, and are distributed statewide in Tennessee, where they can be found from the lowest Mississippi River bottomlands to the peaks and balds of the Blue Ridge Mountains. However, populations are extremely localized throughout this range and individuals are infrequently encountered except in certain well-preserved areas.

Habitat: True to their name, Timber Rattlesnakes prefer mature hardwood and mixed forests, pine forests, bottomland hardwood forests, and even swamplands of large river systems. They are most often encountered basking in sunny spots where light penetrates the forest floor, or crossing trails and occasionally roads. They can also be found beneath rocks or other large cover objects such as boards, especially in areas frequented by rodents. Timber Rattlesnakes

Left: Timber Rattlesnake adult, Wilson County. (Photo by Matthew L. Niemiller) *Right:* Timber Rattlesnake adult, Coffee County. (Photo by Matthew L. Niemiller)

are sometimes, although certainly not always, heard before they are seen. They den and hibernate in sandstone escarpments, tablelands, rocky outcrops, steep bluffs, boulder fields, and talus slopes flanking the largest river systems to the smallest creeks and canebrakes on the Cumberland Plateau and Blue Ridge Mountains.

Natural History: Timber Rattlesnakes are mainly active from spring to autumn and are rarely seen outside of their hibernacula during winter. Like most vipers, Timber Rattlesnakes are largely nocturnal. Radio-tracking studies indicate they rarely travel over 2 km (1.2 miles) from their dens or hibernacula. Although they shed their skin and rattle regularly, adding a new rattle segment with each shed, the number of rattles on the tail cannot be reliably used to determine their age, unless combined with some type of capture-mark-recapture study where the age of individuals can be accurately determined at the outset. Reproductive movements begin in the autumn months, and they are regularly seen crossing roads to move between the sandstone bluffs and boulder fields of the Cumberland Plateau, especially in the Big South Fork area. Reproductive activity largely occurs in a den, which may also serve as their hibernaculum. Mating occurs during autumn when males can be observed in "combat," a sort of dance where they rear up and push each other around until one gives up. Females are ovoviviparous, meaning they retain soft, unshelled eggs in the oviduct until they hatch; thus they give birth to live young. Litter sizes range widely with elevation but are generally 8–12 with a maximum recorded of 30 (in Georgia). The young are birthed in late autumn and usually stay near the den or track the scent trails of adults back to the hibernaculum. Females become reproductively mature in 6–10 years but generally only breed every other year or every 3–4 years, depending on elevation (longer at higher elevations). Timber Rattlesnakes are classic "sit-and-wait" or ambush predators, remaining in one spot for hours or days waiting to ambush prey. They take mostly small mammals, particularly nocturnal rodents such as mice, rats, and voles. They also prey on diurnal species, including rabbits, squirrels, chipmunks, and the young of larger mammals, numerous species of forest-dwelling songbirds, and more rarely, ectotherms such as lizards and insects.

They are vulnerable, especially when young, to a myriad of other species, including coyotes, bobcats, hawks, owls, kingsnakes, feral hogs, cats, dogs, and, of course, humans. Timber Rattlesnakes usually are not inclined to bite when encountered, usually biting only when physically bothered. Bites from this species should be considered particularly dangerous to humans and immediate medical attention should be sought if bitten. Although rarely fatal, bites are very painful and can become septic if not properly treated.

Conservation Status: Although destruction and fragmentation of forested habitat are the primary reasons for its endangerment, human disdain for snakes in general, and vipers in particular, has also contributed to the downfall of this beautiful beast. As a result, Timber Rattlesnakes have been extirpated from many isolated forest patches and parks and are absent from entire regions within their range. The USFWS lists the species as a "Species of Special Concern" largely because of its disappearance from around certain large human population centers, but in Tennessee, TWRA has considered Timber Rattlesnakes "Threatened" since 2009. Ultimately, it will prove very difficult to prevent the demise of this species, especially in the face of rapid human expansion into wilderness, fragmentation of forested habitat, and so much hatred for all things venomous.

Comments: This large, dangerous snake should be avoided when encountered and never handled. Venom is delivered with fangs up to 10 mm (0.4 in) long, and is a strong hemotoxin that can cause extensive tissue damage, loss of body parts, and potentially death. Despite their dangerous bite, these snakes are important elements of the natural world, and should be respected not only for their venom but also for their beauty, for their role as predators, and for their incredible array of evolutionary adaptations. The paired loreal pits are probably the most interesting feature of these snakes; they can sense temperature differentials of less than a tenth of a degree. Because they are paired, the pits provide depth of field to these heat sensors, allowing them to calculate distance, and possibly even generate an image. Thus, an entire group of snakes evolved a novel structure to detect an entirely different range of wavelengths in the electromagnetic spectrum.

Todd Campbell

Pygmy Rattlesnake adult, Trigg County, Kentucky. (Photo by David F. Frymire)

Pygmy Rattlesnake

Sistrurus miliarius

Description: Pygmy Rattlesnakes are small rattlesnakes with tiny rattles and a thin tail. Most adults are 36–50 cm (14–20 in) TL, but they can attain lengths of at least 64 cm (25 in). Body scales are keeled and the anal plate is undivided. Nine large scales cover the crown of the head. The ground color of the dorsum typically is tan to grayish brown to gray, and a dorsal series of dark blotches with lighter borders extends from the neck to the tail. A gold to orange to reddish stripe often is present along the midline of the dorsum. Dark blotches or spots are also present in one to three rows on the flanks. The venter usually is light colored and often mottled with darker pigment. The head is dis-

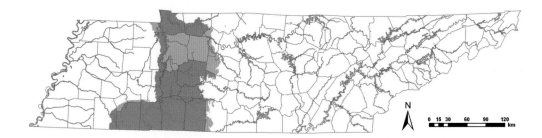

tinctly wider than the neck and appears arrowhead shaped. A dark stripe runs from the eye to the corner of the mouth on each side of the head. Two irregular dark stripes are present on top of the head. The pupils are vertically elliptical. Loreal (heat-sensory) pits are present on each side of the head between the eye and nostril. The rattle is small and the individual segments often break off so that some individuals have few to no segments, only a basal button. The rattle produces a soft, high-pitched sound that is insect-like and is usually only audible at short distances. Males have a slightly longer tail than females. Young are about 12–18 cm (5–7 in) TL at birth. Both young and juveniles resemble adults but have yellow to green coloration on the tail. Three subspecies are recognized, but only the Western Pygmy Rattlesnake (*S. m. streckeri*) is present in Tennessee.

Etymology: The specific epithet, *miliarius* [mill-lee-AIR-ree-us], is Latin for "millet" or "millet-like," which likely refers to the blotches on the dorsum. The subspecific epithet, *streckeri* [STREH-ker-i], honors John Strecker Jr., a prolific Texas natural historian.

Similar Species: Pygmy Rattlesnakes are most similar in appearance to other vipers, such as Timber Rattlesnakes (*Crotalus horridus*), Copperheads (*Agkistrodon contortrix*), and Cottonmouths (*A. piscivorus*). Both Copperheads and Cottonmouths lack rattles on the tail. Timber Rattlesnakes attain a much larger size, usually have a larger rattle that generates more noise, and have several smaller scales on the crown of the head rather than just nine large scales as in Pygmy Rattlesnakes. Pygmy Rattlesnakes also resemble Eastern Hog-Nosed

Pygmy Rattlesnake adult, Carroll County, Arkansas. (Photo by Greg Sievert)

Snakes (*Heterodon platirhinos*), but Eastern Hog-Nosed Snakes have round pupils and lack a rattle on the tail and loreal pits on the head.

Distribution: Pygmy Rattlesnakes occur throughout the southeastern United States, particularly in the Coastal Plain from southern North Carolina south into Florida, and west into eastern Texas, eastern Oklahoma, and southern Missouri. This species apparently is absent from the Mississippi River Valley. In Tennessee, Western Pygmy Rattlesnakes have been documented from a few scattered localities in the Western Highland Rim, including Benton, Decatur, Hardin, Hickman, Lawrence, Lewis, Perry, Stewart, Tipton, and Wayne counties, and two localities in the southern Southeastern Plains and Hills in Hardeman and McNairy counties. Although not frequently encountered in Tennessee, they are common in some small areas.

Habitat: Pygmy Rattlesnakes occur in a number of habitats, particularly those with rocky substrates and near water, including upland hardwood and pine forest, cropland, floodplains, marshes, swamps, and wet prairies. During warm days of summer, this species becomes active in the late afternoon and can be found basking on roads at night, as they absorb heat from the ground. Individuals are known to climb several feet above ground in bushes and vines when their habitat becomes flooded.

Natural History: Little is known about the natural history of Tennessee populations. Pygmy Rattlesnakes likely are active on the surface from March into October, and they hibernate during cold weather. Mating typically occurs in late summer and early autumn but also can occur during spring. Females give birth to 5–10 (but up to 32) live young in the summer following autumn mating. Litter size is positively correlated with female body size. Only one litter is produced during a year, but individuals usually reproduce every year. Sexual maturity is reached in 2–3 years. Adults capture prey primarily by ambush but will also actively hunt and stalk some prey. Adults are opportunistic with respect to diet but seem to show a preference for amphibians and reptiles. Prey of adults includes salamanders, frogs, lizards, small woodland snakes, and small mammals, such as shrews and mice. Young feed on invertebrates, including crickets, spiders, and centipedes. Their small size makes Pygmy Rattlesnakes more vulnerable to predation than larger pit vipers. Predators include mammals (opossums, skunks, domestic dogs and cats), birds of prey,

and other snakes, such as Racers (*Coluber constrictor*) and Black Kingsnakes (*Lampropeltis nigra*). Pygmy Rattlesnakes are well camouflaged and very difficult to see in their native habitat. Consequently, their first line of defense is to remain motionless in a coiled position. Unlike other vipers, Pygmy Rattlesnakes are less inclined to bite when bothered, usually biting only when picked up. This species is not considered particularly dangerous to humans because of their small size and commensurate small amount of venom injected. Bites are painful, but nearly all victims who receive proper medical treatment fully recover.

Conservation Status: Because of its limited distribution and few occurrences, Western Pygmy Rattlesnakes are considered imperiled to vulnerable in Tennessee. Accordingly, TWRA lists this species as "Threatened." However, the status of Tennessee populations is largely unknown and certainly warrants study. Presumed threats include loss of habitat, road mortality, persecution, and timber and agricultural practices that decrease suitable Pygmy Rattlesnake habitat. However, this species might be more common than current data suggest, as individuals are well camouflaged and difficult to detect.

Comments: Pygmy Rattlesnakes are often called "ground rattlers."

Matthew L. Niemiller and
R. Graham Reynolds

Field Notes: Reelfoot Records

Perhaps my favorite location in all of Tennessee to enjoy the outdoors and wildlife is the Reelfoot Lake area in the extreme northwestern part of the state. Reelfoot Lake is a shallow but large natural lake that formed when the land in the region subsided during the New Madrid earthquakes of 1811–12. Because the lake is just three miles from the Mississippi River, it's a major stopover and wintering area for several species of waterfowl and other migratory birds, which follow the course of the river. But what draws me back to Reelfoot Lake is the incredible diversity of amphibians and reptiles, particularly snakes and turtles.

Although my graduate work focused on cave-dwelling salamanders and fishes, I was fortunate to visit Reelfoot Lake several times to assist on other field projects. My good friend and roommate Brad "Bones" Glorioso introduced me to the swampy sloughs and backwaters of Reelfoot in 2004, when I was at Middle Tennessee State University. Bones grew up in

Louisiana and had a great deal of experience with the herpetofauna associated with Mississippi Valley swamps and backwaters. At the time, Bones was working toward his master's degree studying the population ecology and feeding behavior of Stinkpot (*Sternotherus odoratus*) turtles.

From late spring into autumn, Bones made weekly treks to Reelfoot to trap turtles and gather data for his thesis, in addition to assisting on a larger project on the conservation status of the lake's turtle populations funded by the Tennessee Wildlife Resources Agency. I was first to offer Bones my services when I wasn't working on my own thesis. I was there to support Bones and his project, but deep down my real reason for volunteering was the opportunity to "kick around" in the swamp for a few days and see some fascinating creatures! While I have many fond memories of Reelfoot, there is one trip in May 2004 that stands above the rest.

Bones and I left Murfreesboro and began the four-hour drive to Reelfoot around 9:00 a.m. During the drive, we discussed our plans for the next several days of fieldwork: How does the weather look? Where are we going to trap for turtles? What baits are we going to try? But in my mind, there was a bigger question: Will we have time to "herp"—that is, to go and look for as many amphibians and reptiles as we can find? I had purchased a digital camera a few months earlier and was eager to get photographs of several species that I had never seen before in the wild in Tennessee, such as the Western Cottonmouth (*Agkistrodon piscivorus*) and Western Mudsnake (*Farancia abacura*). We arrived at the field station in Samburg around 1:00 p.m. and decided that we would set turtle

---

A roadside ditch near Reelfoot Lake is excellent habitat for snakes and turtles. Lake County. (Photo by Brad M. Glorioso)

traps in the lake the day after next, leaving 36 hours to get all the gear together and explore Reelfoot for herps.

Bones and I are both very competitive, and we decided to attempt to break my own personal record of observing 27 different species of amphibians and reptiles in Tennessee within a 24-hour period. For many of you reading this note, this comment might sound like an unusual endeavor, but for amateur and professional herpetologists alike, it is a metric to gauge one's abilities as a herpetological naturalist. The Reelfoot Lake area includes a variety of habitats in addition to the cypress swamps and sloughs—from ephemeral wetlands to bottomland forests to upland hills, offering a range of places in which snakes, lizards, and turtles live and thrive. With such a diversity of environs for amphibians and reptiles, I thought we had a very good shot of reaching our goal.

The next morning started off well, as we searched along an irrigation ditch on the western side of the lake. We found several Yellow-Bellied Watersnakes (*Nerodia erythrogaster flavigaster*), Broad-Banded Watersnakes (*N. fasciata confluens*), and Western Ribbonsnakes (*Thamnophis proximus*) in branches and vines overhanging the water, attempting to catch the morning sun in an effort to warm their bodies. Western Cottonmouths were up as well, but rather than climbing into vegetation, these bulky, venomous serpents coiled up on the bank amid the vegetation. As I walked through the high grass along the bank, I nearly stepped on a dark three-foot adult. "Whoa! Found a 'mouth!" I shouted to Bones, who calmly replied, "Yeah, they are pretty common here. Watch your step." Now he tells me, I thought. The cottonmouth

told me, too, giving me an early morning greeting with its head up and mouth agape, revealing the cotton white lining of the roof of its mouth. Good morning to you, too!

By early evening, Bones and I had caught or observed 27 species of herps, including 20 species of reptiles. My personal record had already been tied. As the sun set in the western sky, we decided to try our luck in finding several species of frogs that might reveal their presence by calling on a warm spring night. While the anurans pushed us over our goal, the highlight of the night was the capture of a reclusive snake not often seen alive in Tennessee. Bones and I were walking along another ditch on the northwestern side of Reelfoot Lake shortly after dark, attempting to locate and capture calling frogs as well as dipnetting for young turtles residing in the dense aquatic vegetation near the bank. Bones was a few yards in front of me when I saw him kneel down next to the water, intensely focused on something that had caught his eye. Suddenly, he reached in the water, pulled up a medium-sized snake, and held it above his head. My first thought was that it was just another watersnake, like the dozens we had already seen that day. But this snake looked different. I thought I saw a flash of red in the light of my headlamp. I began to go through the list of snakes in my head that had red pigmentation but might also be found swimming in a roadside ditch in West Tennessee. As I approached Bones and his catch, I quickly noticed that this snake was dark over most of its body and had very smooth, shiny scales. As Bones flipped over the snake to show me its belly, I realized what we had—a mudsnake! Although not uncommon, this beautiful snake is very secretive and, unfortunately, usually encountered dead on roads, victims of vehicular snake slaughter as they move about their landscape. Finding a mud-snake alive in aquatic habitat was a special treat, and it was the most memorable of the 32 species of amphibians and reptiles we documented in the 24-hour period. Mission accomplished.

Matthew L. Niemiller

Red-Eared Slider hatchling. (Photo by Lisa Powers)

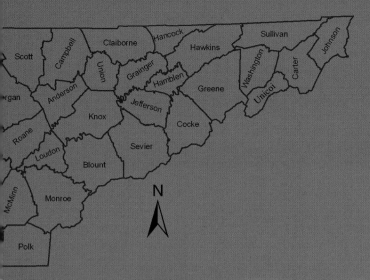

Part 4
Turtles

Eastern Mud Turtle hatchling, Cheatham County (Photo by Lisa Powers)

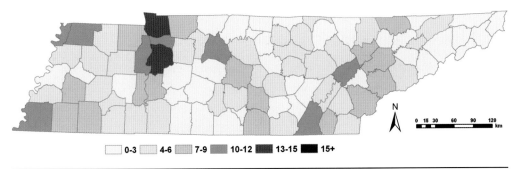

0-3   4-6   7-9   10-12   13-15   15+

Diversity of turtle species in Tennessee based on county records.

1a.   Shell reduced and covered in leathery skin; snout elongate and terminating in tubular proboscis ➜ **2 (Family Trionychidae)**

1b.   Shell not reduced and covered with scutes; snout typically not elongate and not terminating
in a tubular proboscis ➜ **3**

2a.   Nostrils round and lacking horizontal ridges projecting from septum; anterior edge of carapace without tubercles or spines—**Midland Smooth Softshell (*Apalone mutica mutica*)**

2b.   Nostrils crescent shaped and with a horizontal ridge projecting from each side of septum;
anterior edge of carapace with tubercles or spines—**Eastern Spiny Softshell (*Apalone
spinifera spinifera*)**

3a.   Plastron with 10–11 scutes present ➜ **4 (Family Kinosternidae)**

3b.   Plastron with 12 scutes present ➜ **7**

4a.   Pectoral scutes triangular; no skin evident between plastral scutes; two transverse hinges
between pectoral and abdominal scutes ➜ **5**

4b.   Pectoral scutes rectangular; skin usually evident between plastral scutes; a single indistinct
transverse hinge between pectoral and abdominal scutes ➜ **6**

5a.   Two light stripes on each side of the head; from extreme western Tennessee along Mississippi River—**Mississippi Mud Turtle (*Kinosternon subrubrum hippocrepis*)**

5b.   Sides of head without two distinct light stripes but with yellow spotting, mottling, or irregular streaks—**Eastern Mud Turtle (*Kinosternon subrubrum subrubrum*)**

6a.   Two prominent light stripes on each side of head; barbels on both chin and throat—**Eastern Musk Turtle or Stinkpot (*Sternotherus odoratus*)**

6b.   Two prominent light stripes lacking on each side of head; barbels on chin only—**Stripe-Necked Musk Turtle (*Sternotherus minor peltifer*)**

7a.   Plastron reduced and cross shaped; inframarginal scutes present; tail longer than half the carapace length ➔ **8** (Family Chelydridae)

7b.   Plastron oval in shape and not reduced; inframarginal scutes absent; tail less than half the carapace length ➔ **9** (Family Emydidae)

Snapping Turtle

Short Beak

Alligator Snapping Turtle

Fleshy "Eyelashes"

Hooked Beak

Heads of snapping turtles in Tennessee. Alligator Snapping Turtles can be readily distinguished from Snapping Turtles by having an enlarged hooked beak and fleshy projections around the eyes that resemble eyelashes. (Photos by Brad M. Glorioso)

8a.   Carapace with three prominent keels extending the entire length; 3–5 supramarginal scutes between marginals and costals present on each side; upper jaw strongly hooked; eyes lateral and not evident when viewed dorsally—**Alligator Snapping Turtle (*Macrochelys temminckii*)**

8b.   Carapace keels not extending the entire length; supramarginal scutes lacking; upper jaw not strongly hooked; eyes dorsolateral and evident when viewed dorsally—**Snapping Turtle (*Chelydra serpentina*)**

9a.   Plastron with a hinge between pectoral and abdominal scutes ➔ **10**

9b.   Plastron without a hinge between pectoral and abdominal scutes ➜ **11**

10a.   Four toes on the hind feet; statewise except extreme southwest Tennessee—**Eastern Box Turtle (*Terrapene carolina carolina*)**

10b.   Three toes on the hind feet; from extreme southwest Tennessee—**Three-Toed Box Turtle (*Terrapene carolina triunguis*)**

11a.   Carapace without a keel; posterior edge of carapace not serrated but mostly smooth ➜ **12**

11b.   Carapace with or without a keel; posterior edge of carapace serrated ➜ **13**

12a.   Prominent red or yellow middorsal stripe present on carapace—**Southern Painted Turtle (*Chrysemys dorsalis*)**

12b.   Middorsal stripe on carapace indistinct or absent—**Midland Painted Turtle (*Chrysemys picta marginata*)**

13a.   Axillary and inguinal scutes absent or rudimentary; from Blue Ridge Mountains—**Bog Turtle (*Glyptemys muhlenbergii*)**

13b.   Axillary and inguinal scutes present and well developed ➜ **14**

14a.   Crushing surface of upper jaw with a ridge with or without turbercles extending parallel to its margin; apex of lower jaw pointed; margin of lower jaw smooth or serrate ➜ **15**

14b.   Crushing surface of upper jaw without a ridge extending parallel to its margin; apex of lower jaw rounded; margin of lower jaw smooth ➜ **17**

15a.   Crushing surface of upper jaw with a row of tubercles on ridge; margin of lower jaw coarsely serrate; a light C-shaped mark on second pleural scute—**Eastern River Cooter (*Pseudemys concinna concinna*)**

15b.   Crushing surface of upper jaw without a row of tubercles on ridge; margin of lower jaw smooth or finely serrate; light C-shaped mark on second pleural scute lacking ➜ **16**

16a.   Wide red to orange postorbital blotch or stripe present; chin stripes narrow—**Red-Eared Slider (*Trachemys scripta elegans*)**

16b.   Narrow yellow to orange postorbital stripe present; chin stripes broad—**Cumberland Slider (*Trachemys scripta troostii*)**

17a.   Vertebral keel poorly developed without prominent knobs (except in hatchlings); yellow post orbital spot separated from eye by short vertical or diagonal lines—**Northern Map Turtle (*Graptemys geographica*)**

17b.   Vertebral keel well developed with prominent knobs; postorbital markings consisting of large spots or narrow crescent-shaped marks behind and below eye ➜ **18**

Heads of *Graptemys* turtles in Tennessee. Note that identification in this group is largely based on spots and stripes behind and below the eyes. See Key for details. *From left:* False Map Turtle (*G. pseudogeographica pseudogeographica*) adult male, Mississippi Map Turtle (*G. p. kohnii*) juvenile, Ouachita Map Turtle (*G. ouachitensis ouachitensis*) adult.

18a.   Postorbital markings consisting of broad, broken crescent or spots behind eye and below eye; suborbital and mandibular spots large—**Ouachita Map Turtle (*Graptemys ouachitensis ouachitensis*)**

18b.   Postorbital markings consisting of narrow crescent behind and below eye; suborbital and mandibular spots small or absent ➔ **19**

19a.   Curved postorbital stripe elongate preventing neck stripes from contacting the orbit; iris usually white without a bisecting white line—**Mississippi Map Turtle (*Graptemys pseudogeographica kohnii*)**

19b.   Curved postorbital stripe not as elongate allowing 4–7 neck stripes to contact the orbit; iris yellow, greenish yellow, or white with or without a bisecting dark line—**False Map Turtle (*Graptemys pseudogeographica pseudogeographica*)**

17

Family Chelydridae (Snapping Turtles)

Snapping Turtles are large, aquatic turtles with large heads, powerful jaws, broad carapaces, small plastrons, and long, scaly tails. The four members of this family are found in North America and South America. Two genera occur in Tennessee, *Chelydra* [key-LIE-drah], from the Greek *chelys* (turtle) and *hydros* (water), and *Macrochelys* [mack-roh-KELL-ees], from the Greek *macro* (large) and *chelys* (turtle). Fossil Snapping Turtles are among the oldest known turtle fossils, dating to the Late Cretaceous period in North America and the Paleocene in Asia and Europe. This family includes some of the world's largest freshwater turtles, such as the Alligator Snapping Turtle (*Macrochelys temminckii*), which can reach 80 cm (31 in) CL and 113 kg (248.6 lbs). Likewise, some species are known to live 80 years in captivity, perhaps longer in the wild. Snapping Turtles derive their common name from their powerful jaws and cantankerous demeanor when encountered on land. All Snapping Turtles are aquatic and live in a variety of habitats, including ponds, lakes, reservoirs, swamps, sloughs, and quieter sections of major rivers and streams. Because of their large size and tasty flesh, legal commercial harvest of Snapping Turtles occurs in many states; however, data suggest that rates of harvest are unsustainable, and many states are reevaluating their commercial harvest policies. Nonetheless, hatchlings of Snapping Turtles are still harvested (sometimes illegally) and sent overseas, especially to China, and people continue to harvest wild individuals for consumption or sale to Asian markets.

Snapping Turtle

Chelydra serpentina

Description: Snapping Turtles are relatively large aquatic turtles with large heads, long tails, and small plastrons. Most adults are 20–36 cm (8–14 in) CL and weigh 4–16 kg (9–35 lbs) but can reach 49 cm CL and 34 kg (75 lbs). The record size for a captive individual is 39 kg (86 lbs). The CL of >250 adults captured in Murfreesboro, Rutherford County, ranged from 15 to 40 cm (6–16 in), and the mass from 0.8 to 16.9 kg (2–37 lbs). The carapace of adults is various shades of yellow, or golden brown to dark brown and may be monochromatic or streaked with darker pigmentation. However, mud or algae often obscure the color and markings. The margin of the anterior half of the carapace is smooth, whereas the margin of the posterior half is serrated. Three ridges extend the length of the carapace; one ridge is found along the dorsal midline in association with the vertebral scute row, and a ridge is found on the inner part of each of the two costal scute rows. Although prominent in younger turtles, the ridges become

Snapping Turtle adult, Rutherford County. (Photo by Matthew L. Niemiller)

reduced in height as the turtle ages; in older adults, the carapace can be smooth with no trace of ridges remaining. Furthermore, with age the carapace may become flattened, rather than vaulted, and in some individuals the midline is decidedly lower than the sides, so that some old turtles look squeezed in along the midline. The plastron and undersurface of the limbs are yellowish, but the upper surfaces of the limbs are pale to dark brown. The plastron is small and much soft tissue is exposed on the undersurface. The tail has a series of enlarged scales on the midline that extend from base to tip and form a distinct keel. The head is dark brown and the eyes are surprisingly colorful; the iris is a golden brown with dark lines radiating outward from the pupil to the edge of the eye, similar to spokes on a bicycle wheel.

The beak is similarly colored, with streaks of dark brown or black on a golden brown background. The neck is long with several tubercles that are also found on the head. Scattered enlarged scales adorn the anterior surface of the front limbs. Each foot is webbed and bears long, thick claws. Males grow larger than females, have thicker tails, and have a slightly concave plastron. Hatchlings are 20–32 mm (0.8–1.2 in) CL, are dark brown to black, have small white flecks on the undersurface of the chin, limbs and plastron, and they have pale spots on the lower surface of the marginal scales. No subspecies are recognized.

Etymology: The specific epithet, *serpentina* [ser-pen-TINE-ah], is Latin for "snake-like" and refers to either the long neck or the long tail characteristic of this species.

Similar Species: The long tail, large head, and small plastron distinguishes Snapping Turtles and Alligator Snapping Turtles (*Macrochelys temminckii*) from all other turtle species in Tennessee. The beak and ridges on the carapace and tail distinguish these two species from each other. Compared to the Alligator Snapping Turtle, the ridges on the carapace in Snapping Turtles are indistinct in older individuals, the horny beak is shorter, and the middorsal keel of the tail is more prominent. Furthermore, Snapping Turtles lack the supramarginal scute row, which is present between the marginal and costal scutes in Alligator Snapping Turtles.

Distribution: Snapping Turtles are native to the eastern two-thirds of North America and have been introduced into scattered localities in the western United States and Canada. Although records are lacking from some counties, this species occurs throughout most of Tennessee except the highest elevations.

Habitat: This primarily aquatic turtle prefers slow-moving bodies of water with soft bottoms and dense vegetation. Individuals spend much time buried in the soft bottoms, often with just their eyes exposed. Although associated with permanent bodies of water, such as ponds, lakes, streams, rivers, marshes, swamps, and ditches, Snapping Turtles also commonly inhabit wetlands subject to seasonal drying. When these wetlands dry, turtles either move over land to neighboring wetlands or, as at Sinking Pond at Arnold Air Force Base in Middle Tennessee, they burrow into the mud during summer and await the wetland filling the following winter. Unlike many other aquatic turtles, Snapping Turtles are frequently seen on land, particularly females during the nesting season. Females will readily cross roads, although they frequently do not survive, and corpses are a common sight along southeastern roadways in the early summer.

Natural History: Snapping Turtles are primarily aquatic but occasionally bask on land or travel overland to seek new wetlands. Females are often found traveling on land in search of nesting sites. Breeding activities typically occur from early through late spring but continue in decreasing frequency through late summer. During mating, a male sits on top of a female and grips her carapace with his long claws. He wraps his long tail around the females, while twisting it upward so that his vent is aligned with hers. Sperm is then transferred from male to female. Depending on the time of year, eggs may be fertilized soon after mating, or females can store sperm and fertilize eggs in subsequent years. Nesting usually occurs from mid-May through late June, but the exact timing is dependent on temperatures during late winter

Top left: Snapping Turtle hatchling, Cannon County. (Photo by Brian T. Miller) *Top right:* Plastron of a Snapping Turtle, Rutherford County. (Photo by Brad M. Glorioso) *Bottom right:* A Snapping Turtle nest, Henry County. (Photo by David F. Frymire)

and early spring (which affect the timing of breeding), and therefore nesting may occur throughout the summer months. Females select open areas for nesting, generally near the wetland they inhabit, but they will travel great distances (>1.5 km [0.9 miles]) to locate suitable sites. They use their hind limbs to dig a flask-shaped nest in which to deposit 20–30 relatively small, spherical eggs that roughly resemble ping-pong balls. The eggshells are hard and brittle, in sharp contrast to the leathery eggshells of lizards and snakes. The temperature at which the eggs are incubated determines the sex of the young, with females developing at relatively warm (29–33°C [84.2–91.4°F]) temperatures and males at cooler (22–28°C [71.6–82.4°F]) temperatures. Incubation usually takes 10–14 weeks; consequently, hatching occurs from late July through early November. Hatchlings can overwinter in their nest and, thereby, delay emerging from the ground until early spring. Because females often travel far overland to nest, hatchlings might have to travel relatively great distances to reach water. For example, I have found hatchlings in small garden ponds, at least 1 km (0.6 miles) from any body of water capable of supporting a population of Snapping Turtles. Depending on location, males become sexual mature at 3–9 years, females at 4–12 years. Snapping Turtles can live long lives, with many reports of individuals surviving 30–40 years in the wild. Some biologists speculate that Snapping Turtles can live for at least 100 years. Snapping Turtles are omnivores, but dietary preferences change throughout life. Hatchlings and young juveniles are primarily carnivorous and eat a variety of small invertebrate and vertebrate prey. I have watched hatchlings buried in the mud of shallow pools suck American Toad (*Anaxyrus americanus*) tadpoles into their mouths. As they grow, larger prey items are included in the diet, such as any vertebrate or invertebrate animal (insects,

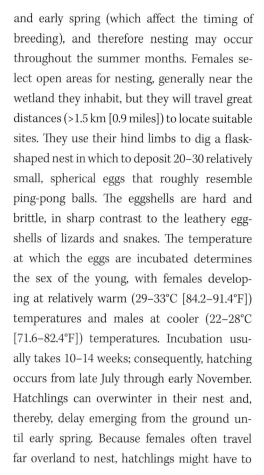

crayfish, fish, amphibians, snakes, other turtles, birds, mammals, and carrion). Older individuals also eat algae. I have witnessed Snapping Turtles foraging on large algal mats in several bodies of water. Predators of eggs and small turtles include other Snapping Turtles, snakes, birds, and many mammals, such as raccoons, foxes, coyotes, and skunks. Relatively few nests survive to hatching, and predation on hatchlings and small turtles is also significant. Predation decreases as a turtle ages (and grows in size), and important predators of adults are limited to humans, other Snapping Turtles, American Alligators, and large carnivores such as raccoons, coyotes, bears, and domestic dogs. Leeches are common and obvious ectoparasites and are often found on exposed flesh and attached to the spaces between scutes on the carapace. The common name refers to the defensive behavior of these turtles when threatened on land. They rear up on their front limbs and rapidly thrust their long neck to the perceived predator with their mouth agape. If contact is made, the jaws snap shut. A turtle can repeat this behavior only a few times before the muscles become fatigued. Snapping Turtles typically hibernate buried in the bottom of body of water. As reported for several other species of turtles, Snapping Turtles can move water in and out of their cloaca. Oxygen in the water moves through the walls of their cloaca into their bloodstream. Some individuals use terrestrial burrows in some situations.

Conservation Status: The Snapping Turtle is the only turtle species in Tennessee that is regulated by TWRA as a sport species, thereby requiring a sport fishing license for legal harvest. Although routinely harvested statewide, the species is widespread, locally abundant, and presumably secure. Unfortunately, some people kill Snapping Turtles because of the myth that this species damages game fish and waterfowl populations.

Comments: Snapping Turtles are relished range-wide for their flesh, and similar to domestic fowl, the coloration of the meat varies depending on location in the body. For instance, the meat of the neck associated with the rapid thrusting of the head during defensive snaps is pale (white) compared to the meat of the legs, which is dark.

Brian T. Miller

Alligator Snapping Turtle adult, Houston County. (Photo by A. Floyd Scott)

Alligator Snapping Turtle

Macrochelys temminckii

Description: Alligator Snapping Turtles are the largest turtles in Tennessee; indeed, they are the largest freshwater turtle in North America. Adult males can reach 50–60 cm (20–24 in) CL and 80 kg (176 lbs), although record individuals have reached 80 cm (31 in) CL and 113 kg (250 lbs). Females are considerably smaller, although still sizeable with a CL around 35–50 cm (14–20 in), and usually reaching only 22 kg (48.5 lbs). The head is extremely large with a strongly hooked beak, is covered with fleshy barbels or tubercles, and appears triangular in shape when viewed from above. The carapace has three rows of pointed scutes (keels) as well as pointed marginal scutes that give the shell a jagged appearance, especially when young. Additionally, there are one to five (usually two or three) supramarginal scutes. The carapace is generally brown, with occasional lighter patches on the tips of the domed scutes or is obscured with algae. The plastron is very small

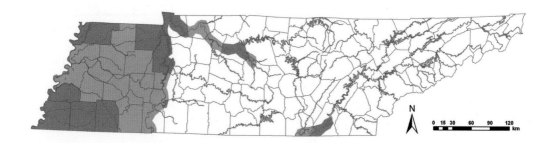

relative to the body, and much exposed flesh is present on the undersides. The small eyes are located on the sides of the head and are striated with alternating pale and dark brown bands radiating from the iris to the sclera, and each eye is surrounded by fleshy pointed projections. The tongue has a pink fleshy projection that is waved such that it resembles a small worm. The tail is long and has rows of jagged rigid tubercles. Young Alligator Snapping Turtles resemble adults, although with a far more jagged appearance, as adults often have dulled scutes. Additionally, young turtles often have faint stripes present on the neck. Hatchlings are 36–44 mm (1.4–1.7 in) CL. No subspecies are recognized.

Etymology: The specific epithet, *temminckii* [tem-ENK-ee-i], honors Coenraad Temminck, an early-19th-century Dutch scientist.

Similar Species: Alligator Snapping Turtles can only be confused with Snapping Turtles (*Chelydra serpentina*), although distinguishing between the two is quite easy. Snapping Turtles have a ridged carapace, but the ridges are much less conspicuous than those on the carapace of Alligator Snapping Turtles. Snapping Turtles have a longer neck, no hooked beak, a shorter snout, and a middorsal keel on the tail. Alligator Snapping Turtles have fleshy "eyelashes," or projections that encircle each eye, and a fleshy pink projection on the tongue, characters that are lacking in Snapping Turtles. Furthermore, additional scutes on the carapace called the supramarginal scutes are present between the marginal and costal scutes in Alligator Snapping Turtles but absent in Snapping Turtles.

Distribution: Alligator Snapping Turtles are native to the Coastal Plain of the Gulf Coast and the lower Mississippi River basin extending up the Mississippi River as far north as Iowa and Illinois. Their range extends east to the eastern Florida Panhandle and west to eastern Texas and Oklahoma. In Tennessee, Alligator Snapping Turtles are found in scattered locations in the central and western part of the state. In West Tennessee, records exist along the Mississippi River in the Wolf and Hatchie rivers of Shelby, Fayette, and McNairy counties as well as Reelfoot Lake in Lake and Obion counties. Furthermore, during the 1990s, TWRA initiated a restocking program in West Tennessee. In central Tennessee, populations exist in tributaries of the north-flowing Tennessee River and in Lake Barkley. A record also exists from the Stones River near Percy Priest Dam in Davidson County. Interestingly, a record exists for Hamilton County just west of Chattanooga as well.

Habitat: This entirely aquatic turtle prefers slow-moving bodies of water with soft bottoms and dense vegetation. Individuals spend much of their time buried in the soft bottoms, often with just their heads exposed. Alligator Snapping Turtles are associated with permanent bodies of water, such as ponds, lakes, streams, rivers, marshes, and cypress swamps. Within these habitats, Alligator Snapping Turtles are found beneath or within logjams, rock piles, and deep holes as well as underneath undercut banks. They are not often encountered in our state, although they can occasionally be found by spotlighting at night around the margins of lakes and rivers. Fishermen also occasionally catch this species when using live or dead bait.

Natural History: Alligator Snapping Turtles are almost entirely aquatic, although females do have to leave the water to lay eggs. Mating typically occurs underwater in the winter and

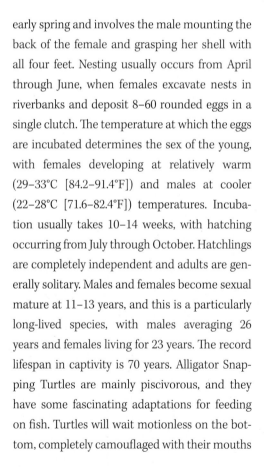

Left: Head of an Alligator Snapping Turtle showing strongly curved beak. (Photo by Brad M. Glorioso) *Right:* Alligator Snapping Turtle adult, Calloway County, Kentucky. (Photo by David F. Frymire)

early spring and involves the male mounting the back of the female and grasping her shell with all four feet. Nesting usually occurs from April through June, when females excavate nests in riverbanks and deposit 8–60 rounded eggs in a single clutch. The temperature at which the eggs are incubated determines the sex of the young, with females developing at relatively warm (29–33°C [84.2–91.4°F]) and males at cooler (22–28°C [71.6–82.4°F]) temperatures. Incubation usually takes 10–14 weeks, with hatching occurring from July through October. Hatchlings are completely independent and adults are generally solitary. Males and females become sexual mature at 11–13 years, and this is a particularly long-lived species, with males averaging 26 years and females living for 23 years. The record lifespan in captivity is 70 years. Alligator Snapping Turtles are mainly piscivorous, and they have some fascinating adaptations for feeding on fish. Turtles will wait motionless on the bottom, completely camouflaged with their mouths open. A fleshy pink projection on the tongue that remarkably resembles a small worm is the only part of the body that moves, and it is wiggled every few seconds. Fish swim near to investigate and are captured by a lightning-fast snapping of the jaws, which either impales the fish on the hooked beak or crushes it in the jaws. Alligator Snapping Turtles will also consume other vertebrate prey including reptiles, such as snakes and other turtles, amphibians, small- and medium-sized mammals, and birds. Invertebrates, such as crayfish and mollusks, also are eaten. Some individuals have been habituated to accept hand feeding, although this is a reckless behavior at best! Predators of eggs and young include large fish, raccoons, wading birds, and otters; adults have no predators other than humans. Nesting success is highly variable, and many nests are lost to predators. Alligator Snapping Turtles have a formidable defense. They will open their mouths when disturbed and snap them shut when the threat nears the mouth. If contact is made, the jaws must be pried apart if they are to be opened. These jaws are capable of delivering a very serious bite, potentially severing digits, and this species should be handled only with extreme caution.

Conservation Status: Alligator Snapping Turtles were heavily exploited for meat in the past century, and populations have declined significantly across the range of this species.

This has been compounded by malicious killing and habitat loss. Alligator Snapping Turtles are infrequently encountered in Tennessee. Hence this species is listed as "Deemed in Need of Management" in Tennessee and is a state listed by NatureServe, meaning that it is rare and vulnerable to extinction in the state.

Comments: Alligator Snapping Turtles are occasionally referred to as "alligator turtles" or "loggerheads." The genus name was recently changed from *Macroclemys* to *Macrochelys*.

R. Graham Reynolds and
Matthew L. Niemiller

Turtles on Base

Twenty years ago, I conducted an inventory of the amphibians and reptiles of Arnold Air Force Base (AAFB) in the Barrens of the Eastern Highland Rim in Middle Tennessee. The herpetofauna of this region had been little studied and the types and number of species known to occur in the region in general and at AAFB in particular were unknown. However, the area was known to have a great and unusual diversity of plants, so with great anticipation, and permission from the base, I began a herpetological inventory.

My student workers and I initially used general field collecting techniques (searching beneath logs, rock, stumps, etc.) to great success. Indeed, on the first day we found individuals of more than 20 species of amphibians and reptiles, but mainly amphibians. We were excited with this diversity of species, as we had found more species in one day than previously had been reported from the base. However, my students were a little disappointed by the lack of reptiles, both number of species and number of individuals. That was to change later in the

month when my "turtle traps" arrived and we began trapping for aquatic turtles.

We set traps in several of the larger bodies of water on base, including Sinking Pond, a few streams, and an embayment of the Woods Reservoir. We baited the traps with beef liver and fish parts and then set them in relatively shallow water so that the captured turtles would be able to surface and breathe. We set the traps in the afternoon and checked them that following morning. The forest canopy above the water's edge was nearly complete and little sunlight penetrated to the water's surface.

Several field assistants accompanied me the following day, but only one decided to wade through the nearly knee-deep mud to help me retrieve and check the traps. As we approached the first trap, I noticed some commotion in the water just outside the trap. This is not unusual, as turtles attracted to the traps often have to search for some time before locating the entrance that allows them to get close to the bait. Using my feet as prods in the now murky water around the trap, I was able to locate and then carefully grab two Snapping Turtles, one weighing about 13 pounds, the other approaching 20 pounds. I took the turtles to shore for the rest of the crew to examine and then returned to the trap, which my helper and I tried to lift out of the water. We were surprised by its weight; rather than lift and carry the trap, we had to drag it and the captured turtles to shore. We suspected at first that we had a few large Snapping Turtles—which we did—but were surprised to see more than 30 Red-Eared Sliders and about a dozen Eastern Musk Turtles in the trap! The second trap was equally overburdened with

Snapping Turtle captured in a trap. (Photo by Collin Jaeger)

sliders, snappers, and musk turtles. From these two traps we counted more than 70 turtles, and countless others were likely in the embayment around the traps. Never before or since have I seen so many turtles attracted to baited traps. I doubt that the students will ever forget the experience, and they will long remember that adult turtles often dominate local aquatic communities. A possible reason why this occurs was witnessed that June, when I was teaching a class on vertebrate zoology.

My summer vertebrate zoology class lasted for five hours Monday through Thursday, which gave me ample time to conduct field trips to diverse habitats in Middle Tennessee, including AAFB. I again had set traps in the embayment, but the relatively few turtles we captured disappointed me. I later realized that the turtles likely had dispersed following courtship activities. My class and I began to walk the shoreline to see if we could spot turtles basking. Although we saw a few turtle heads bobbing in the water, we were most struck by an approximately 50-meter section of ravaged shoreline that obviously was used as a nesting area by many turtles. We counted more than 100 turtle nests that had been ransacked by various predators, such as raccoons, skunks, opossums, and foxes. Literally thousands of cracked and dried eggs were strewn on the ground in destroyed nests. Although we searched for intact nests and eggs that had survived the carnage, we found none. The ordeal was a bit depressing for the students, but the lesson learned will never be forgotten: Turtle nests are subject to extreme predation pressure; few of them survive the scrutiny of predators.

The situation we witnessed at Woods Reservoir possibly is unusual for freshwater turtle species (unlike marine turtles, freshwater species do not necessarily cluster nests and predators won't necessarily unearth one clutch of eggs while digging in a separate nest, as apparently happened here). Unfortunately, most studies indicate that hatchling turtles don't fare much better than the eggs. Which brings home an important lesson: The high densities of adult turtles in wetlands often are critical for survival of the populations. Turtles are renowned for their long lives, and these long lives are necessary for the populations to survive. People are often surprised when they learn of the large population sizes of some species of turtles and the large proportion of adults. However, these attributes of healthy turtle populations underscore their natural histories: low survival of eggs and young but long lives in those individuals that reach adulthood. The longer a female turtle lives, the greater the likelihood that she will lay a clutch of eggs that escapes predation. Removal of adults from a population, either intentionally for food or pets, or inadvertently by automobile traffic, can have dire consequences on population sustainability. Although it often takes decades to notice, populations suddenly "crash" when the recruitment of young into adulthood falls below a critical number.

Brian T. Miller

18
Family Emydidae (Common Water Turtles)

Common Water Turtles are one of the more diverse and speciose families of turtles. Approximately 15% of all turtle species and 12% of all turtle genera are emydids (approximately 50 of 327 species, or 12 of approximately 97 genera). Emydids are found in North America, South America, Europe, and northern Africa. These are small- to medium-sized turtles, seldom exceeding 20 cm (8 in) CL. Most species have a large plastron and a moderately domed, oval carapace that is either smooth dorsally, or keeled along the midline. Sometimes the keel has a series of distinct knobs, which make the keel serrated. Emydid turtles are well represented in the fossil record, with fossils identified as member of this family dating back to the Upper Cretaceous period of North America. Modern emydids are found in a variety of habitats, including lakes, reservoirs, rivers streams, marshes, swamps, and on land. In the past, this family has been referred to either as the Box and Water Turtles or the semiaquatic Pond and Marsh Turtles; however, this family inhabits a much broader array of habitats than these common names imply. Moreover, many species are quite abundant and easily observed basking on logs or fallen trees. Emydids are well represented in Tennessee, with six genera comprising nine species, including our state reptile, the Eastern Box Turtle (*Terrapene carolina carolina*). Common Water Turtles are the most commonly sold turtles in the pet trade. Tens of thousands of individuals of several species are harvested each year in the United States, either for the pet trade or for other commercial enterprises. Some species, such as the Pond Slider (*Trachemys scripta*), have expanded their ranges because of accidental or intentional introductions in the United States and other countries.

The family name Emydidae [eh-MY-deh-dee] is from the Greek *emydos* (freshwater turtle) and *eidos* (similar to). The etymology of the genera found in Tennessee is as follows. The genus name *Chrysemys* [kry-sem-MEES] is from the Greek *khrysos* (golden or pale green) and *emys* (turtle), which refer to the colored stripes on the neck of species within the genus. The genus name *Glyptemys* [glip-tem-MEES], formed by combining the Greek roots *glyptos* (carved) and *emys* (turtle), presumably refers to the woodlike markings on the carapace of the Wood Turtle (*G. insculpta*). The genus name *Graptemys* [grap-tem-MEES] is from the Greek *graptos* (inscribed) and *emys* (turtle), and refers to the markings on the carapace. The genus name *Pseudemys* [SOO-deh-mees] is from the Greek *pseudes* (false) and *emys* (turtle), and refers to the distinction of this group from the European pond turtles of the genus *Emys*. The genus name *Terrapene* [tear-REH-pen-ee] is from the Algonquin word for turtle. Finally, the genus name *Trachemys* [TRAY-kem-ees] is from the Greek *trachys* (rough) and *emys* (turtle), but the reasoning for using this name to describe members of this genus is unclear.

Southern Painted Turtle adult, Lake County. (Photo by Brad M. Glorioso)

Southern Painted Turtle

Chrysemys dorsalis

Description: Southern Painted Turtles are small basking turtles that usually range 10–12 cm (4–5 in) CL, but they can reach up to 15.6 cm (6 in). Their colorful carapace lacks a keel and is low domed, attributes that help streamline the shell. The color of the carapace is variable, ranging from dark green to olive or black. A red, orange, or yellow stripe is prominent along the midline. The vertebral and parietal scutes of-ten have olive, yellow, or cream edges. Otherwise, the carapace looks relatively plain when dry, but faint yellow and red lines or blotches are often revealed when the shell is wet. The marginal scutes are only faintly patterned with yellow and red dorsally but exhibit large, irregular red blotches ventrally. The carapace always posses alternating vertebral and costal scutes; five vertebral scutes are flanked by four pleural and 12 marginal scutes on each side. The posterior edge of the carapace is smooth or wavy (not indented or scalloped). The plastron lacks

a hinge and is generally plain yellow, but it may have a few paired, dark smudges. The head, neck, and legs are covered with alternating yellow and green or black pinstripes, and the yellow stripes gradually blend into red stripes on the lower neck and front legs. The eye is yellow with a dark stripe running through the pupil. The feet are strongly webbed for swimming. Females have short claws on their front feet, and their tail is so short that their vent does not extend past the margin of their carapace. In contrast, mature males have long claws, especially on the front feet, and a very long tail with the vent extending well past the margin of the carapace. Mature females grow to larger sizes than males. Hatchlings are small at about 25 mm (1 in) CL and brightly colored. Their slightly keeled carapace is bright green and has a prominent, relatively broad red, orange, or yellow stripe extending down the midline. The plastron is unpatterned yellow or pale orange. No subspecies are recognized, but the taxonomy of this species is currently under review and remains highly controversial (see **Comments** below and in the Painted Turtle account).

Etymology: The specific epithet, *dorsalis* [door-SAH-lis], is Latin for "back," which refers to the dorsal stripe on the carapace.

Similar Species: Painted Turtles (*C. picta*) are most similar in appearance to Southern Painted Turtles, and the two species are known to hybridize. Painted Turtles are much larger as adults, they lack a distinct red or orange stripe down the midline of the carapace as adults, and the stripes on their neck are either broad or expanded to form two conspicuous yellow blobs. Hatchlings can be differentiated by the width of the carapace stripe, which is merely a faint pinstripe in Painted Turtles. Hybrids might possess characteristics of both species. From a distance, Painted Turtles might be confused with River Cooters (*Pseudemys concinna*), but River Cooters have distinct C shapes on their second costal scutes. They also might be confused with Pond Sliders (*Trachemys scripta*), especially melanistic ones with worn shells, but Pond Sliders have one large yellow, orange, or red patch or a wide stripe on each side of the head behind the eye that is much wider than adjacent stripes. Hatchling Painted Turtles might be confused with several species of Map Turtles (genus *Graptemys*) from a distance because they have a slight keel to their carapace, but hatchling Map Turtles have a much more prominent, usually saw-toothed, keel.

Distribution: Southern Painted Turtles occur in the southern Mississippi River Basin, ranging from extreme southern Illinois, extreme southeastern Missouri, and western Tennessee southward throughout most of Arkansas, Louisiana, and Mississippi, and extreme eastern Oklahoma and Texas. They occur throughout the western third of Tennessee, essentially west of the north-flowing Tennessee River, but they hybridize with Painted Turtles in an intergradation zone extending northwest to southeast along a line from just west of Land Between the Lakes to the Lawrenceburg-Pulaski-Fayetteville area.

Habitat: Southern Painted Turtles prefer the soft-bottomed, vegetated, slow-moving waters of lakes, ponds, reservoirs, rivers, oxbows, and ditches. They bask in great numbers on floating logs and leaning tree boles at the waters' edge. Southern Painted Turtles can be found in virtually any type of aquatic habitat, from pristine river systems to highly altered systems such as reservoirs, farm ponds, and roadside ditches, and are occasionally observed on

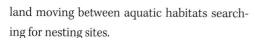

Top left: A female Southern Painted Turtle nesting, Lake County. (Photo by Brad M. Glorioso) *Top right:* Plastron of a Southern Painted Turtle adult, Lake County. (Photo by Brad M. Glorioso) *Bottom right:* Plastron of a Southern Painted Turtle hatchling, Obion County. (Photo by Matthew L. Niemiller)

land moving between aquatic habitats searching for nesting sites.

Natural History: Although highly abundant, Southern Painted Turtles usually do not reach the extreme densities of their northern counterparts, Painted Turtles, possibly because of the presence of other southern species, such as American Alligators (*Alligator mississippiensis*) and much larger aquatic turtles, such as River Cooters and the more aggressive Pond Sliders. After the sun rises, they haul out of the water to bask, and numerous individuals of all age classes will crowd together even on the smallest of logs. Southern Painted Turtles copulate in the water during autumn, but females store sperm over the winter and fertilize developing eggs in the spring. The male initiates the mating ritual by using his long claws to tickle the female's head and neck, enticing her to mate. Nests are constructed during summer months in open, sandy areas adjacent to, but sometimes very far from, their resident water body. In Tennessee and other locations in the northern parts of their range, females lay a single, relatively large clutch of up to 20 eggs that hatch by autumn, but hatchlings from late clutches may overwinter in the nest and emerge during spring. Southern Painted Turtles exhibit temperature-dependent sex determination, in which males are produced at temperatures below 27°C (80.6°F), and females result from temperatures above 30°C (86°F). Mixed nests result from temperatures between 27°C and 30°C. Because of temperature-dependent sex determination, an individual nest may be biased toward either males or females, but population-wide the hatchling sex ratio is usually about 1:1. Males reach sexual maturity in 2–3 years at ca. 6 cm (2.4 in) CL, but females take about twice as long to mature at ca. 10 cm (4 in) CL. Southern Painted Turtles are omnivorous throughout their lives, but their diet shifts with age. The young are carnivorous on aquatic insects (particularly dragonfly naiads), mollusks, fish, tadpoles, and frogs, whereas older individuals are more herbivorous and include algae and duckweed in their diet. Predation occurs at all stages of their life cycle—the egg stage is most vulnerable to raccoons and other mammals, and juveniles are consumed by anything that can capture and ingest them, including large fish, snakes, Snapping Turtles (*Chelydra serpentina*), and minks. Adults are less vulnerable to predation because of their swimming prowess. When captured, Southern Painted Turtles are known to bite.

Conservation Status: Southern Painted Turtles rank among the most abundant of all the turtles in North America and are not considered a species of conservation concern. However, even the most abundant, seemingly tolerant species are potentially vulnerable to the effects of wetland and riparian habitat alteration, chemical pollution, overharvesting, and roadkill. Widespread unsustainable agricultural practices and wetland losses in the southern Mississippi River Basin have resulted in local population declines. Although Southern Painted Turtles are currently "healthy" in terms of population size, it will be important to monitor their populations, especially at the interface between human habitation and natural areas, to ensure the "Least Concern" designation by IUCN is warranted long in the future.

Comments: The systematics and taxonomy of the Painted Turtle "complex" has proven highly controversial. The complex has long been composed of four fairly recognizable subspecies: the Eastern Painted Turtle (*C. p. pica*), Midland Painted Turtle (*C. p. marginata*), Western Painted Turtle (*C. p. bellii*), and Southern Painted Turtle (*C. p. dorsalis*), with all but the western subspecies occurring in Tennessee. In 2003, eight turtle experts published the results of a study of mitochondrial DNA throughout their range and proposed that the Painted Turtle complex was better described as two separate species: (1) the Southern Painted Turtle, which was elevated to full species (*C. dorsalis*) with the same distribution pattern as the original southern subspecies, and (2) the Painted Turtle (*C. picta*), which consists of the three other original subspecies, once thought to be an example of speciation by postglacial hybridization (with *C. p. marginata* occupying the central hybrid zone).

Todd Campbell

Painted Turtle adult, LaGrange County, Indiana. (Photo by Todd Pierson)

Painted Turtle

Chrysemys picta

Description: Painted Turtles are medium-sized basking turtles, generally 11–16 (4.3–6.3 in) CL, but they are known to reach over 19 cm (7.5 in) CL. These beautiful turtles have a streamlined, low-domed, nonkeeled carapace with a background color of dark green or olive and various patterns of yellow and red lines and blotches with black borders. The anterior edges of the anterior-most parietal and vertebral scutes often have wide yellow or cream edges. The patterns on the upper and lower surfaces of the marginal scutes are particularly striking, including hairpin or fingerprint-like shapes above, and irregular red blotches that nearly fill the marginal scutes below. The posterior edge of the carapace is smooth or wavy, not indented or scalloped. The carapace has five vertebral scutes flanked by four pleural and 12 marginal scutes on each side. The plastron is hingeless

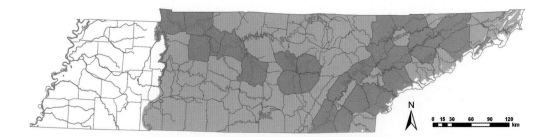

and highly variable in color and pattern but is usually yellow or cream colored with one or two irregular dark or black blotches that appear like images in a kaleidoscope. The head, neck, and legs are covered with alternating yellow, green, and black stripes, and the yellow stripes on the neck usually blend into red stripes on the lower neck and front legs. Two of the parallel yellow stripes on the neck are much wider behind the eye. The eye is yellow with a dark stripe running through the pupil. The feet are strongly webbed for swimming. Mature females are larger but have a shorter tail than do males, and their vent does not extend past their carapace. Mature males have long claws, especially on the front feet, and a relatively long tail with the vent extending well past the carapace. Hatchlings are 25–30 mm (1–1.1 in) CL and have a bright green, slightly keeled carapace with pale borders on most scutes, a pale pinstripe along the midline in some individuals, and a plain yellow or pale orange plastron. Three subspecies are currently recognized with just the Midland Painted Turtle (*C. p. marginata*) occurring in Tennessee, but the taxonomy of this species is currently under review and remains highly controversial (see **Comments** below). Regardless, Painted Turtles are highly variable morphologically. East of the Appalachian Mountains in general, and east of the eastern escarpment of the Cumberland Plateau in Tennessee, Painted Turtles are unique among all other turtles in North America in that the parietal and vertebral scutes line up in four even or nearly even rows across the carapace. The yellow scute borders also line up, forming three narrow stripes across the carapace, and the two stripes on the neck are expanded to oval spots; one behind the eye and one on the neck.

Etymology: The specific epithet, *picta* [PICK-tah], is Latin for "painted" or "colored," and the subspecific epithet, *maginata* [marr-jin-AY-tah], is Latin for "bordered," in reference to the coloration on the marginal scutes.

Similar Species: Southern Painted Turtles (*C. dorsalis*) are most similar to Painted Turtles and the two species are known to hybridize in Tennessee. Adult Southern Painted Turtles are much smaller and always have a distinct red, orange, or yellow stripe down the middle of the carapace and narrower, wavier yellow stripes on the neck. Hatchlings can be differentiated by the width of the carapace stripe, which is always broad and obvious in Southern Painted Turtles. Hybrids may possess characteristics of both species. From a distance, Painted Turtles (*Chrysemys sp.*) might be confused with River Cooters (*Pseudemys concinna*), but River Cooters have distinct C shapes on their second costal scutes. They also might be confused with Pond Sliders (*Trachemys scripta*), especially melanistic ones with worn-down shells, but Pond Sliders have only one large yellow, orange, or red patch or a wide stripe on each side of the head behind the eye that is much wider than adjacent stripes. Hatchling Painted Turtles might be confused with several species of Map Turtles (genus *Graptemys*) from a distance because they have a slight keel to their carapace, but hatchling Map Turtles have a much more prominent, usually saw-toothed, keel.

Distribution: Painted Turtles are the most widespread turtle in North America, ranging from Maine to Georgia in the east, to Washington and Oregon in the west, and extending north into Canada and south into Mexico. They occur throughout the eastern two-thirds of Tennessee, mainly east of the north-flowing

Tennessee River. However, records for Middle Tennessee are scattered within just a few counties, and those records are historic. Despite intense sampling in the Barrens and Central Basin, Painted Turtles have not been documented in many decades. Painted Turtles are known to hybridize with Southern Painted Turtles (*C. dorsalis*) around the Tennessee River in an intergrade zone extending from northwest to southeast along a line from just west of Land Between the Lakes to the Lawrenceburg-Pulaski-Fayetteville area.

Painted Turtle basking, Marion County, Indiana. (Photo by Todd Pierson)

Habitat: These mainly aquatic turtles prefer mud-bottomed lakes, ponds, reservoirs, large rivers, oxbows, and ditches, where they often bask on floating logs and leaning tree boles in huge numbers. Painted Turtles can be found in virtually any type of aquatic habitat, from pristine river systems to highly altered systems such as reservoirs, farm ponds, and roadside ditches. They are occasionally observed on land moving between aquatic habitats or searching for nesting sites.

Natural History: Painted Turtles reach the highest densities of any freshwater turtle species—dozens often will be present on any given basking log on large bodies of water, and small, well-vegetated, highly productive lakes and ponds may harbor 500–800 or more individuals per hectare. They become active just after sunrise, and very early in the year as well, and they may even be seen under the ice of frozen lakes or waterways! In late winter and spring, males court females by tickling them with the long claws on their front feet. Copulation occurs in the water. Female Painted Turtles generally store sperm for one summer's breeding effort of two to five clutches but are known to store sperm successfully for up to three years. During summer, females haul themselves out of the water to lay a clutch of up to 11 eggs in a sand or soil bank. Painted Turtles exhibit temperature-dependent sex determination, in which males are produced at temperatures below 27°C (80.6°F), females result from temperatures above 30°C (86°F), and mixed nests of both sexes result from temperatures in between. Thus any given nest may be biased toward males or females, but population-wide, the hatchling sex ratio is usually about 1:1. If a clutch is laid too late in the year, hatchlings will sometimes overwinter in the nest. Painted Turtles are omnivorous throughout their lives but their diet shifts with age, from carnivorous on aquatic insects (particularly dragonfly naiads), mollusks, fish, and tadpoles to omnivorous, adding plant material and carrion to the juvenile diet. Predation occurs at all stages of their life cycle; the egg stage is most vulnerable to raccoons and other mammals; juveniles are consumed by anything that can capture and ingest them, including large fish, snakes, Snapping Turtles (*Chelydra serpentina*), and minks; adults are less vulnerable due to their swimming prowess. When captured, Painted Turtles are known to bite.

Conservation Status: Painted Turtles are the most widespread and abundant of all the turtles in North America, and as a result, are not of conservation concern. However, even the most abundant, seemingly tolerant species are potentially vulnerable to the effects of wetland and riparian habitat alteration, unsustainable agricultural practices, chemical pollution, overharvesting, and roadkill, often leading to significant and rapid local population declines. Introduced predators, such as largemouth bass and bullfrogs have reduced some local populations in western states. It will be important to monitor their populations, especially at the interface between human habitation and natural areas, to ensure the "Least Concern" designation by IUCN is warranted long in the future. For instance, road-kill prevention programs and "critter crossings" have been established in western states. Painted Turtles themselves have been introduced to California, Arizona, Florida, Germany, Spain, Indonesia, and the Philippines, and could become pests like the Pond Slider (see Pond Slider account) if care is not taken to prevent their sale and transport by the pet industry.

Comments: The systematics and taxonomy of the Painted Turtle "complex" has proven highly controversial. The complex has long been composed of four fairly recognizable subspecies: the Eastern Painted Turtle (*C. p. pica*), Midland Painted Turtle (*C. p. marginata*), Western Painted Turtle (*C. p. bellii*), and Southern Painted Turtle (*C. p. dorsalis*), with all but the western subspecies occurring in Tennessee. In 2003, eight turtle experts published the results of a study of mitochondrial DNA throughout their range and proposed that the Painted Turtle complex was better described as two separate species: (1) the Southern Painted Turtle, which was elevated to full species (*C. dorsalis*) with the same distribution pattern as the original southern subspecies, and (2) the Painted Turtle (*C. picta*), which consists of the three other original subspecies, once thought to be an example of speciation by postglacial hybridization (with *C. p. marginata* occupying the central hybrid zone). Despite the physical differences of the three subspecies, and despite the fact that they freely interbreed in all the hybrid zones, this taxon (*C. picta*) exhibits little genetic differentiation (also termed "geographic genetic structure") in the mitochondrial genes studied across its massive range. Although few authors appeared to support this new organization, the findings of the 2003 study have yet to be refuted in the scientific literature.

Todd Campbell

Bog Turtle adult, Wilkes County, North Carolina. (Photo by Greg Sievert)

Bog Turtle

Glyptemys muhlenbergii

Description: Bog Turtles are Tennessee's smallest turtle. Most adults are 7.5–9 cm (3.0–3.5 in) CL, but some reach at least 11.4 cm (4.4 in). The most conspicuous feature of this mostly brown turtle is the large, irregular orange or yellow blotch on each side of the head behind each eye. The adult carapace is generally dark brown or black, but each pleural and vertebral scute has a pale brown center. In contrast to adults, the carapace of juveniles is usually uniformly pale brown. The low-domed carapace has scalloped posterior edges and prominent growth rings on each scute, features that give the turtles a tortoise-like appearance. Younger individuals have a keeled carapace, and older individuals often have a smooth carapace with worn-down posterior scalloping. The carapace has five vertebral scutes and 12 marginal and four pleural scutes on each side. The plastron

lacks a hinge and is usually black or very dark with pale irregular blotches on each scute that are connected along the midline. However, the plastron of juveniles is a negative image of the adult plastron—yellow or cream colored with a large, irregular black botch in the center. The feet are only slightly webbed. Males have a concave plastron to facilitate mating, and a relatively long tail with the cloaca extending beyond the posterior margin of the shell by about 20 percent of its total length. Females have a flat plastron, a higher-domed shell, and a relatively short tail, with the cloaca extending past the carapace by only about 10 percent of its total length. Hatchlings are 23–34 mm (0.9–1.2 in) CL and resemble adults. No subspecies are recognized.

Etymology: The specific epithet, *muhlenbergii* [moo-len-BERG-ee-i], honors the botanist Gotthilf Heinrich Ernst Mühlenberg, who discovered this species in Pennsylvania.

Similar Species: In Tennessee, Bog Turtles might be confused with Eastern Box Turtles (*Terrapene carolina carolina*). However, Eastern Box Turtles have a colorful, high-domed carapace with radiating black and yellow lines on each scute and a hinged plastron. Furthermore, the carapace of juvenile and small adult box turtles (which are about the same size as adult Bog Turtles) is smooth and contains interlaced patterns of black, green, and yellow stripes. Red-Eared Sliders (*T. scripta elegans*) have a carapace shaped similar to that of Bog Turtles but have a bright red or yellow stripe rather than an irregular spot on either side of its head.

Distribution: Bog Turtles occur from sea level to 1,220 m (4,000 ft) in two disjunct population centers in eastern North America, one of which barely ranges into extreme northeastern

Tennessee. The "northern population" consists of multiple isolated populations in several northeastern states (New York, Pennsylvania, Maryland, and adjacent states). The "southern population" consists of multiple enclaves on the eastern slope of the Blue Ridge Mountains in western North Carolina, southwest Virginia, the northwest tip of South Carolina, extreme northeast Georgia, and extreme northeastern Tennessee. Fewer than 100 separate enclaves or "colonies" (see below) have been located in the southern population. In Tennessee, this species is restricted to a handful of bogs and other wetlands of Johnson County, our easternmost county.

Habitat: True to their name, these semi-aquatic bog specialists are mainly found in acidic sphagnum bogs or fens; however, Bog Turtles also occur in wet meadows, freshwater marshes, seepage springs, small streams, and other shallow, ephemeral wetlands, as long as the habitat is open and sunny, with some shrub, but minimal forest, cover. Bog Turtles can be found on the surface, often basking on matted vegetation from early morning into the late afternoon if surface temperatures are not too warm. If surface conditions become inhospitable, Bog Turtles will either burrow into the mud or use pathways dug by rodents beneath vegetation to remain active. Their home ranges vary from less than 0.5 hectares to about 1.3 hectares (1.2–3.2 acres), but they utilize many different microhabitats within their home wetland. Bog Turtles will move to wet areas, springs, and even stream banks outside their home wetland during dry conditions. This species, particularly juveniles, can be difficult to locate.

Natural History: Bog Turtles are a secretive and reclusive species, often spending

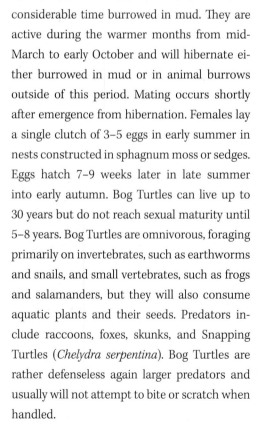

Left: Head of an adult Bog Turtle, Johnson County. (Photo by Phil Colclough) *Right:* Juvenile Bog Turtles in a captive breeding program at the Knoxville Zoo. (Photo by Phil Colclough)

considerable time burrowed in mud. They are active during the warmer months from mid-March to early October and will hibernate either burrowed in mud or in animal burrows outside of this period. Mating occurs shortly after emergence from hibernation. Females lay a single clutch of 3–5 eggs in early summer in nests constructed in sphagnum moss or sedges. Eggs hatch 7–9 weeks later in late summer into early autumn. Bog Turtles can live up to 30 years but do not reach sexual maturity until 5–8 years. Bog Turtles are omnivorous, foraging primarily on invertebrates, such as earthworms and snails, and small vertebrates, such as frogs and salamanders, but they will also consume aquatic plants and their seeds. Predators include raccoons, foxes, skunks, and Snapping Turtles (*Chelydra serpentina*). Bog Turtles are rather defenseless again larger predators and usually will not attempt to bite or scratch when handled.

Conservation Status: Bog Turtles are either extremely rare or extirpated from locations where they once were abundant, mostly due to habitat destruction and collection by the pet trade industry. The southern population likely consists of between only 4,000 and 6,000 individuals distributed among multiple, isolated subpopulations. Bog Turtles often occur in very small, isolated colonies, and their genetic diversity is surprisingly low, apparently suffering from a severe bottleneck a long time ago. Thus Bog Turtles might be robust to the negative genetic effects of further fragmentation of their populations. The species' decline is symptomatic of species with fairly specific habitat requirements and mirrors the decline of its habitat, as bogs and fens have been drained and filled until very few such habitats remain, and the remaining bogs often suffer from hydrological alteration, livestock grazing, and invasion by nonnative plants. Moreover, the small adult size, long lifespan, low reproductive rate, shy disposition, and rarity of this species make it a popular pet species and thereby contribute to its endangerment. As a result, this species is listed as "Critically Endangered" on the IUCN Red List of Threatened Species. Different conservation issues in the northern and southern populations resulted in different federal listing status for each population; in 1997, the USFWS listed the northern population as "Threatened" and the southern population (including Tennessee) as "Threatened Due to Similarity of Appearance." A draft recovery plan was issued by USFWS in 2000, a species action plan was issued in 2009, and a species recovery plan has been recently issued for the northern, but not the southern, population. TWRA lists the species as "Threatened" in Tennessee, and TDEC considers it "Very Rare and Imperiled." In Shady Valley, a 700-acre preserve in Johnson County, the Nature Conservancy is working to preserve and restore high-elevation bog and other wetland habitats for Bog Turtles, wild cranberries, and red spruce. A number of captive husbandry, breeding, and release programs have been implemented for Bog Turtles, including the Southern Appalachian Bog exhibit at Knoxville Zoo. Sadly, Bern Tryon, the director of herpetology at the Knoxville Zoo and one of the Bog Turtle's greatest advocates for 25 years (including awards for his Bog Turtle captive breeding and release program), passed away in May 2011, marking the end of an era in Bog Turtle conservation. Local populations, and possibly the species itself, could go extinct if immediate action is not taken to reduce or prevent the development, draining, and grazing of the bog habitats on which it depends. This species is also at risk of continued exploitation by the nefarious side of the pet industry. Long-term monitoring of existing populations and surveys for new populations are badly needed, as is continued assessment of the value of captive breeding and release programs in Tennessee.

Comments: Since the early 1800s, this species has been considered a member of the genus *Clemmys* (meaning "tortoise") with the Wood Turtle (*C. insculpta*), Spotted Turtle (*C. guttata*), and Western Pond Turtle (*C. marmorata*). However, Bog Turtles and Wood Turtles were sequestered to the genus *Glyptemys* based on the results of molecular genetics studies in 2001. Wood Turtles are considered the Bog Turtle's sister species and are the most similar morphologically; however, this species does not occur in Tennessee.

Todd Campbell

Northern Map Turtle adult basking, Carroll County, Illinois. (Photo by Andrew Durso)

Northern Map Turtle

Graptemys geographica

Description: Northern Map Turtles are medium-sized riverine turtles. Adult females are 18–27 cm (7–11 in) CL, and adult males are smaller at 9–16 cm (3.5–6.3 in) CL. In each sex, the carapace is olive green and moderately domed with a shallow midline keel that lacks prominent knobs or spines. The carapace is adorned with a reticulate pattern of narrow, yellow lines, which resembles the lines on a topographic map (hence the origin of the common name of members of this genus). The pattern on the carapace often is obscured in adults, often by algae. The plastron is pale yellow and unpatterned. The lower half of each marginal scute has thin, green concentric markings. The skin of the head, legs, and tail is various shades of dark green, greenish brown, or gray, and is cov-

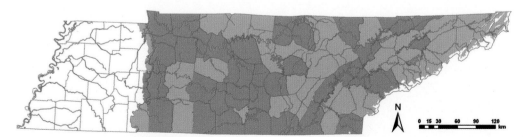

ered with a series of yellow lines. A postorbital spot is present that resembles an elongated triangle. The jaws are moderately large with thick surfaces for crushing prey. The feet are webbed. In addition to growing to larger sizes than adult males, the heads of adult females are larger than those of males, and the maplike pattern on the carapace is more difficult to discern in older females than in older males. Males have thicker tails than females. Hatchlings are circular when viewed from above and about 30 mm (1.1 in) CL. Colors of their carapace, head, and limbs are bright colors, and the pattern of their carapace is bold. Also, the keel on the carapace in hatchlings is pronounced. Dark lines that border seams between plastral scutes are prominent in hatchlings, but these lines fade with age. No subspecies are recognized.

Etymology: The specific epithet, *geographica* [gee-oh-GRAF-eh-kah], is from the Greek words *geo* (world), *graph* (map), and *ikos* (belonging to), which refer to the lines on the carapace that resemble contour lines on a topographic map.

Similar Species: Northern Map Turtles are most similar to other map turtles (*Graptemys sp.*) and emydid turtles. Both Ouachita Map Turtles (*G. ouachitensis ouachitensis*) and False Map Turtles (*G. pseudogeographica pseudogeographica*) have prominent middorsal keels with knobs as adults. Ouachita Map Turtles have a postorbital spot, but they also have a pair of spots below the eye: one directly below the eye and another on the lower jaw. False Map Turtles lack a postorbital spot but have either a short vertical bar or crescent behind the eye. Additionally, young of both Ouachita Map Turtles and False Map Turtles have smudging or lines on the plastron, but these markings often fade

with age. Northern Map Turtles differ from Eastern Painted Turtles (*Chrysemys picta*) and Southern Painted Turtles (*C. dorsalis*) by having a serrated posterior margin to the carapace. River Cooters (*Pseudemys concinna*) and Pond Sliders (*Trachemys scripta*) have a ridged crushing surface in the roof of the mouth and lack a prominent middorsal keel, whereas Northern Map Turtles have a smooth crushing surface and possess at least a low middorsal keel. River Cooters have prominent yellow to yellow-orange bars on the marginal scutes and typically lack a postorbital spot or blotch. Pond Sliders have a broad postorbital stripe that is red to orange to yellow, not a spot.

Distribution: Northern Map Turtles are found from southern Quebec in Canada southward into Tennessee and northern Alabama and west to central Minnesota, eastern Iowa, central Missouri, and central Arkansas. This species is apparently absent throughout much of the Mississippi River Valley. In Tennessee, Northern Map Turtles occur throughout the Interior Plateau, Ridge and Valley, and lower elevations of the Blue Ridge Mountains within the Tennessee and Cumberland River drainages. However, records are lacking from much of northeast Tennessee. A questionable record exists from Reelfoot Lake in northwest Tennessee.

Habitat: Northern Map Turtles can be found in reservoirs, lakes, rivers, and streams that typically have clear, flowing water, gravel substrates, and aquatic vegetation. This species has a preference for larger bodies of water (e.g., larger rivers and streams versus smaller streams). Adults and juveniles can been seen basking on logs and rocks, often in large aggregations, with larger individuals preferring to bask farther from the shore and in areas with

Left: Venter of a Northern Map Turtle, Rutherford County. (Photo by Matthew L. Niemiller) *Right:* Head of a Northern Map Turtle adult, Rutherford County. (Photo by Matthew L. Niemiller)

faster currents compared to smaller individuals. Like other map turtles, this species is very wary and is quick to be alarmed and slide into the water. Consequently, Northern Map Turtles are difficult to capture and uncommonly observed. Juveniles often will hide under large, flat rocks when disturbed. Northern Map Turtles overwinter in deeper pools with slow current.

Natural History: Little is known about the natural history of Northern Map Turtles compared to other emydid species in Tennessee. This species is often one of the first species to emerge in spring and last to enter hibernation in autumn. Northern Map Turtles are most active during the day, usually feeding during the early morning and late afternoon. Adults, particularly males, are known to move between summer home ranges and overwintering sites. Mating occurs in both spring and autumn. Unlike other emydid turtles, male Northern Map Turtles lack elongate claws used to stroke the female's face during courtship. Females dig nests in open areas with loose soil or sand, often a considerable distance from water. Females lay one to three clutches per year, each consisting of 6–15 eggs. Eggs are elongate and sex of developing embryos is determined by incubation temperature. Temperatures above 30°C (86°F) result in mostly females, whereas temperatures around 25°C (77°F) result in mostly males. Eggs hatch after 9–11 weeks, typically during late summer. However, eggs laid late in the season may not hatch until the following spring; similarly, hatchlings may remain in the nest until the following spring. Males probably reach sexual maturity around 4–6 years of age, females around 10 years of age (about 19 cm [7.5 in] CL). The diet consists primarily of snails, mussels, and crayfish, although insect larvae and carrion are also consumed. Likely predators of adults include predatory mammals, such as coyotes, raccoons, river otters, and foxes. Hatchlings are eaten by predatory fishes and several birds, including herons, gulls, crows, grackles, and red-winged blackbirds. Raccoons, skunks, and other mammals are the primary predators of nests. Northern Map Turtles are very shy and rarely attempt to bite when handled.

Conservation Status: Northern Map Turtles likely are more common and widely distributed in the state than current records indicate. This species is difficult to capture and document because of its wary disposition. Northern Map Turtles are threatened by decreased water quality, particularly where mussel populations have declined as a result of damming and channelization. Like other turtles, this species is also threatened by the construction of roads between aquatic habitats and nesting areas that females must cross to lay eggs.

Matthew L. Niemiller and
R. Graham Reynolds

Ouachita Map Turtle adult, Trigg County, Kentucky. (Photo by Greg Sievert)

Ouachita Map Turtle

Graptemys ouachitensis

Description: Ouachita Map Turtles are medium-sized riverine turtles. Adult females are 13–26 cm (5–10 in) CL, and adult males are considerably smaller at 8–14 cm (3–5.5 in) CL. In each sex the carapace is various shades of olive, greenish brown, or brown, and it has a middorsal keel with blunt knobs or spines on the second and third vertebral scutes. The posterior margin of the carapace is serrated. The

carapace has a reticulate pattern of narrow yellow to white lines; however, this pattern often is obscured in adult females and older males. An interrupted middorsal black stripe often is present. The plastron is cream or yellowish brown. Darker markings are present on the plastron in young individuals, but these markings become less prominent with age. The lower half of the marginal scutes have concentric green markings. The ground color of the skin of the head and neck is dark green, but a series of yellow lines, some of which may reach the eye, break up the overall green color. A large postorbital spot or blotch is present behind each eye. These large blotches come close to meeting at the midline but are separated from each other by at least one line on the top of the head. Yellow spots are also present directly below the eye and on the lower jaw. These jaw spots occasionally touch the large postorbital markings. The jaws are moderately large with thick surfaces for crushing prey. The feet are webbed. The iris of the eye is white to yellow with a black horizontal bar. In addition to growing to larger sizes than males, females also have larger heads. However, compared to adult females, the claws on the forelimbs of adult males are greatly elongated, and their tails are thicker. Hatchlings are circular in outline when viewed from above and about 25–35 mm (1–1.1 in) CL. The green and yellow of their shell and head are bright and their pattern is bold. Their middorsal keel is obvious and has prominent knobs, which give the hatchings a "saw-backed" appearance. The plastron has a complex design of darker lines and markings that fades with age. Two subspecies are recognized, but only the Ouachita Map Turtle (*G. o. ouachitensis*) occurs in Tennessee.

Etymology: The specific and subspecific epithet, *ouachitensis* [waw-sheh-TEN-sis], means "belonging to the Ouachita Mountains" in Arkansas, where the type specimen was collected.

Similar Species: Ouachita Map Turtles are most similar to other map (*Graptemys sp.*) and emydid turtles. False Map Turtles (*G. pseudogeographica pseudogeographica*) also have prominent middorsal keels with knobs and postorbital markings, but individuals of this species lack the pair of spots below the eye (one directly below the eye and another on the lower jaw) found in Ouachita Map Turtles. Northern Map Turtles (*G. geographica*) also have a postorbital spot but typically lack darker markings on the plastron, and they have a low middorsal keel on the carapace. Ouachita Map Turtles are known to hybridize with Northern Map Turtles, confounding attempts at accurate identification by physical features. Ouachita Map Turtles differ from Eastern Painted Turtles (*Chrysemys picta*) and Southern Painted Turtles (*C. dorsalis*) by having a serrated posterior margin on the carapace. Both River Cooters (*Pseudemys concinna*) and Pond Sliders (*Trachemys scripta*) have a ridged crushing surface in the roof of the mouth and lack a prominent middorsal keel, whereas False Map Turtles have a smooth crushing surface and possess a middorsal keel. River Cooters typically lack a postorbital spot or blotch, and their marginal scutes have prominent yellow to yellow-orange bars. Pond Sliders have a broad postorbital stripe that is red to orange to yellow, not a spot.

Distribution: Ouachita Map Turtles have a wide distribution throughout the Mississippi, Missouri, and Arkansas River drainages from North Dakota, Iowa, and Minnesota south to eastern Texas, Louisiana, and Mississippi. In Tennessee, Ouachita Map Turtles have been found in the Tennessee River, Mississippi River, Reelfoot Lake, and direct tributaries to the Mississippi River. Ouachita Map Turtles also are known from the Tennessee River in eastern Tennessee and the Clinch River drainage in Claiborne County. Records are lacking from the Cumberland River drainage in Tennessee, although this species has been documented from the Upper Cumberland River in Kentucky. Scott and Redmond (2008) reported questionable literature records of False Map Turtles from the upper Tennessee River drainage in Hamblen County and from the middle Cumberland River in Clay County. However, these records likely represent Ouachita Map Turtles, rather than False Map Turtles.

Habitat: Ouachita Map Turtles are associated with riverine habitats with faster currents than those inhabited by False Map Turtles, including larger rivers, oxbows, lakes, and reservoirs with aquatic vegetation and sand to silt bottoms. However, this species is occasionally observed in backwater sloughs and floodplain swamps. Adults and juveniles can been seen basking on logs and rocks, often in large aggregations. Like other map turtles, Ouachita Map Turtles are extremely wary and quick to be alarmed and slide into the water. Basking sites are rarely near the shore. This species is seldom found on land, except during the nesting season, when females travel overland to nesting sites.

Natural History: The behavior of Ouachita Map Turtles is similar to that of False Map Turtles. Adults and juveniles hibernate in underwater rock piles, bank burrows, muskrat lodges, or

Ouachita Map Turtle adult, Clark County, Arkansas. (Photo by Greg Sievert)

soft sediments. Ouachita Map Turtles are most active during the day, usually feeding during the early morning and late afternoon. During the rest of the day, Ouachita Map Turtles spend much of their time basking. Mating occurs in both spring and autumn. During courtship, a male will vibrate his elongated forelimb claws against the female's head and bob his head up and down. A male also may bite the neck, limbs, and tail of a female. Females dig nests in open areas with sand or loose soil in June, usually during the morning hours. Up to three clutches are laid per year, each with 6–17 eggs. Eggs are elongate and sex of developing embryos is determined by incubation temperature. Eggs incubated at temperatures above 30°C (86°F) result in mostly females, whereas eggs incubated at temperatures around 28°C (82°F) result in mostly males. Eggs hatch in 9–12 weeks during late summer, although eggs laid late in the season may not hatch until the following spring. Similarly, hatchlings occasionally remain in the nest until the following spring. Males reach sexual maturity during their second or third year, whereas females mature in six to seven years. Ouachita Map Turtles can live up to 15 years in captivity. Their diet consists of a variety of prey, including freshwater sponges, bryozoans, clams, snails, earthworms, crayfish, aquatic insects, fish, and even fledgling birds. Fruits, seeds, and other plant material also are consumed. Likely predators of adults include predatory mammals, such as coyotes, raccoons, river otters, and foxes. Nest predators include foxes, raccoons, river otters, and fly maggots. Predatory fishes, herons, gulls, crows, grackles, and other birds eat hatchlings. Ouachita Map Turtles are very shy and rarely attempt to bite when handled.

Conservation Status: Little is known about Tennessee populations, but Ouachita Map Turtles do not appear to be of immediate conservation concern. Regardless, populations should likely be monitored. Tragically, basking turtles, including Ouachita Map Turtles, are used for target practice by misguided people.

All too often turtles with obvious gunshot injuries are found along the shores of our rivers and streams, either dead or dying.

Comments: The taxonomy of Ouachita Map Turtles has been in a state of flux for many years. In the past, *G. ouachitensis* has been recognized as a subspecies of *G. pseudogeographica*. The confusing taxonomy is associated with, in part, morphological variation in these taxa and potential hybridization with False Map Turtles and Northern Map Turtles. We adhere to the taxonomy followed by other recent authors, recognizing *G. ouachitensis* and *G. pseudogeographica* as distinct species. Furthermore, *G. pseudogeographica* includes two subspecies, *G. p. kohnii* and *G. p. pseudogeographica*. However, this taxonomy might change again after future morphological and molecular studies.

Matthew L. Niemiller and
R. Graham Reynolds

False Map Turtle adult, Cape Girardeau County, Missouri. (Photo by Matthew L. Niemiller)

False Map Turtle

Graptemys pseudogeographica

Description: False Map Turtles are medium-sized riverine turtles. As with other species of map turtles, females grow to large sizes and have larger heads than males. Adult females are 13–28 cm (5–11 in) CL, and adult males are considerably smaller at 9–15 cm (3.5–6.0 in) CL. In each sex the carapace is various shades of olive, greenish brown, or brown, and it has a middorsal keel with small knobs. The posterior margin of the carapace is serrated. The carapace has with a reticulate pattern of narrow yellow to white lines; however, this pattern often is obscured in adult females and older males. A middorsal black stripe often is also present. The plastron is yellowish brown. Darker marks,

Left: Mississippi Map Turtle hatchling, Lake County. (Photo by Matthew L. Niemiller) *Right:* Venter of a Mississippi Map Turtle hatchling, Lake County. (Photo by Brad M. Glorioso)

smudges, or parallel lines are present on the plastron of young individuals, but these markings become less obvious with age. The lower half of the marginal scutes have concentric green markings. The ground color of the skin of the head and neck is green, but a series of yellow lines, some of which reach the eye, break up the overall green color. On each side of the head, one of these lines is enlarged to form a prominent postorbital bar or crescent. The jaws are moderately large with thick surfaces for crushing prey. The feet are webbed. Compared to females, the claws on the forelimbs of adult males are greatly elongated, and their tails are thicker. Hatchlings are circular in outline when viewed from above, and about 25–35 mm (1–1.4 in) CL. The green and yellow of their shell and head are bright and their color pattern is bold. Their middorsal keel has prominent knobs giving them a "saw-backed" appearance. The plastron has a complex design of darker lines and markings that fades with age. Two subspecies are recognized, and both occur in Tennessee. False Map Turtles (*G. p. pseudogeographica*) have a short, postorbital vertical bar that allows at least some yellow lines to reach the orbit of the eye. The iris of this subspecies is yellowish brown with a black horizontal bar. The Mississippi Map Turtle (*G. p. kohnii*) has a postorbital crescent that keeps all yellow lines on the neck from reaching the orbit of the eye. The postorbital crescent sometimes is broken. Additionally, the iris of this subspecies is white.

Etymology: The specific and subspecific epithet, *pseudogeographica* [su-do-gee-oh-GRAF-eh-kah], is from the Greek *pseudo* (false) and the words *geo* (world), *graph* (map), and *ikos* (belonging to), which describe the lines on the carapace that resemble contour lines on a topographic map. This name indicates the similarity of this species to the Northern Map Turtle (*G. geographica*). The subspecific epithet, *kohnii* [KOH-nee-i], honors Joseph Kohn of New Orleans, Louisiana, who collected the type specimen.

Similar Species: False Map Turtles are most similar to other map turtles (*Graptemys sp.*) and emydid turtles. Ouachita Map Turtles (*G. o. ouachitensis*) have prominent middorsal keels with knobs and postorbital markings. However,

Left: Head of a False Map Turtle adult. (Photo by Matthew L. Niemiller) *Right:* Venter of a False Map Turtle hatchling, Lake County. (Photo by Brad M. Glorioso)

Ouachita Map Turtles also have a pair of spots below the eye—one pair occurs directly below the eye and another on the lower jaw. Northern Map Turtles (*G. geographica*) have a postorbital spot, but it is not in the form of a short vertical bar or crescent. Additionally, Northern Map Turtles typically lack darker markings on the plastron and have a lower middorsal keel on the carapace. False Map Turtles differ from Eastern Painted Turtles (*Chrysemys picta*) and Southern Painted Turtles (*C. dorsalis*) by having a serrated posterior margin to the carapace. Both River Cooters (*Pseudemys concinna*) and Pond Sliders (*Trachemys scripta*) have a ridged crushing surface in the roof of the mouth and lack a prominent middorsal keel, whereas False Map Turtles have a smooth crushing surface and possess a middorsal keel. River Cooters have prominent yellow to yellow-orange bars on the marginal scutes and typically lack a postorbital spot or blotch. Pond Sliders have a broad postorbital stripe that is red to orange to yellow, not a spot.

Distribution: False Map Turtles have a wide distribution throughout the Mississippi, Missouri, and Arkansas River drainages from North Dakota and Minnesota south to eastern Texas and Louisiana. In Tennessee, both subspecies are largely confined to more riverine habitats in West Tennessee. Mississippi Map Turtles have been found in the Tennessee River, Mississippi River, Reelfoot Lake, and direct tributaries to the Mississippi River; False Map Turtles are associated only with the Mississippi River and its direct tributaries. However, intergrades between these two subspecies exist in Tennessee and the range of each subspecies has yet to be accurately determined. Scott and Redmond (2008) reported questionable literature records from the upper Tennessee River drainage in Hamblen County and from the middle Cumberland River in Clay County. These records likely represent Ouachita Map Turtles rather than False Map Turtles.

Habitat: False Map Turtles are associated with riverine habitats with slower currents; however, both subspecies can be found in floodplain swamps, bayous, sloughs, oxbows, and lakes that are frequently connected to larger rivers during floods. These habitats typically have abundant aquatic vegetation and

Mississippi Map Turtle adult, Lake County. (Photo by Brad M. Glorioso)

fallen timber that provide basking sites. Adults and juveniles can be seen basking on logs and rocks, often in large aggregations. Like other map turtles, False Map Turtles are extremely wary, quick to be alarmed, and slide into the water at the slightest disturbance. Basking sites are rarely near the shore, and turtles often will climb well above the water to bask (up to 2 m). This species is seldom found on land; however, females travel overland to nesting sites.

Natural History: False Map Turtles emerge from hibernation when water temperatures are above 4–7° C (40–45°F). Adults and juveniles hibernate in underwater rock piles, bank burrows, muskrat lodges, or soft sediments. False Map Turtles are most active during the day, usually feeding during the early morning and late afternoon. During the rest of the day, False Map Turtles spend much of their time basking. Mating occurs during the spring and autumn. During courtship, a male will vibrate his elongated forelimb claws against the female's head, and he will bob his head up and down. Females dig nests in open areas with sand or loose soil in June, usually during the morning hours. Females are known to use the same nest sites year after year. Although females have been reported to lay as many as four clutches of eggs per year, most lay fewer clutches. Clutch size is variable (2–22 eggs), perhaps associated with differences in sizes of females and variability in number of clutches. Eggs are elongate, and the sex of developing embryos is determined by incubation temperature. Eggs incubated at temperatures above 30° C (86°F) result in mostly females, whereas eggs incubated at temperatures around 25° C (77°F) result in mostly males. Eggs hatch in 9–12 weeks in late summer, although eggs laid late in the season may not hatch until the following spring. Similarly, hatchlings may remain in the nest until the following spring. Males probably reach sexual maturity during their fourth to sixth year, whereas females do not become sexually mature until their eighth to twelfth year. False Map Turtles can live up to 35 years in captivity. Their diet consists of a variety of prey, including freshwater sponges, bryozoans, clams, earthworms, crayfish, aquatic insects, fish, and even fledgling birds. Fruits, seeds, and other plant material also are consumed. Likely predators of adults include predatory mammals, such as coyotes, raccoons, river otters, and foxes. Nest predators include foxes, raccoons, river otters, skunks, coyotes, and fly maggots. Catfishes, basses, rice rats, gulls, crows, grackles, and other birds eat hatchlings. False Map Turtles are very shy and rarely attempt to bite when handled.

Conservation Status: Throughout the range of False Map Turtles, pollution and channel and floodplain alterations for flood control have been mentioned as causes of population declines. Other factors of declines include siltation, unlawful shooting, commercial harvest, and loss of nesting habitat. Little is known about Tennessee populations, but this species does not appear to be of immediate conservation concern.

Comments: The taxonomy of False Map Turtles, including Ouachita Map Turtles (*G. o. ouachitensis*), has been in a state of flux for many years. At one time or another, both subspecies of *G. pseudogeographica* have been recognized as distinct species. Additionally, *G. ouachitensis* has been recognized as a subspecies of *G. pseudogeographica*. The confusing taxonomy is, in part, owing to morphological variation in these taxa and potential hybridization. We follow the taxonomy followed by other recent authors, recognizing *G. ouachitensis* and *G. pseudogeographica* as distinct species with *G. pseudogeographica* comprising the subspecies *G. p. kohnii* and *G. p. pseudogeographica*. However, this taxonomy might change again after future morphological and molecular studies.

Matthew L. Niemiller and
R. Graham Reynolds

River Cooter female, Lake County. (Photo by Brad M. Glorioso)

River Cooter

Pseudemys concinna

Description: River Cooters are large basking turtles. Most adults are 23–33 cm (9–13 in) CL, but old adults have exceeded 37 cm (14.5 in) CL. These turtles have a high-domed, nonkeeled carapace that often appears solid dark green, brown, or black from a distance. However, close inspection reveals that the carapace is beautifully patterned with fine red, orange, and black reticulations that look much like fingerprints. An obvious C-shaped marking is present on the posterior margin of each second pleural scute, and

Left: Plastron of a male River Cooter, Bradley County. (Photo by Matthew L. Niemiller) *Right:* River Cooter basking, Lake County. (Photo by Matthew L. Niemiller)

similar markings often can be found in the same location on some or all of the other pleural scutes. Actually, the C-shape is only part of a complete circle that continues on the adjacent, posterior costal scute. These patterns are often very difficult to see if the shell is dry from basking but are obvious when the turtle is either under water or has just hauled itself out to bask. The seams between each carapace scute are often much darker than the background color, but this too may not be visible if the shell is dry. The upper surfaces of the marginal scutes are similar in coloration to the larger pleural and vertebral scutes, but they lack the C marking, and the lower sur-

face of each has a single black or sometimes red blotch or doughnut-shaped marking. The plastron is hingeless and mostly yellow or cream colored, with dark irregular spots that usually are confined to the anterior-most scutes and nearly form mirror images on either side of the midline. The shells of mature adults are somewhat elongated and narrower, or "pinched," at the bridge between the shell openings, giving the shell a waistline or slight dumbbell shape from above. The posterior edge of the adult carapace is wavy or scalloped due to indentations between the marginal scutes, but this feature may be worn down in older turtles. River Cooters have five vertebral scutes flanked by 4 costal scutes and 12 marginal scutes on each side. The head and neck are festooned with yellow or cream-colored stripes, some of which are wide near the jaw, and

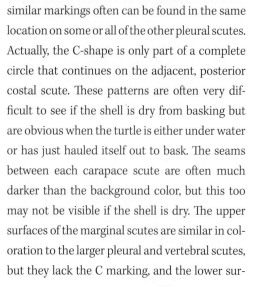

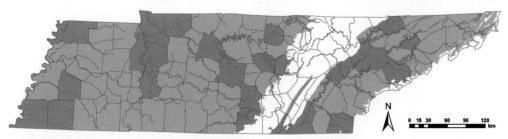

Left: River Cooter hatchling, Shelby County. (Photo by Matthew L. Niemiller) *Right:* Plastron of a hatchling River Cooter, Shelby County. (Photo by Matthew L. Niemiller)

five of these stripes run continuously between the eyes on the dorsal surface of the head. The underside of the chin appears relatively flat when viewed from an anterior position. The feet are strongly webbed for swimming. Mature females are larger but have a shorter tail than males, and their vent does not extend past their carapace. Mature males have long claws, especially on the front feet, and a long tail with the vent extending well past the carapace. Hatchlings are usually 30–40 mm (1.2–1.6 in) CL and look much like the adults, except the patterns on the carapace are more doughnut-like and the C shape on the second costal scutes may be unformed or difficult to discern. Two subspecies currently are recognized, but only the Eastern River Cooter (*P. c. concinna*) occurs in Tennessee. However, up to five subspecies were recognized as recently as 2008. The taxonomy of this species is currently under review and remains highly controversial (see **Comments** below).

Etymology: The specific epithet, *concinna* [kon-SIN-na], is Latin for "well arranged," which is a reference to the intricate pattern on the carapace.

Similar Species: From a distance, River Cooters might be confused with Pond Sliders (*Trachemys scripta*), especially melanistic ones with worn-down shells. However, Pond Sliders lack the C-shape on the second costal scutes and instead have a wide banded pattern on the carapace (if visible) and a large yellow, orange, or red patch or a wide stripe on each side of the head behind the eye that is much wider than adjacent stripes. River Cooters could possibly be confused with either of the Painted Turtles (*Chrysemys dorsalis* and *C. picta*). The range of River Cooters overlaps with Southern Painted Turtles extensively, but the latter have fairly low domed, very colorful shells with black, red and yellow patterns, wide yellow bands or blobs on their neck behind the eye, and Southern Painted Turtles of all ages have a red, orange, or yellow middorsal stripe on the carapace. Additionally, both species of Painted Turtles have a smooth edge to the posterior margin of the carapace.

Distribution: River Cooters are mostly confined to the Old South, ranging from Virginia and Florida westward to Kansas, Oklahoma, and Texas, but they are absent from much of the

Blue Ridge and Cumberland Mountains. In Tennessee, River Cooters are likely found statewide in larger rivers and bodies of water at lower elevations and are absent from the Cumberland Plateau and much of the Blue Ridge Mountains. However, records are missing from much of northeast Tennessee.

Habitat: True to their name, River Cooters are mainly denizens of large, slow-moving, muddy river systems, such as the Mississippi, Tennessee, and Cumberland rivers and their reservoirs, but they are also abundant in smaller, faster, rockier systems, such as the Big South Fork of the Cumberland River and the Conasauga River, where they are found in riparian habitats from deep pools to shallow, rocky shoals. They can regularly be observed foraging on algae and snails in the shallow water over the rocky shoals below the Leatherwood Ford Bridge in the Big South Fork National River and Recreation Area. They are abundant in the still waters of Reelfoot Lake and in oxbow lakes, which are formed by large river systems meandering out of their banks over time, but can occasionally be found in smaller lakes and even farm ponds, as long as the water is relatively clear.

Natural History: River Cooters are classic basking turtles, spending much of the day hauled out on logs and rocks, or less often, riverbanks. They are mostly active during the summer months, but even the occasional warm winter day in Tennessee will entice them to forage and bask. River Cooters are rarely seen overland, and their home ranges are generally small relative to the large rivers systems they inhabit; evidence from radio-tracking studies in West Virginia even suggests they rarely move between adjacent pools. Mating occurs during spring, and they nest during summer months, usually prior to July. Nests are constructed in sandy sites within the riparian zone of the river they inhabit. In central Arkansas, females deposit several clutches of 9–18 eggs, but reproductive output may be different in Tennessee. Eggs hatch 60–90 days later, but like many aquatic turtles in Tennessee, late arrivals overwinter in the nest. Male River Cooters mature in about six years at just under 20 cm (8 in) CL, whereas females mature at 10 or more years between 23 and 26 cm (9.0–10 in) CL. River Cooters undergo a classic shift in their diet from consuming aquatic insects, snails, and worms as juveniles to being largely herbivorous as adults, feeding on various emergent wetland plants and algae in Alabama, but adults may rarely feed on snails and even carrion. Predators are largely undocumented, but experts suggest that raccoons rob nests and minks, Snapping Turtles (*Chelydra serpentina*), large fish, and even snakes may eat hatchlings.

Conservation Status: River Cooters are widespread and abundant where they occur and are not of conservation concern in Tennessee. However, the large river systems of Tennessee, especially the Tennessee River, have been severely altered by the construction of the highest concentration (by river mile) of hydroelectric dams on Earth. This converted much of its length to a series of oligotrophic lakes with low nutrient content and bathtub-like banks with little emergent vegetation or basking and nesting sites, interrupted by only short stretches of physically intact but hydrologically altered riparian zones with scoured bottoms and clear, flowing water. While River Cooters maintain a foothold in many TVA reservoirs, the long-term effects of the recent, disturbing increase in residential development along the banks of these once restricted shorelines remains to be seen. Although River Cooters have long been harvested for sale to Asian markets in the Carolinas and possibly Georgia, the extent of this activity is unknown in Tennessee, and clearly the harvesting of our native turtles for overseas markets should not be allowed. Ultimately, even the most abundant species are vulnerable to the effects of wetland and riparian habitat alteration, unsustainable agricultural practices, chemical pollution, overharvesting, and roadkill, often leading to significant and rapid local population declines. This large, beautiful species should be monitored carefully, especially in systems where development is allowed, to ensure they will be part of the Tennessee landscape for many years to come.

Comments: Five subspecies of River Cooters were long recognized and are still recognized by some authors, and the taxonomy of this group is confusing and controversial, but clarity is on the horizon. Currently only two subspecies are recognized. The Hieroglyphic River Cooter (*P. c. hieroglyphica*) and Missouri River Cooter (*P. c. metteri*) were recently considered clinal variants of the Eastern River Cooter (*P. c. concinna*). Both the Suwannee River Cooter (*P. suwanniensis*) and Rio Grande River Cooter (*P. gorzugi*) are now considered distinct species.

Todd Campbell

Eastern Box Turtle adult, Cannon County. (Photo by Brian T. Miller)

Eastern Box Turtle

Terrapene carolina

Description: Eastern Box Turtles are small- to medium-sized terrestrial turtles that generally reach 10–15 cm (4–6 in) CL but are known to reach nearly 20 cm (8 in). The high-domed cara-

pace is highly variable but generally has a dark brown background with orange or yellow spots that often appear to radiate from one spot on each scute. The plastron is also variable, with broad areas of dark brown or black on a yellow or cream background on each scute. This turtle is best known for its ability to withdraw

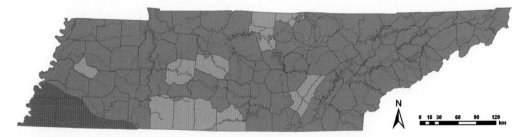

Two subspecies of Eastern Box Turtles are recognized in Tennessee: the Eastern Box Turtle (*Terrapene carolina carolina*), whose range appears in orange, and intergrades with the Three-Toed Box Turtle (*T. c. triunguis*), whose range appears in purple.

its head, legs, and tail entirely within its shell, which can be closed tightly like a box, even forming a watertight seal between the carapace and plastron that cannot be pried open without hurting the turtle. This is possible via two hinges on the plastron. One hinge lies between the humeral and pectoral scutes and the other is a pair of hinges, each between the marginal and abdominal scutes where the bridge would be (a true bridge is absent in box turtles). The carapace has five vertebral scutes flanked by four pleural scutes and 12 marginal scutes on each side. A keel is present along the midline of the carapace in most individuals but may wear down with age. The posterior edge of the carapace is flanged outward. The front feet are only slightly webbed and have strong claws, and the rear feet are "elephantine" and are not webbed, much like that of a tortoise. Some individuals have only three toes on their hind feet (they can even have three on one foot and four on the other), and in one subspecies described below, all have three toes. The jaw has a prominent beak. Males have a concave plastron with a broad dent to accommodate the female's carapace during mating, and a relatively long tail with the cloacal opening extending well beyond the carapace. Females have a flat or slightly convex plastron, and a tiny tail with the cloacal opening close to its origin under the carapace. Males usually have red eyes, and females have pale brown or orange eyes. Hatchlings are tiny (25–35 mm [1–1.4 in] CL) versions of the adults in color and shape but have a more prominent keel on the carapace. Four subspecies are recognized, two of which occur in Tennessee: the Eastern Box Turtle (*T. c. carolina*), which is described above and the Three-Toed Box Turtle (*T. c. triunguis*), which has three toes and claws on the rear feet and a solid olive, light brown, or

even orange or yellowish carapace. Three-Toed Box Turtles also sometimes have radiating yellow spots, but if present, they will be subtle.

Etymology: The specific and subspecific epithet, *carolina* [ka-roh-LIE-nah], indicates the general region where scientists first collected this species. The subspecific epithet, *triunguis* [try-UN-gwis], is from the Latin *tri* (three) and *unguis* (claws), in reference to the three toes on the hind limbs of this subspecies.

Similar Species: Eastern Box Turtles are hard to mistake for any other species because they possess a high-domed, keeled, posteriorly flanged carapace, a multihinged plastron, and tortoise-like feet. These turtles are usually found in terrestrial locations, and no other box turtle species occurs in Tennessee. Eastern Mud Turtles (*Kinosternon subrubrum subrubrum*) have high-domed shells and plastral hinges but are much smaller, are usually entirely black or very dark with a few stripes, and have webbed feet.

Distribution: Eastern Box Turtle are widespread throughout the eastern United States, from southern New England, the mid-Atlantic, and the southern Great Lakes region southward to eastern Texas, the Gulf Coast, and Florida. In Tennessee, this species occurs statewide from the highest peaks of the Blue Ridge Mountains to the lowest river bottomlands of West Tennessee. The Eastern Box Turtle ranges throughout most of the state but intergrades with the Three-Toed Box Turtle in the extreme southwest corner of Tennessee.

Habitat: These predominantly terrestrial turtles are common in woodland habitats of nearly any type, from the driest upland pine forests, tablelands, and cedar glades to mixed pine-oak forests to moist bottomland hardwood forests and even swamps. Like other members of the family Emydidae, Eastern Box Turtles can be

Left: Head of a male Eastern Box Turtle, Wilson County. (Photo by Matthew L. Niemiller) *Right:* Eastern Box Turtle hatchling, Grundy County. (Photo by Brad M. Glorioso)

found around water, and seem to be most common in riparian zones and broad river valleys, where they can be found (sometimes in groups) sitting at the water's edge, walking along shallow creeks, or buried in the mud under shallow puddles, roadside ditches, and puddles made by off-road vehicles.

Natural History: Eastern Box Turtles are largely terrestrial and diurnal, moving slowly through the forest or basking in a sun fleck in the morning and evening hours and resting during the hot part of the day, often in a mud puddle. They often hide, and hibernate, by digging themselves into the leaf litter just enough to bury their shell, which explains why some of them have scars from past fires on the very top of their carapaces. Amazingly, Eastern Box Turtles can reach densities of nearly 30 turtles per hectare. Mating occurs during spring, and females dig nests during the summer months, depositing 2–7 eggs that hatch in August and September. They grow rapidly and mature at around 10 cm (4 in) CL (females mature at a slightly smaller size), and their lifespan is measured in decades, potentially outliving their human caretakers in captivity. Eastern Box Turtles

are the ultimate omnivores throughout their lives, consuming many kinds of plants (dead or alive, but fruits and berries are favorites), shelf fungi and mushrooms, insects and their larvae (especially beetle grubs), earthworms, slugs and snails, slow-moving soil vertebrates (such as salamanders), and even carrion (including roadkills). Predators include any terrestrial mammal that can figure out how to open their shell, or crush it—neither being an easy task. Adults are most vulnerable to large woodland predators, such as raccoons, bobcats, foxes, hawks, and crows. Hatchlings are small and vulnerable to several other predators, including many species of snakes, small mammals, and feral cats, dogs, and hogs. When threatened, Eastern Box Turtles will retreat into their shell and close it tightly but will rarely try and bite when handled.

Conservation Status: Because of their calm temperament, beautiful shell, ability to close up, terrestrial lifestyle, ease of capture, suitability for captivity, and longevity, Eastern Box Turtles are highly persecuted by commercial collectors and the public alike. They probably rank second only to Pond Sliders in the

 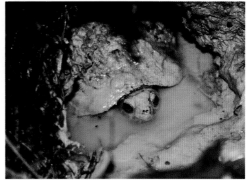

Left: Plastron of an Eastern Box Turtle hatchling, Grundy County. (Photo by Brad M. Glorioso) *Right:* An adult Eastern Box Turtle burrowing into a streambed, Cannon County. (Photo by Brian T. Miller)

number of individuals kept as household pets. However, they are most vulnerable to the loss and fragmentation of forested habitat. Eastern Box Turtles are also killed on roadways in great numbers each year, especially after spring rains or after they come out of hibernation and start looking for food and mates in a landscape highly fragmented by roads. Urban sprawl around Memphis will likely result in rapid and severe population declines of Three-Toed Box Turtles in Tennessee. The IUCN considered Eastern Box Turtles to be "Near Threatened" until 2010, when the designation was changed to "Vulnerable." Most states, including Tennessee, prohibit, or at least regulate, the collection of Eastern Box Turtles.

Comments: The empty shells of Eastern Box Turtles are regularly encountered in the woods, and often the pieces fit together quite nicely. True to their namesake, Native Americans often used their empty shells as containers, and their shells are regularly found in archaeological digs in Virginia. Eastern Box Turtles are frequently called "terrapins."

Todd Campbell

Adult male Red-Eared Slider, Lake County. (Photo by Brad M. Glorioso)

Pond Slider

Trachemys scripta

Description: Pond Sliders are medium- to large-sized basking turtles. Most adults are 13–20 cm (5–8 in) CL, but adults up to of 29 cm (11.4 in) CL have been reported. This is one of the most widely ranging and morphologically variable species of freshwater turtles. Adults have yellow, green, and black pinstripes on their carapace, legs, tail, and head, but they are best known for the broad yellow, orange, or red stripe or patch behind the eye. The carapace has a "rough" appearance due to raised annuli on each scute and numerous alternating yellow, green, and black pinstripes on the

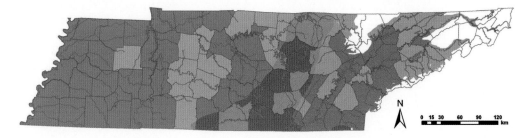

Two subspecies of Pond Sliders are recognized in Tennessee: the Red-Eared Slider (*Trachemys scripta elegans*), whose range appears in blue, and the Cumberland Slider (*T. s. troostii*), whose range appears in orange. Intergrades occur in Middle Tennessee, shown in purple.

scutes that form a pattern much like human fingerprints. The posterior marginal scutes are notched between, giving the posterior edge of the carapace a scalloped or wavy appearance. The plastron is typically yellow with smudges or eyelike spots on the scutes that may become dark and mottled, particularly in older males, or may be absent. All four feet are strongly webbed for swimming. Multiple yellow stripes are present on the dorsal and anterior surfaces of the front legs. When looking at one head on, its lower jaw and chin appear rounded. Pond Sliders are extremely variable in color and pattern over their lifespan. Juveniles and young adults are vividly colored as described above, but large, old individuals, particularly males, usually have a very dark shell with wide dark or black bars due to the infusion of melanin in their carapace scutes. Females are generally larger than males, have much shorter front claws than males when mature, and their relatively tiny tails are only about five percent of the CL and the cloaca barely extends beyond the carapace, whereas males have long front claws as adults, their tails reach about 10 percent of the CL, and their cloaca extends beyond the carapace about half the length of the entire tail. Hatchlings are 25–35 mm (1.0–1.4 in) CL and are more vividly patterned than adults. At least 16 subspecies are recognized in North, Central, and South America. Two subspecies occur in Tennessee: the Red-Eared Slider (*T. s. elegans*) and Cumberland Slider (*T. s. troostii*). As their name indicates, Red-Eared Sliders have a broad red stripe behind their eye, but this marking is sometimes orange or yellow. Juvenile Red-Eared Sliders have yellow plastrons with circular or doughnut patterns on each scute (not confined to the anterior scutes). Cumberland Sliders

look very similar to Red-Eared Sliders but have a narrower, yellow "ear" stripe behind the eye, much broader stripes on most skin surfaces, and smaller dark spots on the ventral surfaces of the marginal scutes. Intergrades between these two subspecies are known where their respective ranges contact. Intergrades with Yellow-Bellied Sliders (*T. s. scripta*) may exist in the southeastern part of the state, particularly in the Conasauga River drainage. Yellow-Bellied Sliders have vertical "striped-pants" patterns on the posterior surface of the hind legs. Juvenile Yellow-Bellied Sliders have a mostly yellow plastron with small, dark circles or doughnuts in each of the anterior scutes.

Etymology: The specific epithet, *scripta* [script-TAH], is Latin for "writing," in reference to the pinstripe patterns on the carapace. The subspecific epithet, *elegans* [EH-leh-gans], is Latin for "elegant," and *troostii* [TROO-stee-i] honors Gerald Troost, a Dutch-born naturalist from Nashville, Tennessee.

Similar Species: Red-Eared Sliders are hard to confuse with any other turtle when they exhibit the classic red ear stripe. However, some individuals, including Cumberland Sliders, have yellow or orange stripes behind the eye, which renders their head and neck patterns similar to River Cooters (*Pseudemys concinna*) and Painted Turtles (*Chrysemys picta* and *C. dorsalis*). However, River Cooters have a flat lower jaw or chin when viewed head on and a C-shaped marking on the second costal scute. Painted Turtles have a smooth edge to the posterior margin of the carapace that lacks pinstripes. Yellow-Bellied Sliders or intergrade individuals could potentially be mistaken for Chicken Turtles (*Deirochelys reticularia*), as both have networks of stripes on their heads

Left: Plastrons of male *(left)* and female (*right*) Red-Eared Sliders, Rutherford County. (Photo by Matthew L. Niemiller) *Right:* Hatchling Red-Eared Slider, Lake County. (Photo by Brad M. Glorioso)

and "striped-pants" patterns on the posterior surface (thigh) of their rear legs, but the latter has one broad yellow stripe on its front legs, a longer neck, and a longer, narrower carapace.

Distribution: Pond Sliders are distributed throughout much of eastern North America from Indiana and Illinois southwestward into Mexico and east along the Gulf Coast and up the Atlantic Coast into Virginia. This species occurs throughout much of Tennessee; however, the ranges of the two subspecies in Tennessee split the state approximately in half, with Red-Eared Sliders occurring in the western half of the state primarily west of the Cumberland Plateau and Cumberland Sliders from the Cumberland Plateau eastward to the lower elevations of the Blue Ridge Mountains. A broad intergrade zone exists where the two subspecies contact. Complicating matters, the range of Yellow-Bellied Sliders might extend up into southeastern Tennessee.

Habitat: Pond Sliders can be found in virtually any lake, pond, reservoir, large river, or slow-water habitat with muddy bottoms and vegetated littoral zones, and they will even inhabit small creeks as long as a series of slow-water zones or deep pools are present. These turtles are usually the most numerous of the basking turtles in any given location, with many individuals from a wide range of age classes often observed basking on large logs in any water depth, but they are rarely seen on the banks. Pond Sliders are absent or extremely rare in the cool waters of fast-flowing mountain streams.

Natural History: Pond Sliders are diurnal and active in Tennessee from March into October. However, individuals have been observed basking on warm, sunny days during winter and even moving underneath ice in ponds. This species, especially males, are prone to going on walkabouts over land to disperse between lakes or river systems and are often seen walking across yards and roads in even the most urbanized areas. Mating occurs during autumn, winter, and occasionally spring, and nesting occurs during late spring and summer. They mate in the water, where males flex or wiggle their long front claws to attract the female's attention, and even "tickle" the sides of the female's face. Nest sites vary widely, from river banks to pond edges to parks and yards long distances from water bod-

Left: Plastron of a hatchling Red-Eared Slider, Lake County. (Photo by Brad M. Glorioso) *Right:* Head of a Cumberland Slider, Roane County. (Photo by Matthew L. Niemiller)

ies (up to 150 m [492 ft]), as long as the substrate is sandy. Clutch size is highly variable, from 2 to 30 eggs, depending on female body size, but usually about 10 elliptical eggs are laid in each of two or three nesting events per year. Males become sexually mature at about 10 cm (4 in) CL (about four years of age), and the larger females mature at about 15 cm (6 in) CL (about eight years of age). Pond Sliders are classic omnivores, but they undergo a dietary shift from arthropods to plants as they age, probably associated with, or a result of, their shift from shallow littoral zone habitats to deeper water zones with age. Despite the general diet shift toward plants, these turtles will consume most anything as adults, including fish, amphibians, reptiles, and even carrion. Predators include large, predatory fish, such as catfish, gar, and bass, raccoons, river otters, and wading birds. Hatchlings are the most vulnerable life stage. When captured, Pond Sliders are known to bite. Many mammals prey on nests, including but not limited to dogs, foxes, skunks, raccoons, opossums, and various rodents.

Conservation Status: Pond Sliders are generally very common in water bodies where they occur throughout their range. However, they face myriad threats, such as residential and industrial development in the riparian zone, river channelization, chemical pollution, and exploitation, resulting in loss of nesting, basking, and foraging habitats and leading to local population declines. Because Pond Sliders are so common, local population decline may eventually be reversed by immigration of individuals from other populations in adjacent watersheds. Some of the commercially harvested populations in Louisiana exhibit a body size distribution that is indicative of exploited populations in that large individuals are much less prevalent. This large, colorful species is wonderful to behold in the wilds of Tennessee in part because it is so despised where it has been introduced. It is considered to be one of 100 of the world's worst invasive exotic species and shares this distinction with only one other reptile, the notorious Brown Treesnake. Extensive traffic in hatchlings of this popular pet species, along with their sale for meat, has resulted in their becoming established in the Northeast, Midwest, and western United States, as well as

Japan, China, Australia, Brazil, the Caribbean, and many other countries worldwide. The impacts of Pond Sliders include competition with, and predation on, other aquatic turtles.

Comments: This is probably the most studied turtle species on Earth, with hundreds of published laboratory studies of its anatomy, physiology, and genetics; nearly as many published studies of its natural history, ecology, behavior, and evolution in its native range; and a rapidly increasing number of studies of the distribution, abundance, impacts, and management of this species where it has been introduced worldwide. A search of the scientific literature will return literally thousands of published papers. The taxonomy of this genus of turtles (previously known as *Chrysemys*) has proven difficult, largely due to the fact that this genus ranges throughout the New World, and taxonomy may change over time as new information, especially the results of molecular genetics studies throughout its range, becomes available.

Todd Campbell

19
Family Kinosternidae (Mud and Musk Turtles)

Mud and Musk Turtles are primarily small, aquatic turtles that have a moderate to high-domed carapace and a plastron that is often hinged. Most species are less than 15 cm (6 in) CL, but those in the genus *Staurotypus* can reach about 30 cm (12 in). Members of this family are found in temperate and tropical regions of North America, Central America, and South America, with the center of diversity in Mexico. The taxonomy of this family has been in a state of flux for years, and the number of species and subspecies recognized varies among sources. At present, most authorities recognize 25 species in four genera (*Claudius, Kinosternon, Staurotypus,* and *Sternotherus*). Three species occur in Tennessee. Fossil kinosternids are known from the Oligocene of North America. Mud and Musk Turtles are found in a variety of aquatic habitats, although most species generally prefer slower flow with soft bottoms consisting of sand or mud and abundant vegetation.

Kinosternids derive their common name from their ability to release a foul-smelling musk from glands located under the margin of the carapace. Kinosternids are primarily carnivorous but also scavenge on carrion. Although aquatic, this group is known to bask, occasionally well up into branches overhanging water. Furthermore, some species of mud turtles spend much of each year on land, either on terrestrial forays or aestivating in soil.

The family name Kinosternidae is from the Greek *kino* (move) and *sternon* (chest) and describes the ability of this group to move the plastron at hinged junctions. This family is represented by two genera in Tennessee: *Kinosternon* [ki-no-STER-non] and *Sternotherus* [ster-no-THER-us], the latter of which is from the Greek *sterno* (chest) and *thairos* (hinged).

Eastern Mud Turtle

Kinosternon subrubrum

Description: Eastern Mud Turtles are relatively small semiaquatic turtles with a domed, oval carapace and a large plastron. Both males and females range from 7.5 to 10 cm (3–4 in) CL but can reach 12.5 cm (5 in). The carapace is uniformly colored with various shades of dull yellow, brown, or black. The plastron is brown or black, but it often has paler yellow blotches, which are particularly prominent on the inner section of each scute and represent the remnant scute present at hatching. Similarly, the lower half of each marginal scute is dark brown with yellowish spots expanding outward from the remnant hatchling section. The plastron has two transverse hinges, which are aligned with the front and rear surface of each bridge of the shell. The hinges allow the turtles to tightly seal the plastron to the carapace and thereby protect the retracted head and limbs. Growth rings are noticeable on the carapace but are more prominent on the plastron. The head and neck are brown to black with scattered small yellowish

Eastern Mud Turtle adult, Coffee County. (Photo by Brad M. Glorioso)

spots and variously shaped yellowish reticulations. Each foot is webbed. Sexual dimorphism is subtle, but males have enlarged scales on the surface of their front limbs, a slightly concave plastron, and a stout tail. In comparison, females lack enlarged scales on their front limbs, their plastron is flat, and their tail is thin. Hatchlings, about 16–25 mm (0.5–1.0 in) CL, resemble adults but have a prominent middorsal keel and two lateral keels on the carapace.

The marginal scutes have pale yellow to orange spots; otherwise, the hatchling carapace is dark brown or black. The plastron has irregularly shaped orange-red patches, and the hinges are difficult to discern. Three subspecies are recognized and two occur in Tennessee: the Eastern Mud Turtle (*K. s. subrubrum*) found throughout Tennessee and the Mississippi Mud Turtle (*K. s. hippocrepis*) found along the Mississippi River. Mississippi Mud Turtles differ from Eastern

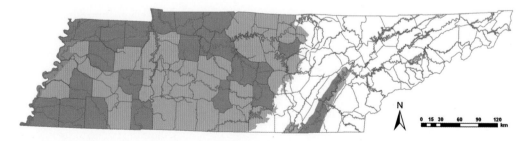

Two subspecies of Eastern Mud Turtles are recognized in Tennessee: the Eastern Mud Turtle (*Kinosternon subrubrum subrubrum*), whose range appears in orange, and the Mississippi Mud Turtle (*K. s. hippocrepis*), whose range appears in blue. Populations around the Mississippi River are likely intergrades, shown in purple, though those around Reelfoot Lake are likely Mississippi Mud Turtles.

Mud Turtles by possessing two pale stripes on each side of the head rather than pale spotting, mottling, or reticulations. These two subspecies intergrade in West Tennessee, but presumably "pure" populations of Mississippi Mud Turtles exist around Reelfoot Lake in Lake and Obion counties.

Etymology: The specific and subspecific epithet, *subrubrum* [sub-ROOB-rum], is from the Latin *sub* (below) and *rubrum* (red) and refers to the reddish coloration of the plastron in hatchlings. The subspecific epithet, *hippocrepis* [hip-oh-CREH-piss], is from the Greek *hippos* (horse) and *krepis* (shoe) and refers to the shape of the pale stripes on the head of this subspecies.

Similar Species: Eastern Mud Turtles most closely resemble other turtles with a domed carapace, including Eastern Musk Turtles (*Sternotherus odoratus*), Loggerhead Musk Turtles (*S. minor*), and young Eastern Box Turtles (*Terrapene carolina carolina*). The relatively large, double-hinged plastron readily distinguishes Eastern Mud Turtles from both species of Musk Turtles, which have a smaller plastron that lacks a double hinge. Furthermore, the uniformly colored carapace and double-hinged plastron of Eastern Mud Turtles distinguishes them from Eastern Box Turtles, which have a carapace adorned with a series of cream to yellow or orange spots, lines, or reticulations and a plastron lacking a true bridge connection to the carapace.

Distribution: Eastern Mud Turtles are found throughout much of the southeastern United States. Their distribution extends eastward from eastern Texas and Oklahoma through most of Arkansas, Louisiana, Mississippi, Alabama, Georgia, and South Carolina. Populations also occur in southeast Missouri, extreme northern Florida, Indiana, and lower

elevations in western Kentucky, eastern North Carolina, Virginia, Delaware, Maryland, New Jersey, and New York. The Tennessee distribution is based on records scattered across the western and central regions of the state. Few records exist east of the Cumberland Plateau in eastern Tennessee. Mississippi Mud Turtles occur primarily around Reelfoot Lake in Lake and Obion counties but intergrade with Eastern Mud Turtles in the Mississippi Alluvial Plain. Eastern Mud Turtles are found throughout the remainder of the state where the species occurs.

Habitat: Eastern Mud Turtles are primarily associated with shallow, slow-moving bodies of water with emergent aquatic vegetation. However, they also inhabit streams with gravel and slab bottoms. These turtles are often found walking submerged on the bottom of the wetland, or basking in the shallows with the top of their carapace exposed. They also are found wandering on land during the summer, particularly following rainstorms or when moving to a more permanent body of water after their previous wetland has dried up. Eastern Mud Turtles are rarely observed basking out of water.

Natural History: Eastern Mud Turtles are truly an amphibious species of turtle with a relatively short period of activity each year. The active season begins during late winter or early spring when the turtles emerge from their hibernation sites, which are chambers in the soil just beneath the frost zone that they constructed and entered the previous summer or fall. After emerging from hibernation, Eastern Mud Turtles migrate to a nearby wetland (e.g., marsh, pond, stream, river, or ditch). Turtles mate shortly after arriving in the wetland. Females become terrestrial a few weeks after mating and dig a nest on land to deposit their

Left: Mississippi Mud Turtle adult, Lake County. (Photo by Brad M. Glorioso) *Right:* Eastern Mud Turtle hatchling, Cheatham County. (Photo by Lisa Powers)

eggs. Typically, females lay 2–4 (range 1–9) small, hard and brittle, oval eggs. Females that breed early in the year in more southern areas are capable of producing a second and possibly a third clutch of eggs; however, clutch size decreases with subsequent clutches. Eggs laid during mid-spring typically hatch in late summer or early autumn, after an 8–14 week incubation period. Eggs laid during late spring or summer often do not hatch until the following spring. The sex of the hatchlings is determined by incubation temperature. Hatchlings grow rapidly and attain the coloration and appearance of adults during their second or third year. They become sexually mature during their third to seventh year. The differences in age at sexually maturity occur because males mature a year or so earlier than females, and individuals from more southern populations grow faster and mature earlier than those from more northern populations. Depending on the time of year the clutch is laid, females may become terrestrial and wander through fields or forests or they may burrow in the soil and aestivate. Males remain in and around the wetlands long

after females have departed. Some males remain in the wetland for much of the year, but others leave a few weeks after the females. Terrestrial individuals seek refuge from summer heat by burrowing into the soil. They often remain in their excavated summer cavities until a heavy rain, which might prompt their eruption to the surface. They then wander and forage on land but eventually burrow again as ground surfaces dry. Many individuals aestivate during much of the summer, and some remain in these chambers through winter, thus using aestivation chambers as hibernation sites. Growth rings on scutes of the plastron can be used to estimate the age of individuals. Based on these growth rings, Eastern Mud Turtles can live for >15 years. However, more than one growth ring can be added annually, and rings become obscured by wear in older individuals. Thus the age estimates might be off by a few years in some individuals. Eastern Mud Turtles feed in water but are also known to feed when on their terrestrial excursions. Although omnivores, their diet consists largely of a variety of small invertebrates. However, they are also known

Plastron of an Eastern Mud Turtle, Coffee County. (Photo by Brad M. Glorioso)

to eat small frogs and fish and to scavenge carcasses of larger vertebrates. Predators of hatchlings include frogs, snakes, birds, and mammals. Predators of adult turtles are restricted to larger carnivorous mammals.

Conservation Status: Although secretive and rarely seen, Eastern Mud Turtles are thought to be secure and are not listed as a species of concern in Tennessee. However, the status of this species is difficult to assess because of the unusual life history. Although found in some urban areas (downtown Murfreesboro), habitat loss associated with historic draining of wetlands, and habitat fragmentation associated with roadways negatively impact populations. Also, because of their habit of aestivating in fields, mud turtles of several species are injured and killed when croplands are cultivated or disked.

Brian T. Miller

Stripe-Necked Musk Turtle adult, Houston County. (Photo by A. Floyd Scott)

Loggerhead Musk Turtle

Sternotherus minor

Description: Loggerhead Musk Turtles are small, moderately domed, riverine turtles. Adults are 7.5–10 cm (3–4 in) CL but can reach at least 11.7 cm (4.6 in). The carapace of hatchlings has three keels: a central keel that extends down the midline of the carapace in the vertebral scute row and a pair of lateral keels that extend down the length of each of the costal scute rows. The keels become less pronounced with age, and the two lateral keels are nearly absent in the subspecies that occurs in Tennessee. The carapace is various shades of dark brown, gray, or light tan, with a series of darker streaks and spots. Darker pigment typically borders the seams of scutes. The vertebral scutes overlap. The plastron is small and, therefore, does not completely protect the limbs, and it has a single inconspicuous hinge. The plastron is various shades of pink, yellow, orange, or tan, and

usually lacks dark spots. A single gular scute is present. The large head has a short tubular or protruding snout and is marked with dark spots or vermiculations on a light brown, gray, or olive green background. Two fleshy barbels are present on the chin only. All four feet are webbed. Males have thicker and longer tails than females. Males also have two or three keratinized transverse ridges on the on upper surface of their front limbs and on the undersurface of their hind limbs. Furthermore, in males the plastral scutes are widely separated, whereas in females they are relatively closer together. Lastly, older males have extremely enlarged heads. Hatchings are circular when viewed from above, with a CL of 22–29 mm (0.8–1.1 in). Hatchlings and young juveniles have much bolder patterning on the skin and carapace, pink plastrons, and more pronounced keels on the carapace. Two subspecies are recognized, but only the Stripe-Necked Musk Turtle (*S. m. peltifer*) occurs in Tennessee. This subspecies has a series of broken yellowish stripes on the head and neck, an orange to brown plastron, and lacks the pronounced lateral keels of the carapace found in the Loggerhead Musk Turtle (*S. m. minor*).

Etymology: The specific epithet, *minor* [MY-nor], is Latin for "smaller" or "lesser," which describes the smaller size of this species relative to the congener *S. carinatus*. The subspecific epithet, *peltifer,* is from the Latin *pelta* (crescent-shaped shield) and *ifer* (bearing), which refers to the small bridge scutes between the plastron and carapace.

Similar Species: Loggerhead Musk Turtles most close resemble other kinosternid turtles, including Eastern Musk Turtles (*S. odoratus*) and Eastern Mud Turtles (*Kinosternon subrubrum subrubrum*). Eastern Musk Turtles are very similar in appearance to Loggerhead Musk Turtles, but they have two pale yellow lines on each side of the head and neck and have pairs of barbels on both the chin and throat, as opposed to just the chin as in Loggerhead Musk Turtles. Eastern Mud Turtles have a smoother carapace that lacks pronounced keels, and a larger, two-hinged plastron that completely covers the limbs and tails when closed against the carapace.

Distribution: Loggerhead Musk Turtles are distributed from central Florida westward along the Gulf Coast to southeastern Louisiana and north into portions of Tennessee, Kentucky, and Virginia. Loggerhead Musk Turtles are known from the Tennessee River and its major tributaries in the Ridge and Valley and Blue Ridge Mountains of eastern Tennessee, and from scattered larger tributaries of the north-flowing Tennessee River in Houston, Humphreys, Lewis, and Wayne counties. Other scattered, questionable records exist for Cumberland, Overton, and Fentress counties. This species has recently been discovered in the Little Sequatchie River.

Habitat: Loggerhead Musk Turtles are frequently found in rivers, streams, and creeks with faster flow and cobble substrate. However, this species also inhabits oxbows, ponds, and swamps with soft bottoms. Individuals are often found in shallow water in and around snags and fallen trees, but they have also been found underneath rocks and boulders. Hatchlings and juveniles are known to hide in crayfish burrows. Although highly aquatic, Loggerhead Musk Turtles can occasionally be seen basking high above the water on the limbs of overhanging trees. This species is encountered infrequently

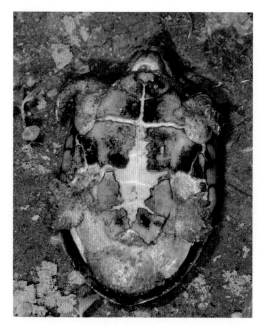

Left: Plastron of a Stripe-Necked Musk Turtle, Polk County. (Photo by Matthew L. Niemiller) *Right:* Stripe-Necked Musk Turtle juvenile, Houston County. (Photo by A. Floyd Scott)

on land. Loggerhead Musk Turtles overwinter in submerged rock crevices or within softer sediments in deeper pools.

Natural History: Adults are most active in the morning but also may be active just after dark. Mating occurs underwater during spring and autumn. Several males may attempt to mate with a lone female at one time. Females are known to lay two or more clutches, each with 1–5 oblong, hard eggs, in shallow nests in loose soil at the bases of trees or around fallen logs. The sex of developing embryos is determined by incubation temperature. The incubation period lasts 9–17 weeks, although some eggs and hatchlings might overwinter in the nest. Males reach sexual maturity at 5.5–7.0 cm (2–2.7 in) CL in 3–5 years, whereas males reach sexually maturity at 7.0–9.0 cm (2.7–3.5 in) CL in 6–8 years. Loggerhead Musk Turtles are known to live more than 20 years in the wild. The diet is broad, including snails, clams, aquatic insects, crayfish, fish, carrion, and some aquatic plant matter. The diet shifts from one dominated by insects to one dominated by mollusks as the individual ages. Known predators of adults include Alligator Snapping Turtles (*Macrochelys temminckii*) and American Alligators (*Alligator mississippiensis*), but larger carnivorous mammals such as raccoons are likely predators. Nest predators include raccoons, crows, reptiles, and other mammals. Individuals of this species are quick to bite when handled, and they will emit a pungent musk from their anal glands.

Conservation Status: Loggerhead Musk Turtles are common where they occur in eastern Tennessee. However, this species is susceptible to pollution from road runoff, coal mining, and chemicals within the habitats where they live, particularly pollution that results in declines in mollusk populations. Habitat destruction and modification, such as construction of reservoirs, likely has affected this species in Tennessee.

Matthew L. Niemiller and
R. Graham Reynolds

A Slough of Stinkpots!

As a young graduate student pondering a good and successful thesis project, I had a dream opportunity to study the turtle populations at renowned Reelfoot Lake in northwestern Tennessee. As part of a larger three-year project, I needed to design a study for my thesis that I could complete in a single field season. I thought about studying feeding behavior in River Cooters (*Pseudemys concinna*), and for a while this was my project—until a fateful day in late September 2004.

It was only my second trip to Reelfoot Lake to trap turtles. Liz Young, a field technician and friend, and Vince Cobb, my advisor, came along to assist in turtle trapping. We spent Friday evening setting hoop nets in Reelfoot Lake. On Saturday morning, the plan was to use the crawfish nets I have had since I was a young boy to try to capture turtles in a roadside slough near the lake. I hardly slept that night, as I was nervous about the upcoming day. I really wanted to impress my advisor by catching a slew of turtles in the roadside slough (pun intended).

Brad "Bones" Glorioso with a huge pile of his beloved "stinkies," Lake County. (Photo by Vince Cobb)

As a young boy growing up in the New Orleans area, my late uncle and I used those crawfish nets during many weekends of crawfishing in the ditches and bayous of southern Louisiana. The crawfish simply fed on the bait attached to the center of the open net and thus were not trapped. In order to capture the crawfish, the net had to be lifted from the water swiftly by grabbing the floating cork attached to the net with a long pole. Those boiled crawdads sure seemed to taste better when you personally put in the sweat and effort to catch them! But I digress. I remember many times my uncle wanted to change crawfishing spots because we started catching turtles in the nets, usually Eastern Musk Turtles (*Sternotherus odoratus*), also known as Stinkpots, and Eastern Mud Turtles (*Kinosternon subrubrum*). He once said, "Where there are a bunch of turtles, there are not many crawfish."

As I got older, my crawfish nets sat idle, but I always took them with me as I moved around, in the hopes of one day putting them to use again. I ended up at Middle Tennessee State University for graduate school, and I remember telling Matt Niemiller early in 2004 about my crawfishing days and the turtles my uncle and I would catch, and he asked me if I had ever purposely used them to catch turtles. I had never even thought of doing that, but I felt it might work in the right habitat. Matt and I tried out the nets during an amphibian and reptile survey of the Stones River National Battlefield (SRNB) months before my Reelfoot work, and they performed fairly well, catching mostly Red-Eared Sliders (*Trachemys scripta elegans*) and Stinkpots but also a handful of Snapping Turtles (*Chelydra serpentina*), Spiny Softshells (*Apalone spinifera*), and Northern Map Turtles (*Graptemys geographica*).

This prior work at the SRNB, and my earlier crawfishing days, led me to talk up this technique to my advisor in the days leading to our trip to Reelfoot Lake. This would be the first time Vince would see the nets and technique in action. While I was optimistic, I also knew we might not catch anything. We set out 16 baited nets at 9:00 a.m. that Saturday morning in a slough adjacent to the lake and waited. I planned on checking these nets three times an hour. Those first 20 minutes before checking the nets for the first time went by slowly, as I was so anxious to see if the nets were going to produce. I was so excited when I pulled up the first net and found a Stinkpot. Then another! And another! I was grinning from ear to ear with excitement and pride.

It soon became difficult to process all the turtles captured in the short breaks we had in between checking the nets. We began to use several five-gallon buckets to hold the turtles captured from each round of checking the nets. However, we quickly ran out of buckets. Eventually, Vince and I went and checked the nets while Liz stayed back to continue processing turtles. We needed a place to put the turtles until trapping was complete for the day. As we did not have enough containers, we had to improvise. We grabbed a large trashcan from the nearby picnic area, removed the trash bag, and began placing turtles in the can after processing. In our last couple hours at the slough, Vince, being the voice of reason, tried to make me realize that we were so far behind in processing the turtles that we needed to pick up the nets and finish processing those we had before we ran out of daylight. That was hard for me to do as the captures were still coming, and I continued capturing turtles for at least an hour.

We eventually stopped at 5:00 p.m., but we were at the slough until well after dark processing all the turtles we had captured. In all, we captured 125 turtles in eight hours, with 111 being Stinkpots! We also captured eight Southern Painted Turtles (*Chrysemys dorsalis*), five Red-Eared Sliders, and one Snapping Turtle. Thus a new thesis was born—population demography and feeding activity of Stinkpots in this slough. Ever since that day, with their feisty personalities and odiferous scents, "stinkies," as I affectionately call them, have been my favorite turtle species. What an amazing day of turtle catching!

Brad M. Glorioso

Eastern Musk Turtle adult, Hardeman County. (Photo by Matthew L. Niemiller)

Eastern Musk Turtle

Sternotherus odoratus

Description: Eastern Musk Turtles, also called Stinkpots, are one of the smallest turtles found in North America, with adults typically 8–12 cm (3–5 in) CL but occasionally reaching at least 15 cm (6 in). Sexual size dimorphism is generally not found in this species, especially in the middle of its geographic range, with males and females attaining similar maximum sizes. The color of the carapace ranges from olive green to black, and sometimes possesses irregular spots or dark streaks, especially juveniles. The posterior marginal scutes of the carapace are non-overlapping, giving a smooth edge appearance to the dorsal shell. Many individuals have algae on the carapace, especially around the edges but sometimes covering the entire shell. The relatively small plastron has a single gular scute and an indistinct, slightly movable hinge on the anterior lobe, and it varies in coloration from

Left: Eastern Musk Turtle basking, Lake County. (Photo by Brad M. Glorioso) *Right:* Eastern Musk Turtle hatchling, Lake County. (Photo by Brad M. Glorioso)

dull yellow to brown. Exposed skin can be seen between the plastral scutes, especially in adults, and more so in males than females. Glands located at the anterior and posterior edge of the bridge secrete a yellow or orange foul-smelling substance, giving the turtle its common and scientific name. The ground color of the head, legs, and tail is gray to black. The head typically contains two prominent yellowish lines that begin at the snout and extend posteriorly past the eye (above and below) and to the neck. These stripes are highly variable and, depending on the individual, can be distinct, faded, broken, or completely absent. Fleshy barbels are present on both the chin and the neck. The feet are webbed. Adult males possess a small patch of enlarged and hardened scales on the inner surface of each hind limb that aids in reproduction. Furthermore, the tail of adult males terminates in a blunt nail, and is thicker and longer than that of females, which lacks the terminal nail. Leeches are commonly observed on both the skin and the shell of Eastern Musk Turtles. Hatchlings are typically 20–25 mm (0.8–1 in) CL and possess both prominent head stripes and a prominent vertebral keel on the midline

of the carapace and smaller keels on each side, which become less pronounced or absent during maturation. There is little molecular variation among Eastern Musk Turtle populations and no subspecies are recognized despite its extensive geographic range.

Etymology: The specific epithet, *odoratus* [oh-dor-RAY-tus], is Latin meaning "to have an odor," in reference to the foul-smelling substance produced by musk glands.

Similar Species: Eastern Musk Turtles most close resemble other kinosternid turtles in Tennessee, including Loggerhead Musk Turtles (*S. minor*) and Eastern Mud Turtles (*Kinosternon subrubrum subrubrum*). Eastern Musk Turtles can be distinguished from Eastern Mud Turtles by examining the plastron. In Eastern Mud Turtles, the plastron is larger and covers the limbs more so than the plastron of the Eastern Musk Turtle. The plastron of the Eastern Mud Turtle has two transverse hinges as opposed to one in the Eastern Musk Turtle, and the pectoral scute is triangular in Eastern Mud Turtles, whereas it is squarish in Eastern Musk Turtles. Eastern Musk Turtles can be distinguished from Loggerhead Musk Turtles by examining the head

Plastrons of female (*left*) and male (*right*) Eastern Musk Turtles, Lake County. (Photo by Brad M. Glorioso)

and throat. In Loggerhead Musk Turtles, there are usually multiple alternating dark and pale stripes of similar width down the head and neck, whereas in Eastern Musk Turtles there are generally two defined pale lines beginning at the snout that run above and below the eye and down the head and neck. Also, Eastern Musk Turtles have barbels on both the chin and the throat, whereas Loggerhead Musk Turtles have barbels only on the chin. Eastern Musk Turtles are easily differentiated from juvenile Snapping Turtles (*Chelydra serpentina*) by examining the posterior marginal scutes and tail. Unlike Eastern Musk Turtles, Snapping Turtles possess serrated posterior marginal scutes and a long tail equipped with saw-toothed projections extending well beyond the carapace.

Distribution: Eastern Musk Turtles are found from southern Ontario, Canada, and the New England states south to Florida and west to central Texas and southern Wisconsin. In Tennessee, Eastern Musk Turtles occur nearly statewide, but they are absent at high elevations of the Blue Ridge Mountains in eastern Tennessee.

Habitat: Eastern Musk Turtles are highly aquatic, seldom venturing far from water, and can be found in nearly any permanent water, including reservoirs, lakes, rivers, oxbows, swamps, sloughs, canals, streams, and ponds. Although Eastern Musk Turtles can be found in relatively swift gravel-bottomed streams and rivers, they prefer and reach their highest abundances in sluggish waters with soft bottoms and abundant aquatic vegetation and shelter. Although occasionally found in deep waters, this species typically occupies shallow littoral zones, where individuals are often found basking, sometimes with part of their shell exposed. Eastern Musk Turtles do not aerially bask with the frequency of most emydid turtles, but they occasionally climb atop floating vegetation or logs to bask and sometimes ascend to relatively great heights on slanted trees or branches overhanging water; startled Eastern Musk Turtles basking on these slanted trees have been known to drop into canoes, surprising passing canoeists!

Natural History: In Tennessee, Eastern Musk Turtles are likely active from early spring to late autumn. Adults and juveniles will become inactive at very cold water temperatures and will overwinter in the soft substrate, under debris, or in mammal burrows until the water becomes warmer. They are most active at dusk and dawn, usually feeding during the few hours after sunrise and before sunset. Courtship and mating occur sporadically throughout the year, with peaks in both spring and autumn. Females lay eggs in excavated nests beneath stumps or logs, under leaf litter, in the nests of Snapping Turtles and American Alligators, in the walls of muskrat and beaver lodges, or even on open ground. Seldom are the eggs laid more than 50 m (165 ft) from the water's edge. Communal nests are occasionally found. The specifics of reproduction vary greatly by latitude in this wide-ranging species, and little is known about Tennessee populations. Males typically mature in 2–7 years, and females in 4–8 years; both genders usually mature at a CL of 65–85 mm (2.5–3.3 in). Depending on latitude, females lay 1–6 clutches per year, with two or three clutches being likely for Tennessee. Each nest consists of 1–10 eggs, with clutch size usually being positively correlated with female size. In West Tennessee, clutch sizes of 1–6 eggs are known to occur. Eggs are elliptical with a white brittle shell, and the sex of developing embryos is determined by incubation temperature. Eggs incubated at temperatures of 28°C (82.4°F) or higher result in females, whereas eggs incubated at temperatures around 25°C (77°F) produce mostly males. Intermediate temperatures produce similar numbers of both sexes. Eggs hatch in 56–132 days, with an average of 80 days, in late summer or early autumn. The longevity of wild Eastern Musk Turtles has been estimated to be at least 28 years in Pennsylvania, and 15–19 years in South Carolina. A captive individual lived nearly 55 years at the Philadelphia Zoo. Eastern Musk Turtle populations can be quite large, despite being an inconspicuous turtle. Density and biomass estimates have varied greatly among the habitats and populations studied, with the largest estimates being from a roadside slough near Reelfoot Lake in West Tennessee. Most studies of movement in Eastern Musk Turtles report very small home ranges, usually less than two hectares (five acres), with males having larger home ranges than females. They are omnivorous, and will consume nearly any living or dead aquatic animal or plant. Eastern Musk Turtles have also been noted to feed upon invertebrates, such as slugs, on land near the water. Known predators of adults and juveniles include river otters, hawks, eagles, kites, grackles, bass, bullfrogs, snakes, alligators, and other larger turtles. Raccoons, skunks, foxes, muskrats, herons, crows, snakes, and other turtles are known predators of eggs. When captured or threatened, many Eastern Musk Turtles will secrete musk and attempt to bite repeatedly by extending their long necks over and under their shell, searching for something to grab.

Conservation Status: Records of Eastern Musk Turtles in the state are relatively sparse, especially for Middle and East Tennessee. However, this is an artifact of localized collecting, usually around college towns. Also, the usual sight records of basking turtles and salvage of road-killed specimens are infrequent with Eastern Musk Turtles, as they aerially bask infrequently and seldom move far from the water onto roads. Eastern Musk Turtles are more abundant and widely distributed in the state than current records indicate, and their populations appear stable throughout Tennessee.

Brad M. Glorioso

20
Family Trionychidae (Softshell Turtles)

Softshell Turtles have soft, leathery shells that are devoid of scutes or scales. Their body is dorso-ventrally compressed (flattened from top to bottom) and circular or ovoid in outline when viewed from above. Because of their body shape, they are reminiscent of pancakes, a term often associated with colloquial names of members of this family. These turtles are highly aquatic and powerful swimmers, with paddle-like limbs and webbed feet. Softshell Turtles are represented by 13 genera and about 31 species found in North America, Africa, and Southeast Asia, although they have also been introduced to Hawaii. Fossil trionychids date to the early Cretaceous period of China and Mongolia and also suggest a broader distribution in the past, which included South America and Europe. Two species in the genus *Apalone* [ah-pah-LOW-nee], from the Greek *apalos* (soft), occur in Tennessee. Softshell Turtles can be found in a variety of habitats, although the two species found in Tennessee are more often associated with riverine habitats. Although they occasionally bask, Softshell Turtles also frequently lie buried in the sand or mud in shallow water, where they wait in ambush of passing prey. They use their long neck and elongate, tubular snout to reach the surface to breathe.

Some species, such as the Florida Softshell (*Apalone ferox*), are quite large, reaching 63 cm (25 in) CL. Females grow to larger sizes than males, but males have longer and thicker tails than females. Because of their long necks, powerful beaklike mandibles, sharp claws, and agility, Softshell Turtles should be handled with caution. Softshell Turtles are eaten in most areas of their range, especially Southeast Asia. Accordingly, some species are farmed for food, while others are harvested from the wild and shipped overseas. Although historically regulated as a game species in Tennessee, Softshell Turtles are protected and can no longer be legally harvested for food.

Smooth Softshell

Apalone mutica

Description: Smooth Softshells are medium-sized, dorso-ventrally flattened turtles with a soft, leathery shell that lacks scales or scutes. Females are 16–25 cm (6–10) CL but reach over 35 cm (14 in), whereas males are smaller at 11–18 cm (4.3–7.0 in) CL. The carapace is tan, olive brown or brown, and it often is marked with small, darker spots or streaks in juveniles and adult males, or larger, irregular dark blotches in mature females. A lighter-colored border sometimes is present along the edge of the carapace. The carapace is smooth and lacks spiny projections along the anterior edge. The plastron is white or cream colored, and the underlying bones sometimes are evident through the skin. Narrow postorbital stripes, pale white to yellow and bordered by dark pigment, are present on the head. Similar but poorly defined stripes often extend from the tip of the snout to the eyes. The snout is tubular in appearance. The nostrils are round and lack transverse ridges in their interior. The feet are webbed and lack conspicuous

Smooth Softshell adult, Crawford County, Arkansas. (Photo by Greg Sievert)

streaks or spots. Adult males are smaller than adult females but have larger, thicker tails so that the vent opens beyond the rear edge of carapace. Hatchlings vary from 31 to 50 mm (1.2–2 in) CL and resemble miniature adults. Two subspecies are recognized, but only the Midland Smooth Softshell (*A. m. mutica*) occurs in Tennessee.

Etymology: The specific and subspecific epithet, *mutica* [MEW-ti-ka], is from the Latin

muticus for "curtailed" or "unarmed," likely in reference to the lack of hard protective scutes on the shell.

Similar Species: Smooth Softshells can only be confused with the related Spiny Softshell (*A. spinifera*) in Tennessee. These two species can be distinguished by examining the anterior edge of carapace just behind the base of the neck and the nostrils. Smooth Softshells lack spines or tubercles along the anterior edge

Left: Head of a Smooth Softshell adult, Houston County. (Photo by A. Floyd Scott) *Right:* Smooth Softshell subadult, Crawford County, Arkansas. (Photo by Greg Sievert)

of the carapace and feel smooth to the touch. The snout has a rounded tip, and the interior of the nostrils lack transverse ridges. Spiny Softshells have numerous spiny tubercles and projections along the anterior edge of the carapace, and the carapace often feels like sandpaper. The interior of the nostrils of Spiny Softshells have transverse ridges. Additionally, the feet of Spiny Softshells are strongly patterned with streaking and spotting, whereas the feet of Smooth Softshells lack dark patterns of streaks or spots.

Distribution: Smooth Softshells are found in the central United States throughout much of the Ohio River, Missouri River, Mississippi River, and some Gulf Coast drainages. This species is most common in medium to large rivers. In Tennessee, distributional records are sparse. Most records in the state are from the Reelfoot Lake area in Lake and Obion counties. Smooth Softshells also have been documented from the lower Tennessee River drainage in Benton, Henry, Houston, Humphreys, and Stewart counties and from the Cumberland River drainage in Davidson and Pickett counties.

Habitat: Smooth Softshells are primarily found in medium- to large-sized rivers and streams but also can be found on occasion in larger lakes and reservoirs with sandy or muddy bottoms. Riverine habitats usually have appreciable flow and gravelly, sandy, or muddy substrates. Adults bask on sand bars, mud bars, or, less frequently, large logs and rocks; however, they always bask at the water's edge and will quickly dash into the water at the slightest disturbance. Sandbars also are important sites for nesting.

Natural History: Smooth Softshells are active from April through late September or early October. They overwinter buried in the soft bottoms of deeper pools of rivers or lakes. Their metabolism is slowed during the winter and they extract oxygen from water through the walls of their cloaca, which decreases their need to breathe air. During warm seasons, these turtles are diurnal. They sleep at night hidden buried in the soft bottoms of shallow water or beneath submerged trees. During the day, Smooth Softshells often rest buried in the sand in shallow water, extending their necks and using their long piglike noses as snorkels to breathe. Mating occurs during spring and, occasionally, extends into late summer. Females lay 3–26 spherical eggs (usually 10–16) in excavated nests on sunlit sandbars above normal flood levels. The number of eggs laid is positively correlated with female body size. Eggs hatch in 9–12 weeks during August or September. The sex of developing embryos is determined at the time of fertilization rather than by incubation temperature, as in most other turtle species in Tennessee. Smooth Softshells primarily eat invertebrates, including aquatic insects, worms, snails, clams, isopods, crayfish, and spiders. However, small fish, tadpoles, frogs, salamanders, young birds, and small mammals have been reported in the diet. Algae, seeds, and berries also are consumed. Females usually patrol deeper water while foraging, whereas males typically forage in shallow water. Predators of adults are few but include American Alligators (*Alligator mississippiensis*) and Alligator Gar, whereas smaller juveniles and hatchlings are prey for larger fish, Snapping Turtles (*Chelydra serpentina*), wading birds, predatory mammals, and some aquatic snakes, such as Cottonmouths (*Agkistrodon piscivorus*). Nest predators include raccoons, skunks, foxes, and several species of birds. Flooding is a significant cause of egg mortality. Smooth Softshells rely on their agility and speed in water to escape predators. Although less aggressive than Spiny Softshells, larger individuals can inflict a painful bite and should be handled with caution.

Conservation Status: Although not tracked by state agencies, Smooth Softshells are not nearly as common as Spiny Softshells, and they are thought to be more vulnerable than Spiny Softshells to activities that adversely affect their aquatic habitats, such as pollution, agricultural runoff, siltation, and impoundments. This species also might be a victim of by-catch from turtle and fish trapping for other legally harvested species.

Comments: Smooth Softshells are sometimes called "spineless softshells."

Lisa Powers and
Matthew L. Niemiller

Eastern Spiny Softshell adult female, Lake County. (Photo by Brad M. Glorioso)

Spiny Softshell

Apalone spinifera

Description: Spiny Softshells are medium-sized, dorso-ventrally flattened turtles with a soft, leathery shell that lacks scales and scutes. Females are 17–43 cm (6.7–17 in) CL but reach 52 cm (20 in), whereas males are small at 13–22 cm (5–8.7 in) CL. The carapace is grayish green, olive green, tan or brown with brown to black eyelike spots, called ocelli, which are variable in size. These spots are prominent in juveniles and males but become obscured in mature females, which have an irregular mottled pattern. Although appearing smooth, the surface of the carapace is replete with tiny projections—the

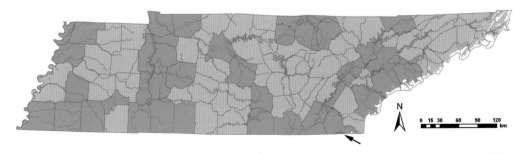

Two subspecies of Spiny Softshells are recognized in Tennessee: the Eastern Spiny Softshell (*Apalone spinifera spinifera*) whose range appears in orange, and the Gulf Coast Spiny Softshell (*A. s. aspera*) whose range intergrades with the Eastern Spring Softshell and appears in purple in extreme southeast Tennessee.

Left: Adult female Gulf Coast Spiny Softshell, Bradley County. (Photo by David Hedrick) *Right:* Eastern Spiny Softshell adult male, Roane County. (Photo by Matthew L. Niemiller)

sensation of rubbing your hand over the carapace, particularly of males, is comparable to rubbing your hand over sandpaper. Furthermore, a series of spiny projections or tubercles are present along the anterior edge (hence "spiny" softshell). The plastron is white to cream in color, and the underlying bones are sometimes evident through the skin. In adults of both sexes, prominent white to pale yellow stripes bordered with darker pigment extend from each side of the head onto the neck. Furthermore, a postorbital stripe begins just behind the eye, and another stripe runs in parallel with the postorbital stripe and begins under the tubular, piglike snout. The snout has a transverse ridge in each nostril. The feet are webbed and strongly streaked or spotted. Adult males are smaller than adult females but have larger, thicker tails such that the vent opens beyond the rear edge of carapace. Hatchlings are 30–45 mm (1.2–1.8 in) CL and, except for prominent ocelli on the carapace, resemble adults. Five subspecies are recognized, but only the Eastern Spiny Softshell (*A. s. spinifera*) and the Gulf Coast Spiny Softshell (*A. s. aspera*) occur in Tennessee, with the latter subspecies found only in the Conasauga River drainage in south-

ern Polk and Bradley counties. The Gulf Coast Spiny Softshell differs from the Eastern Spiny Softshell in having two or more dark lines that may be broken running parallel to the rear edge of the carapace. Additionally, the two stripes on the head usually unite on the side of the head in Gulf Coast Spiny Softshells but remain separated in Eastern Spiny Softshells.

Etymology: The specific and subspecific epithet, *spinifera* [spy-NIFF-er-ah], is from the Latin *spina* (thorn or spine) and *ifer* (bearing), referring to the tubercles present on the anterior carapace. The subspecific epithet, *aspera* [ass-PEAR-ah], is from the Latin *asper* (rough).

Similar Species: In Tennessee, Spiny Softshells can only be confused with the related Smooth Softshell (*A. mutica*). These two species can be distinguished by examining the anterior edge of carapace just behind the base of the neck, also the nostrils, and the feet. Smooth Softshells lack spines or tubercles along the anterior edge of the carapace and feel smooth to the touch. The snout has a rounded tip, and the nostrils lack transverse ridges. Spiny Softshells have numerous spiny tubercles and projections along the anterior edge of the carapace and the

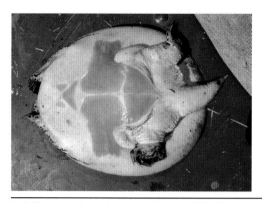 

Left: Plastron of a male Eastern Spiny Softshell, Lake County. (Photo by Brad M. Glorioso) *Right:* Eastern Spiny Softshell juvenile, Rutherford County. (Photo by Brad M. Glorioso)

carapace often feels like sandpaper. The nostrils of Spiny Softshells have transverse ridges. Additionally, the feet of Spiny Softshells are strongly patterned with streaking and spotting, whereas those of Smooth Softshells are not.

Distribution: Spiny Softshells are found throughout much of eastern North America from the Great Lakes west into Nebraska, southeastern Wyoming, and eastern Colorado southward to the Gulf Coast of the United States and Mexico east to the South Carolina and North Carolina. Several disjunct populations occur in the western United States as well as New York. Spiny Softshells likely occur statewide, except for the higher elevations of the Blue Ridge and Cumberland Mountains. However, confirmed records are lacking from parts of northeastern Tennessee and the Cumberland Plateau. Gulf Coast Spiny Softshells were recently confirmed from the Conasauga River of Bradley County in southeastern Tennessee.

Habitat: Spiny Softshells inhabit a variety of permanent aquatic habitats, including medium- to large-sized rivers and streams with a preference or clean, clear water with sandy bottoms. However, they also can found in ditches, bayous, ponds, reservoirs, and lakes with sandy or muddy bottoms. This species is almost never found in ephemeral wetlands, although they will occasionally move over land. Adults will bask on logs, fallen trees, and rocks as well as on sand or mud bars at the water's edge. Sandbars along the banks of rivers are an important nesting habitat.

Natural History: Spiny Softshells are active from April through early October. These turtles are diurnal and sleep at night underwater buried in sand or mud in the shallows or under submerged trees. Spiny Softshells often rest buried in the sand in shallow water, extending their necks and using their long piglike noses as snorkels to breathe. Mating occurs in deep water in April or early May followed by nesting in May and June. Females lay 3–39 (usually 12–18) spherical eggs in excavated nests on sandbars and sandy riverbanks above normal flood levels. The number of eggs is positively correlated with female body size. Eggs hatch in 9–12 weeks in late August into September. The sex of developing embryos is not determined by incubation temperature, unlike most other turtles in Tennessee. Spiny Softshells are known to live more than 20 years in captivity. Spiny Softshells feed primarily on invertebrates, such as aquatic insects, snails, clams, and crayfish, but will also eat small fish and some plant material, such as aquatic vegetation and berries. Predators likely include American Alligators (*Alligator mississippiensis*) and predatory mammals, such as river otters, raccoons, and foxes. Juveniles and hatchlings are prey for wading birds, raccoons, and large fish. Nest predators include raccoons, skunks, foxes, and several birds. Flooding also is a significant cause of egg mortality. Fishermen often catch this species on hook and line. Spiny Softshells are fast, powerful swimmers and rely on their agility and speed to escape from predators. Spiny Softshells have a long neck and powerful bite, and, therefore, should be handled with caution.

Conservation Status: Spiny Softshells are likely distributed statewide and can be locally abundant. However, this species may be more vulnerable to pollution of their aquatic habitats than hard-shelled species because of the permeability of their leathery shell. Spiny Softshells also are sometimes caught and killed by fishermen. This species also might be a victim of bycatch from turtle and fish trapping for other legally harvested species.

Lisa Powers and
Matthew L. Niemiller

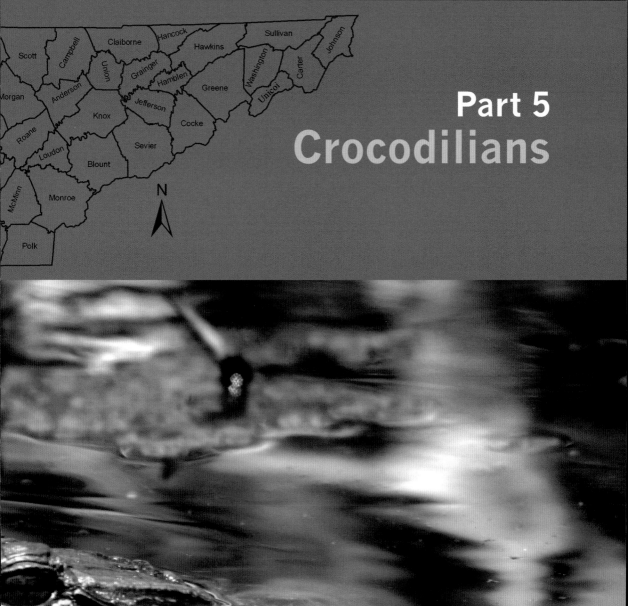

Part 5
Crocodilians

Juvenile American Alligator, Charlton County, Georgia.
(Photo by R. Graham Reynolds)

21
Family Alligatoridae (Alligators)

Alligators are members of the order Crocodylia (crocodilians), a group of large primarily aquatic reptiles that range across southeastern North America, Central America, the Caribbean, South America, Africa, southern Asia, Australia, and the South Pacific. There are 25 recognized species of crocodilians belonging to three families, the Crocodylidae (Crocodiles), Alligatoridae (Alligators and Caimans), and Gavialidae (Gharials). Crocodilians are known from as far back as the Jurassic, 200 million years ago, although the ancestors of the living crocodilians, the Eusuchians, evolved in the Cretaceous, roughly 120 million years ago. Ancestral crocodilians exhibited a variety of body forms, including terrestrial, aquatic, and marine forms, the latter of which possessed flippers instead of clawed feet. Crocodilians are the closest living relatives of the birds. Indeed, it is this group that unites the birds with the rest of the reptiles into a group known as the Sauropsids, which also includes the very distantly related turtles. All modern crocodilians are primarily aquatic and most inhabit fresh or brackish waters, but some species routinely take to long-distance travel across open seas. Crocodilians worldwide are facing a variety of threats, including persecution, habitat loss, and human consumption. Six of the 25 species (24%) are critically endangered, one is endangered, and three are vulnerable; indeed, some species may be lost from the wild in the very near future.

Alligators (*Alligator mississipiensis*) are a large crocodilian, reaching lengths of close to 4.3 m (14.1 ft) from snout to tail tip, but they are not the largest members the family Alligatoridae. This distinction belongs to the Black Caiman of the Amazon, which can reach lengths that exceed 5 m (16 ft). The family and genus names, Alligatoridae and *Alligator,* are English pronunciations of the Spanish *el lagarto* (the lizard), probably a description used by early Spanish explorers in the Western Hemisphere.

American Alligator

Alligator mississippiensis

Description: American Alligators are large aquatic reptiles that reach lengths of 4.3 m (14.1 ft) TL, although none larger than about 2 m (6.6 ft) have been seen in Tennessee and most individuals are considerably smaller. American Alligators have thick black to gray dorsal scales with knobby bumps running in several rows down the dorsum and tail. Known as osteoderms, these are bony plates embedded in the skin that give protection to the dorsal surface of the animal. The venter is pale colored, and the snout is rounded with some teeth visible when the mouth is closed. Hatchlings are around 23 cm (9 in) TL, with large eyes and a short snout. Juveniles and hatchlings are considerably more colorful than adults and possess pale to bright

American Alligator adult, St. Tammany Parish, Louisiana. (Photo by Brad M. Glorioso)

yellow or white crossbands along the body and tail. These bands usually fade after a year or so, when the juveniles take on the adult coloration. Occasional albino and leucistic individuals have been found, although never in Tennessee.

Etymology: The specific epithet, *mississip-piensis* [miss-iss-sip-pee-EN-sis], means "from the Mississippi," which is originally from the Chippewa words *mici zibi* (great river) and the Algonquin *messipi,* and it refers to the presence of this species in the Mississippi River.

Similar Species: American Alligators are frequently confused with both American Crocodiles (*Crocodylus acutus*) and Spectacled Caimans (*Caiman crocodylus*), neither of which are native to Tennessee, although caimans are popular in the pet trade and occasionally turn up in Tennessee as released pets (see Chapter 22, **"Erroneous Species and Species of Possible Occurrence"**). American Crocodiles are native to south Florida, where they co-occur with both Alligators and the introduced Specta-

Left: Head of an American Alligator adult, St. Martin Parish, Louisiana. (Photo by Brad M. Glorioso) *Right:* American Alligator juvenile, Leon County, Florida. (Photo by Ken Wray)

cled Caiman. Crocodiles are distinguished from American Alligators by having a much more narrow snout, with teeth that project from the jaw when the mouth is closed. Caimans are smaller than American Alligators and American Crocodiles and have a bony ridge between their eyes.

Distribution: American Alligators are distributed throughout the southeastern Coastal Plain from eastern Texas to southeastern North Carolina and can be locally common. They are frequently encountered in central Mississippi near the Mississippi River as far north as Tunica County, and they are common in eastern Arkansas, where they have been reintroduced to some areas following decades of persecution, which shrunk their range. Individuals found farther north, including those in Tennessee, Missouri, and Illinois, were considered to be transients from more southern populations, although it is likely that they once occupied this area prior to European colonization. However, American Alligators are thought to be expanding their range along the Mississippi River and are now frequently sighted in Tennessee in Shelby County, particularly around McKellar Lake and T. O. Fuller State Park, as well as the Ghost River

section of the Wolf River in Fayette County. Juveniles have been found in Shelby County and presumably represent a small breeding population in this area. Although they are currently rare and barely established in Tennessee, we include this species as a part of the state's herpetofauna and expect that American Alligators will become more common in the future. Established or potentially established populations occur in Shelby, Fayette, Tipton, and Haywood counties, although Lauderdale, Dyer, Hardeman, and Lake counties might one day harbor populations as well. Alligators released from captivity often turn up in unexpected places, and escaped (or released) individuals have been documented from the following counties: Rhea (Watt's Bar Lake), Dickson (on a road), Bedford (Bedford Lake), and Putnam (in a drain pipe).

Habitat: American Alligators inhabit slow-moving or stagnant waterways—from rivers to lakes, ponds, swamps, sloughs, and borrow pits. They prefer areas with plenty of aquatic vegetation and places to bask that include either logs or banks exposed to sun. Females require secluded banks to nest, where they excavate and construct a large nest of dirt, sticks, and leaves. Alligators, like most reptiles, are ectothermic and absorb heat from their environment. This limits the potential range of the species to the warmer southeastern counties of Tennessee, at least for the near future.

Natural History: American Alligators feed on a wide variety of animals from birds, fish, small mammals, and reptiles to carrion. They are ambush predators, waiting beneath the water, usually with just the tops of their heads exposed, to snatch fish or birds with a sideways snap of the jaws. They also occasionally remain completely submerged, ambushing prey on the water surface or on a bank. Larger individuals are capable of overpowering deer or feral pigs, and these larger meals are often cached for consumption at a later time. Juveniles feed on small invertebrates, reptiles, amphibians, fish, and scraps from their mother's feeding. American Alligators are mostly nocturnal, although they are frequently seen while basking during the day. They are capable of surviving near freezing conditions; as long as the surrounding water does not freeze solid they can remain dormant by maintaining a breathing hole in surface ice. In more extreme conditions, they will use burrows in mud banks to survive the cold, although extended freezing temperatures are often lethal. During periods of drought, Alligators can serve as crucial "environmental engineers" by digging wallows and holes in drying waterways, often maintaining the only water sources available to other aquatic organisms during extreme drought. They are, thus, an important component of aquatic ecosystems. American Alligators breed in the spring, when males gather together to bellow and wrestle over females. Males are territorial during this time and will occasionally fight to the death with other males. Females lay between 30 and 50 eggs in large

(2 m [6.6 ft] in diameter) nests that they construct on banks near water, usually in late spring. Female American Alligators exhibit sophisticated parental care and guard the nest at all times. The eggs begin to hatch about two months later, although the young require the assistance of the mother to emerge from the nest. The young begin to grunt while still in the egg, alerting the mother to excavate the nest and assist the hatchlings with breaking out of their shells. The hatchlings have egg teeth, although occasionally the mother will assist in puncturing the eggs to release the hatchlings. After hatching, the babies are gathered up by the mother and led to shallow, heavily vegetated water. If the water is farther away, the mother will pick the hatchlings up in her mouth and carry them to the water. American Alligators, like some turtles, exhibit temperature-dependent sex determination, such that the temperature of the eggs during incubation determines the sex of the young. Temperatures above 34°C (93.2°F) will produce males, whereas temperatures below 30°C (86°F) will produce females. Temperatures in between will result in mixed sexes, and so females carefully construct nests to produce both sexes. Hatchlings are easy prey for many predators, such as other American Alligators, wading birds, osprey, mammals, and reptiles, and the female guards them fiercely for the first year of life. After about 4 years, juveniles are large enough to venture away from their siblings and are immune to most of the smaller predators. Large American Alligators are virtually immune to predation, although males might be killed in territorial disputes, and adults are subject to slaughter by humans for food, hides, sport, or persecution. In spite of the occasional sensational story, American

Alligators pose relatively little threat to people or their pets when they are left undisturbed in their natural habitat.

Conservation Status: American Alligators are a relatively recent (re)addition to Tennessee's herpetofauna and currently exert a tenuous hold in our state. These populations could either increase as a result of continued range expansion up the Mississippi River drainage from northern Mississippi or they could be wiped out by a series of cold-weather years or poaching and persecution. Currently, Alligators are listed as protected in Tennessee. If you see an American Alligator in Tennessee, officials request that you contact the Region I office of TWRA in Jackson: (800) 372-3928 (subject to change).

Comments: American Alligators are very vocal and communicate using sounds that can be either endearing or alarming to the human observer. Hatchlings and juveniles call continuously to their mother in a series of short grunts, even before they emerge from their eggs, a behavior that helps the mother keep track of them and assist them in hatching. If they are alarmed, juveniles will also produce an alarm call, a series of more frantic, higher-pitched grunts or yelps that will bring a mother rushing toward them. Adults produce a variety of sounds, including a breathy warning hiss or a short "cough" when approached by humans. Males bellow during mating season. The bellow is a very low ethereal guttural roar that carries long distances through water. Each bellow lasts one to several seconds and drops in pitch.

R. Graham Reynolds and
Matthew L. Niemiller

Part 6
Other Reptiles

22

Erroneous Species and Species of Possible Occurrence

Several species of reptiles with distributions outside of Tennessee have been erroneously reported as a component of the herpetofauna of Tennessee because of either misidentification or error in cataloging of specimens. The occurrences in Tennessee of species discussed below are considered either *erroneous,* in that the species have previously been reported in the state but the reports lack sufficient verification or originate from questionable records, or *of possible occurrence,* in that they might yet be found in our state. Some of these species have distributions in neighboring states and might have historically occurred in Tennessee, but as of this writing none have established breeding populations in the state. Subsequent surveys might discover breeding populations, and we along with others are currently engaged in investigating unverified reports of several of these species.

Some species of reptiles reported in Tennessee are obviously introduced from elsewhere, and most individuals of these species do not survive more than a few years in our climate. However, some introduced species could become established within our state. Finally, the popularity of many exotic reptiles in the pet trade has led to numerous escapees or intentional releases, and we have included reports of species known to have been released from the pet trade in Tennessee. These are usually represented by one or a few specimens, known as waifs, and are unlikely to found a new population. Exotic reptiles are extremely popular in the pet trade, accounting for hundreds of millions of dollars of business annually in the United States alone. Often the animals being kept and traded are unsuitable for captivity and either escape or are released intentionally when they have outgrown their owner. Verifiable sightings or photographs of the species listed below should be reported to wildlife authorities or professional and state herpetologists, so that we might further our understanding of the state's faunal composition.

Erroneous records of snakes include Smooth Greensnakes (*Opheodrys vernalis*) and Gophersnakes (*Pituophis catenifer*), while erroneous turtles include Ornate Box Turtles (*Terrapene ornata*) and Razor-Backed Musk Turtles (*Sternotherus carinatus*). Lined Snakes (*Tropidoclonion lineatum*), Eastern Diamond-Backed Rattlesnakes (*Crotalus adamanteus*), Gopher Tortoises (*Gopherus polyphemus*), Brown Anoles (*Anolis sagrei*), Texas Horned Lizards (*Phrynosoma cornutum*), Mediterranean House Geckos (*Hemidactylus turcicus*) and Indo-Pacific Geckos (*Hemidactylus garnotii*) are all known from one or more specimens collected in Tennessee, although none appear to be established. Brahminy Blind Snakes (*Ramphotyphlops braminus*) could possibly turn up as waifs in our state.

Other potentially occurring species include Graham's Crayfish Snake (*Regina grahamii*) and the Western Chicken Turtle (*Deirochelys reticularia miaria*), both of which occur across the Mississippi River from Tennessee. Finally, Spectacled Caimans (*Caiman crocodilus*) and Green Iguanas (*Iguana iguana*) are common exotic species in the pet trade that occasionally turn up in our state as escaped (or released) pets.

Family Colubridae: Smooth Greensnake

Opheodrys vernalis

Smooth Greensnakes, small relatives of the Rough Greensnake, inhabit the north-central and northeast United States, but isolated populations are scattered throughout the Rocky Mountain states, southeastern Texas, and Chihuahua, Mexico. Although Endsley (1954) lists Smooth Greensnakes as inhabitants of Tennessee, his records are likely erroneous. Smooth Greensnakes do occur in north-central Virginia and south-central Illinois, but no verified specimens have been taken from Tennessee; consequently, this species likely does not occur here.

Smooth Greensnake, Iron County, Wisconsin. (Photo by Pierson Hill)

Family Colubridae: Gophersnake

Pituophis catenifer

Gophersnakes are large-bodied terrestrial snakes related to the native Northern Pinesnake. They range across much of the central and western United States and are associated with a variety of habitats that have few trees and access to burrows or other subterranean cover. Gentry (1956) reported *P. catenifer* from the "Tennessee River area," although no specific locality information was given. The closest population of Gophersnakes to Tennessee occurs 200 km (125 miles) away in northwestern Reynolds County, Missouri. No specimens exist from Tennessee, and no individuals have been found since Gentry's report; hence we consider the earlier report of Gophersnakes to be based on a misidentification and the inclusion of this species in the herpetofauna of Tennessee to be erroneous.

Gophersnake, Kankakee County, Illinois. (Photo by Todd Pierson)

Family Colubridae: Graham's Crayfish Snake

Regina grahamii

Graham's Crayfish Snakes are medium-sized, semiaquatic snakes that inhabit small bodies of water, such as streams, swamps, and ponds. Individuals of this species are secretive and infrequently encountered, likely because they are shy and enter the water from basking sites before they can be observed. Graham's Crayfish Snakes range from the Gulf Coast north along the Mississippi River through the Missouri Bootheel Region to Nebraska and Iowa, although they appear to occur only west of the Mississippi River. No specimens are reported from Tennessee, although populations exist in Pemiscot County, Missouri, and Mississippi County, Arkansas, so we expect that this species eventually may be discovered in Tennessee.

Graham's Crayfish Snake adult, Harris County, Texas. (Photo by Ken Wray)

Family Colubridae: Lined Snake

Tropidoclonion lineatum

Lined Snakes are not considered to be native to Tennessee, even though a specimen with a locality record of Franklin County, Tennessee, exists in the United States National Museum (USNM 56030). As pointed out by Scott and Redmond (2011) and Johnson (1987), this locality record is likely in error as the nearest population of *T. lineatum* to Franklin County is 530 km (330 miles) away in Jefferson County, Missouri. Ramsey (1953) indicated that this specimen was a part of an estate gift of 5,000 specimens donated to the USNM by the estate of Julius Hurter, but the gift lacked Hurter's original catalogues; hence details such as precise collection localities could not be verified. However, two Lined Snakes were discovered recently in Shelby County (F. Scott, pers. comm.). Presumably, these two individuals represent intentionally or accidentally released captives. Although there is no indication of a reproductive population in Shelby County, continued searches might result in the discovery of more individuals.

Lined Snake adult, Jefferson County, Missouri. (Photo by Matthew L. Niemiller)

Family Emydidae: Ornate Box Turtle

Terrapene ornata

The Ornate Box Turtle ranges primarily across the Midwest, but populations do occur eastward into extreme northwestern Indiana. Ornate Box Turtles are small terrestrial turtles, very similar in appearance to the Eastern Box Turtle. The single record from Tennessee east of Memphis (Parker 1948) almost certainly represents a released individual brought in from elsewhere. The closest populations to Tennessee occur in Craighead and Prairie counties, Arkansas.

Ornate Box Turtle adult, Carroll County, Illinois.
(Photo by Ken Wray)

the western side of the Mississippi River, as no individuals are documented from Tennessee or Illinois to our knowledge. Because records exist for New Madrid County, Missouri, and Mississippi County, Arkansas, just across the river from Tennessee, Western Chicken Turtles could turn up in our state. Western Chicken Turtles can be distinguished from other basking turtles by the presence of a single, distinct yellow stripe on the forelegs, vertical yellow stripes on the tail, and a head and neck that nearly equal the length of the carapace.

Chicken Turtle adult, Liberty County, Florida.
(Photo by Brad M. Glorioso)

Family Emydidae: Chicken Turtle

Deirochelys reticularia

The Western Chicken Turtle (*D. r. miaria*) is a subspecies of the Chicken Turtle, a medium-sized emydid turtle, which ranges from the Gulf Coast north along the Mississippi River drainage to the Missouri Bootheel Region. Interestingly, this species appears confined to

Family Gekkonidae: Indo-Pacific Gecko

Hemidactylus garnotii

Although native to Southeast Asia, the Indo-Pacific Gecko has been introduced into tropical regions throughout the world, including the southeast United States. This species is well established in Florida. This species success at colonizing new regions is facilitated by its adaptability and unisexual mode of reproduction (parthenogenesis). A single individual released into an area can begin to reproduce. The Indo-Pacific Gecko is a nocturnal species and is often found on buildings, fences, and bridge abutments in urban areas where it has been introduced. After populations are established, hitchhiking as stowaways in transported nursery stock and building materials facilitates range expansion of the Indo-Pacific Gecko. Two individuals have been documented from Tennessee—a female and a juvenile found in 2004 near the campus of Middle Tennessee State University in Rutherford County (Niemiller and Glorioso 2010). The presence of this species in Murfreesboro might be due to hitchhiking or individuals might represent escaped pets. Regardless, this species likely cannot survive the harsher winters in Tennessee and the establishment of a permanent population seems unlikely.

Indo-Pacific Gecko subadult, Rutherford County. (Photo by Brad M. Glorioso)

Family Gekkonidae: Mediterranean Gecko

Hemidactylus turcicus

The Mediterranean Gecko is native to southern Europe, around the Mediterranean Sea; however, this species has been introduced all over the world, including the United States. Indeed, breeding populations of the Mediterranean Geckos are established in more than 20 states, with the majority of populations established in the Southeast. Mediterranean Geckos have been reported in several regions of Tennessee, although it is unknown whether breeding populations have yet become established. Mediterranean Geckos are popular in the pet trade, which undoubtedly has facilitated their spread across the United States. These lizards are relatively small and have sticky toe pads that allow them to climb nearly any surface, including glass. They are often seen foraging at night for insects around external houselights. Mediterranean Geckos are similar in appearance to Indo-Pacific Geckos but have warty skin. Detrimental effects of this species on native species of lizards currently are unknown.

Mediterranean Gecko adult, St. Martin Parish, Louisiana. (Photo by Matthew L. Niemiller)

Family Kinosternidae: Razor-Backed Musk Turtle

Sternotherus carinatus

Razor-Backed Musk Turtles occur from central Texas to Arkansas and Mississippi, although they occur east of the Mississippi River only in southwestern Mississippi, where they co-occur with Loggerhead Musk Turtles (*S. minor*). Razor-Backed Musk Turtles are very similar in appearance to Loggerhead Musk Turtles, and the two species are often mistaken for each other. Brimley (1926) presumably misidentified Loggerhead Musk Turtles as Razor-Backed Musk Turtles when he listed the latter species from West Tennessee. No specimens are deposited in museum collections, and no other reports or information exists to corroborate Brimley's assertion.

Razor-Backed Musk Turtle adult, Natchitoches Parish, Louisiana. (Photo by Matthew L. Niemiller)

Family Phrynosomatidae: Texas Horned Lizard

Phrynosoma cornutum

The horned lizards are a group of ant-eating terrestrial lizards primarily adapted to arid regions of the desert southwest. The Texas Horned Lizard ranges from Kansas south to northern Mexico and east to northern Louisiana. Although difficult to keep healthy, horned lizards are popular "pets." Consequently, people who encounter them while traveling through or vacationing in Texas, Oklahoma, or other western states often collect horned lizards. After tiring of the pet, horned lizards are subsequently released elsewhere. Horned lizards are adapted to arid microhabitats of warmer climates, such as sand dunes and coastal scrub, and introduced populations are established in coastal areas of Florida, Georgia, and South Carolina. Gentry (1956) included this species in a state list, and one specimen from Rhea County exists in the Illinois Museum of Natural History (UIMNH 33960). Presumably, the Tennessee records represent escaped or released pets, and there are no records of established (= breeding) populations of horned lizards in the state.

Texas Horned Lizard adult, Union County, New Mexico. (Photo by Todd Pierson)

Family Dactyloidae: Brown Anole

Anolis sagrei

Brown Anoles are small, arboreal lizards native to the Greater Antilles (Cuba) and the northern Bahamas, although they have been inadvertently introduced elsewhere because of their tendency to stowaway on shipping, decorative foliage, and freight. They are firmly established in Florida, where they are displacing the native Green Anole (*A. carolinensis*) along much of the Gulf Coast and southern Georgia. This species is limited in its ability to colonize more northern areas, as it appears much more susceptible to cold weather than the Green Anole. Individuals are documented annually all over Tennessee, almost exclusively in plant nurseries, greenhouses, home improvement store garden centers, and similar areas where plants arrive from Florida and other Gulf Coast areas. This species is not established in our state and likely will not be in the near future.

Brown Anole adult, Seminole County, Florida. (Photo by Matthew L. Niemiller)

Family Testudinidae: Gopher Tortoise

Gopherus polyphemus

Gopher Tortoises are handsome inhabitants of southeastern xeric longleaf pine forests and are a federally protected species. The only species of tortoise in the southeastern United States, they occur naturally in Coastal Plain areas, although they are very patchily distributed due to habitat loss. Tortoises are entirely terrestrial and can be easily identified by their "elephantine feet" and the burrows that they dig in sandy areas underneath vegetation. They are frequently seen crossing southeastern Coastal Plain roads in the morning and evening. A single individual was recently found in the middle of a soccer field in Monroe County, Tennessee, probably an individual that someone picked up on a trip to Florida or Georgia (a highly illegal activity) and then brought to Tennessee. The tortoise was captured and taken to the Knoxville Zoo for safekeeping and possible translocation to a rehabilitation center.

Gopher Tortoise adult, Seminole County, Florida. (Photo by Matthew L. Niemiller)

Family Typhlopidae: Brahminy Blind Snake

Ramphotyphlops braminus

Although not yet officially documented in Tennessee, this circumtropical snake, native to the Indian subcontinent, is likely to show up at some point—most likely in decorative vegetation shipped from Florida. This species is often called the Flowerpot Snake, as it burrows into the soil of flowerpots, a habit that has allowed it to spread almost anywhere tropical plants are shipped. Brahminy Blind Snakes are parthenogenic, so only one female is required to reproduce and found a new population. This species lives almost exclusively underground or underneath surface cover objects such as rocks and boards. Brahminy Blind Snakes are established only in tropical areas and hence this species is unlikely to survive a Tennessee winter.

Brahminy Blind Snake adult, Palm Beach County, Florida. (Photo by Todd Pierson)

Family Viperidae: Eastern Diamond-Backed Rattlesnake

Crotalus adamanteus

Eastern Diamond-Backed Rattlesnakes are beautiful and imperiled natives to the southeastern United States and are the largest venomous snake north of Central America. This species has declined precipitously due to persecution and the repugnant practice of "rattlesnake round-ups," where rattlesnakes are "gassed" out of their holes (frequently Gopher Tortoise burrows). Two individuals of this species have turned up in Tennessee: one in Hardeman County (Norton and Harvey 1975) and one in Grundy County in 2007. These individuals are almost certainly animals that were picked up elsewhere and deposited in our state. The specimen from Grundy County was transported to the Knoxville Zoo, where it was kept for several years.

Eastern Diamond-Backed Rattlesnake adult, Liberty County, Florida. (Photo by Ken Wray)

Family Alligatoridae: Spectacled Caiman

Caiman crocodilus

Caiman are medium-sized members of the family Alligatoridae native to Central and South America. They are frequently, although inexplicably, kept as "pets" in the United States. They require a very large area and a great deal of food and hence are poorly suited for companionship. Caiman occasionally show up in Tennessee when their owners either abandon them because they have outgrown their "cute" phase or they escape from their enclosures, as in 2008, when a 1.2-m (3.9-ft) Caiman was captured in the Little River (Sevier County).

Spectacled Caiman, Trinidad. (Photo by Pierson Hill)

Family Iguanidae: Green Iguana

Iguana iguana

Green Iguanas are popular pets, even though adult males can reach sizes of up to 1.8 m (5.9 ft) in length and inflict a nasty bite or tail lash. These large green lizards are arboreal and have a crest of spines running down the center of the dorsum. Individuals frequently escape or are released, but this species would never survive a typical Tennessee winter. The University of Tennessee Veterinary School has received dozens of iguanas over the years that have been captured on or near the university campus.

Green Iguana, Miami-Dade County, Florida. (Photo by Dick Bartlett)

Checklist of the Reptiles of Tennessee

Lizards

Family Anguidae
☐ Eastern Slender Glass Lizard (*Ophisaurus attenuatus longicaudus*)

Family Phrynosomatidae
☐ Eastern Fence Lizard (*Sceloporus undulatus*)

Family Dactyloidae
☐ Northern Green Anole (*Anolis carolinensis carolinensis*)

Family Scincidae
☐ Northern Coal Skink (*Plestiodon anthracinus anthracinus*)
☐ Southern Coal Skink (*Plestiodon anthracinus pluvialis*)
☐ Common Five-Lined Skink (*Plestiodon fasciatus*)
☐ Southeastern Five-Lined Skink (*Plestiodon inexpectatus*)
☐ Broad-Headed Skink (*Plestiodon laticeps*)
☐ Little Brown Skink (*Scincella lateralis*)

Family Teiidae
☐ Eastern Six-Lined Racerunner (*Aspidoscelis sexlineata sexlineata*)

Snakes

Family Colubridae
☐ Eastern Wormsnake (*Carphophis amoenus amoenus*)
☐ Midwestern Wormsnake (*Carphophis amoenus helenae*)
☐ Northern Scarletsnake (*Cemophora coccinea copei*)
☐ Kirtland's Snake (*Clonophis kirtlandii*)
☐ Northern Black Racer (*Coluber constrictor constrictor*)

☐ Southern Black Racer (*Coluber constrictor priapus*)

☐ Northern Ring-Necked Snake (*Diadophis punctatus edwardsii*)

☐ Mississippi Ring-Necked Snake (*Diadophis punctatus stictogenys*)

☐ Western Mudsnake (*Farancia abacura reinwardtii*)

☐ Eastern Hog-Nosed Snake (*Heterodon platirhinos*)

☐ Prairie Kingsnake (*Lampropeltis calligaster calligaster*)

☐ Mole Kingsnake (*Lampropeltis calligaster rhombomaculata*)

☐ Scarlet Kingsnake (*Lampropeltis elapsoides*)

☐ Eastern Black Kingsnake (*Lampropeltis nigra*)

☐ Red Milksnake (*Lampropeltis triangulum syspila*)

☐ Eastern Milksnake (*Lampropeltis triangulum triangulum*)

☐ Eastern Coachwhip (*Masticophis flagellum flagellum*)

☐ Mississippi Green Watersnake (*Nerodia cyclopion*)

☐ Yellow-Bellied Watersnake (*Nerodia erythrogaster flavigaster*)

☐ Copper-Bellied Watersnake (*Nerodia erythrogaster neglecta*)

☐ Broad-Banded Watersnake (*Nerodia fasciata confluens*)

☐ Diamond-Backed Watersnake (*Nerodia rhombifer*)

☐ Midland Watersnake (*Nerodia sipedon pleuralis*)

☐ Northern Watersnake (*Nerodia sipedon sipedon*)

☐ Northern Rough Greensnake (*Opheodrys aestivus aestivus*)

☐ Red Cornsnake (*Pantherophis guttatus*)

☐ Gray Ratsnake (*Pantherophis spiloides*)

☐ Northern Pinesnake (*Pituophis melanoleucus melanoleucus*)

☐ Queensnake (*Regina septemvittata*)

☐ Northern Brownsnake (*Storeria dekayi dekayi*)

☐ Midland Brownsnake (*Storeria dekayi wrightorum*)

☐ Florida Red-Bellied Snake (*Storeria occipitomaculata obscura*)

☐ Northern Red-Bellied Snake (*Storeria occipitomaculata occipitomaculata*)

☐ Southeastern Crowned Snake (*Tantilla coronata*)

☐ Orange-Striped Ribbonsnake (*Thamnophis proximus proximus*)

☐ Eastern Ribbonsnake (*Thamnophis sauritus sauritus*)

☐ Eastern Gartersnake (*Thamnophis sirtalis sirtalis*)

☐ Rough Earthsnake (*Virginia striatula*)

☐ Western Smooth Earthsnake (*Virginia valeriae elegans*)

☐ Eastern Smooth Earthsnake (*Virginia valeriae valeriae*)

Family Viperidae

☐ Southern Copperhead (*Agkistrodon contortrix contortrix*)

☐ Northern Copperhead (*Agkistrodon contortrix mokasen*)

☐ Western Cottonmouth (*Agkistrodon piscivorus leucostoma*)

☐ Timber Rattlesnake (*Crotalus horridus*)

☐ Western Pygmy Rattlesnake (*Sistrurus miliarius streckeri*)

Turtles

Family Chelydridae

☐ Snapping Turtle (*Chelydra serpentina*)

☐ Alligator Snapping Turtle (*Macrochelys temminckii*)

Family Emydidae

☐ Southern Painted Turtle (*Chrysemys dorsalis*)

☐ Midland Painted Turtle (*Chrysemys picta marginata*)

☐ Bog Turtle (*Glyptemys muhlenbergii*)

☐ Northern Map Turtle (*Graptemys geographica*)

☐ Ouachita Map Turtle (*Graptemys ouachitensis ouachitensis*)

☐ Mississippi Map Turtle (*Graptemys pseudogeographica kohnii*)

☐ False Map Turtle (*Graptemys pseudogeographica pseudogeographica*)

☐ Eastern River Cooter (*Pseudemys concinna concinna*)

☐ Eastern Box Turtle (*Terrapene carolina carolina*)

☐ Three-Toed Box Turtle (*Terrapene carolina triunguis*)

☐ Red-Eared Slider (*Trachemys scripta elegans*)

☐ Cumberland Slider (*Trachemys scripta troostii*)

Family Kinosternidae

☐ Mississippi Mud Turtle (*Kinosternon subrubrum hippocrepis*)

☐ Eastern Mud Turtle (*Kinosternon subrubrum subrubrum*)

☐ Stripe-Necked Musk Turtle (*Sternotherus minor peltifer*)

☐ Eastern Musk Turtle (*Sternotherus odoratus*)

Family Trionychidae

☐ Midland Smooth Softshell (*Apalone mutica mutica*)

☐ Gulf Coast Spiny Softshell (*Apalone spinifera aspera*)

☐ Eastern Spiny Softshell (*Apalone spinifera spinifera*)

Crocodilians

Family Alligatoridae

☐ American Alligator (*Alligator mississippiensis*)

Glossary

Abiotic: Nonliving components of a habitat, community, or ecosystem.

Aestivate: To spend a period of time in an inactive state, often underground during hot or dry weather.

Aestivation: State of dormancy used by some organisms in response to hot, dry conditions.

Albinism: Devoid of pigment.

Allele: A variant of a particular gene, located in a specific position on a specific chromosome.

Allopatric: Two populations or species occurring in geographic isolation from one another; non-overlapping ranges.

Alluvial: Referring to material deposited by flowing water.

Ambient: Surrounding, as in ambient temperature.

Anterior: At or situated near the head of an organism.

Anthropogenic: Produced or caused by humans.

Arboreal: Living in trees.

Arthropod: A group of invertebrates, which have an exoskeleton, segmented body, and jointed appendages. Includes the insects, crustaceans, arachnids, chilopods (centipedes), and diplopods (millipedes).

Autotomy: The loss or breakage of the tail either reflexively or after being captured.

Band: A pigmented area that spans the dorsum of the body from one side to another extending onto, or nearly to, the ventral surface.

Barred: A pattern of stripes on the trunk or appendages.

Batesian mimicry: Imitation of a noxious species by a harmless or palatable species; mimicry of warning colors or signals to avoid predation.

Biomass: The total mass of all or a group of living organisms in a given habitat.

Blotch: A pigmented area that is usually large and round or squarish that differs from the background coloration.

Body length: See **SVL**.

Carapace: Dorsal surface of a turtle or tortoise, formed from the bony shell and covered with scutes or a leathery tissue.

Caudal: At or near the tail.

Channelization: Artificial deepening and straightening of a body of water, typically rivers and streams.

Chert: Fine-grained sedimentary rock that contains quartz microcrystals.

Class: A higher-level taxonomic group below the level of phylum but above the level of order, as in the class Reptilia or Sauropsida.

Climate: A composite of the prevailing weather conditions of an area or region, as measured over years to centuries. Contrast with weather, which is the local condition at any given time.

Cloaca: Chamber or cavity located just anterior to the tail that serves as a reservoir common to the urinary, digestive, and reproductive tracts.

Community: An assemblage of species occurring in a discrete space and time shaped by inter- and intraspecific interactions.

Congeneric: Belonging to the same genus.

Conspecific: Belonging the same species.

Cranial: Referring to the head or skull.

Crepuscular: Active during dawn and/or dusk.

Cryptic species: Two or more related species that are morphologically indistinguishable but are genetically divergent.

Deciduous: Vegetation that loses foliage with the onset of cold weather (usually the autumn) and regrows it during the spring.

Diploid: Possessing two sets of chromosomes in each body cell.

Disjunct population: A population of organisms that is geographically separated from other populations of the same species.

Diurnal: Active during the day.

DNA: Deoxyribonucleic acid; nucleic acid that contains genetic information of an organism.

Dorsal: Referring to the uppermost surface of a bilateral organism; the back.

Dorsolateral: Pertaining to the back and sides of a bilateral organism.

Dorsum: The uppermost surface of a bilateral organism; the back.

Ecoregion: An ecological region that is characterized by similar landscape, climate, soil, vegetation, hydrology, fauna, and other ecological attributes.

Endemic: Restricted to a particular region; found nowhere else.

Ephemeral: Existing for only a short time period, as in an ephemeral wetland.

Eutrophication: The natural or anthropogenic process of sedimentation and nutrient deposition in aquatic habitats leading to changes in community structure.

Extirpation: Local extinction of a population of a species.

Facultative: Refers to an optional process or behavior; contrast with **Obligate**.

Family: A higher-level taxonomic group below the level of order but above the level of genus, as in the family Colubridae.

Fauna: The species of animal in a particular region or time period.

Fecundity: Reproductive capacity of an organism relating to the number of potential offspring.

First-order stream: See **Stream order**.

Floodplain: Low-lying area adjacent to rivers and streams that is subject to periodic flooding during heavy rainfall.

Fossorial: Living or adapted for life underground, typically in the soil or in burrows.

Genus: A lower-level taxonomic group below the level of Family that usually consists of several closely related species, as in the genus *Lampropeltis*. Plural genera.

Gravid: Carrying developing eggs or young.

Hemotoxin: A toxin present in the venoms of some pit vipers that causes blood cells to burst, causing tissue damage and disrupting proper blood clotting.

Herpetofauna: The amphibians and reptiles of a region or time period.

Herpetology: The scientific study of amphibians and reptiles.

Heterogeneity: Variation, in relation to characters or characteristics of an environment, population, and so on.

Hybridization: Reproduction between two different organisms, typically species, that results in the formation of offspring.

Hydrology: Study or characterization of the characteristics, including the distribution, movement, and cycling, of aquatic resources.

Hydroperiod: The period of time in which a wetland contains water.

Intergradation: Reproduction between morphologically distinguishable populations or subspecies.

Introduced species: A species that has been transported by humans, either indirectly or purposely, outside of its native range. See **Invasive species**.

Introgression: Movement of alleles from one species to another through hybridization or mating of two species. Also called introgressive hybridization.

Invasive species: An introduced species that has become established and has spread, often causing detriment to local biotic communities.

Iris: The pigmented part of the eye.

Karst: An area or region characterized by irregular limestone or other carbonate bedrock in which erosion has produced numerous caves, sinkholes, fissures, and subterranean streams.

Lentic: Referring to nonflowing water, such as a lake or pond.

Life history: The history of changes undergone by an organism over its lifetime, from conception to death.

Lineage: A group of individuals, species, or taxa related by a common ancestor.

Linnaean taxonomy: A hierarchical classification system based on the work *Systems of Nature* by Carolus Linneaus in 1758. Organisms are given standard Latin names that reflect the most recent and accepted evolutionary relationships.

Loess: Sediment formed from wind-blown deposits of silt and sand.

Longitudinal: Referring to length or lengthwise.

Loreal scale: Scale found between the preocular and nasal scales in some squamates.

Lotic: Referring to flowing water, such as streams, creeks, and rivers.

Melanistic: Possessing dark pigmentation due to the presence of melanin in the epithelium. Melanistic individuals are either aberrant individuals that are darker than normal members of a species or those that become more darkly pigmented with age.

Mesic: Characterized by being moderately moist, as in a mesic habitat.

Middorsal: Referring to the midline of the dorsum or back.

Mimicry: The resemblance of one organism to another or to an object in its environment, presumably for camouflage or reduced risk of predation.

Morphology: The study of the form and structure of an organism, especially external structure and form. Relating to the external characteristics of an organism.

Mottled: An indistinct pattern of spots or flecks of various sizes.

Nares: External nostril openings. Singular naris.

Natural history: The study of the biology of species including their origins, evolution, behavior, and relationships with other species.

Neurotoxin: A chemical that causes damage to nerves or nerve tissue, frequently similar in structure to neurotransmitters, which allows passage into and disruption of the nervous system.

Nocturnal: Active at night.

Nomenclature: A system of names that organizes organisms based on their relationships to one another.

Noxious: Irritating, painful, or otherwise harmful.

Nymph: The immature stage of certain insect groups, which is often aquatic.

Obligate: A behavior or process that is not optional. Also used to imply that an organism is only able to survive in a specific habitat or environment; contrast with **Facultative**.

Olfaction: The sense of smell.

Ontogenetic: Describing changes in the traits or behaviors of an organism as it ages.

Order: Higher-level taxonomic group below the level of class but above the level of family, as in the order Squamata.

Ova: Undeveloped eggs; singular ovum.

Oviposit: To lay eggs or egg masses.

Parthenogenesis: A form of asexual reproduction in which eggs develop without fertilization.

Pectoral: Referring to the chest.

Perennial: Lasting or active throughout the year or multiple years.

Physiographic province: A landscape region characterized by unique geological history and physical attributes.

Physiography: The study of the physical features of the surface of the Earth.

Physiology: The study of the mechanical, physical, and biochemical functions of organisms.

Plastron: Ventral surface of a turtle or tortoise, formed from the bony shell and covered with scutes or a leathery tissue.

Pleistocene: Geologic time period from 1.7 to 0.1 million years before the present.

Polymorphism: Multiple color or body forms in a species or population.

Population: A group of organisms of the same species living within a continuous area.

Protuberance: A bulging or knoblike feature.

Quaternary: The current geologic period, also known as the most recent period of the Cenozoic era, encompassing the Pleistocene and Holocene and dating from the present to approximately 2.58 million years ago.

Regenerated: The repair and replacement of lost structures, such as the tail of lizards.

Reticulum: A network design or pattern (reticulation) of color that results in a netlike patterning of an organism.

Riparian: Pertaining to the edges of aquatic habitats, such as stream banks and floodplains.

Road cruising: The act of searching for animals by driving slowly along roads. A productive way of finding reptiles, particularly at night after heavy rains.

Scutes: Large scales covering the carapace and plastron in turtles; also another name for larger scales.

Sexual dimorphism: Morphological characters that differ between males and females of the same species.

Siltation: The natural or anthropogenic process of sediment deposition in aquatic habitats.

Species: The least inclusive taxonomic level, below the level of genus and consisting of a group of organisms capable of interbreeding that are reproductively isolated from other such groups.

This is known as the biological species concept, though species are not natural categories and hence many other species concepts exist. Latin species names, always italicized and lowercased, follow italicized and capitalized genus names.

Stream order: A classification system for streams and rivers, ranging from first to twelfth in order of size. First-order streams are the smallest and form the first tributaries; hence they have no other water flowing into them. First- through third-order streams are considered headwaters and usually consist of faster flowing, more highly oxygenated water that contains less silt than higher-order streams.

Subspecies: A taxonomic subdivision of a species typically consisting of a group of populations with unique characteristics and occupying a definable geographic region but capable of interbreeding with other such races. Subspecies are named using a trinomial with a third name following the species binomial, such as in the Mississippi Mud Turtle, *Kinosternon subrubrum hippocrepis.*

Substrate: The soil or material on or in which an organism lives.

SVL: Snout-vent length, measured from the tip of the snout to the posterior opening of the cloaca, usually on the ventral surface of the organism. A common measurement used to assess body size in reptiles.

Sympatric: Populations or species occurring within the same regions with overlapping ranges.

Talus: An aggregation of rocks, particularly at the base of steep slopes or cliffs.

Taxonomy: The science of naming and classifying organisms.

Temperate: A moderate climate, characterized by deciduous forest and distinct seasonality.

Terrestrial: Living on the land.

TL: Total length, measured from the tip of the snout to the tip of the tail.

Topography: Study of the Earth's surface or the characterization of the physical features of the Earth.

Transverse: Crossing from one side to the other.

Urbanization: Referring to the growth of cities into surrounding rural areas.

Vent: The common external body opening of the reproductive, urinary, and digestive systems located on the underside of a reptile.

Venter: The underside of an organism; opposite of dorsum.

Ventral: Referring to the underside of an organism.

Vermiculation: A wormlike pattern.

Vestigial: Degenerate or atrophied, often referring to limbs. Structures, such as limbs, which have evolved to be reduced in size or function.

Voucher: A specimen collected from a particular area and housed in a museum that provides evidence of that species' occurrence in that area. Some museums allow photographic vouchers instead of physical vouchers, reducing the need to collect wild animals.

Xeric: Characterized by being dry, as in a xeric habitat.

Recommended Readings, Organizations, and Websites

Reptile Conservation and Monitoring

Baker, K. J. Reptiles: Biology, behavior and conservation. Nova Science Publishers, Hauppage, NY.

Crump, M. L. 2002. Amphibians, reptiles, and their conservation. Linnet Books, Hamden, CT.

Crump, M. L. 2011. Amphibians and reptiles: An introduction to their natural history and conservation. McDonald & Woodward, Granville, OH.

Dorcas, M., and J. D. Wilson. 2011. Invasive pythons in the United States: Ecology of an introduced predator. Univ. of Georgia Press, Athens.

Gibbons, J. W., D. E. Scott, T. J. Ryan, K. A. Buhlmann, T. D. Tuberville, B. S. Metts, J. L. Greene, T. Mills, Y. Leiden, S. Poppy, and C. T. Winne. 2000. The global decline of reptiles, déjà vu amphibians. BioScience 50:653–666.

Kraus, F. 2009. Alien amphibians and reptiles: A scientific compendium and analysis. Invading Nature: Springer Series in Invasion Ecology. Springer, New York.

McDiarmid, R. W., M. S. Foster, C. Guyer, J. W. Gibbons, and N. Chernoff, editors. 2012. Reptile biodiversity: Standard methods for inventory and monitoring. Univ. of California Press, Berkeley and Los Angeles.

Mullin, S. J., and R. A. Seigel. 2009. Snakes: Ecology and conservation. Cornell Univ. Press, Ithaca, NY.

Southeastern Reptiles

Barbour, T. 1971. Amphibians and reptiles of Kentucky. Univ. Press of Kentucky, Lexington.

Beane, J. C., A. L. Braswell, J. C. Mitchell, and W. M. Palmer. 2010. Amphibians and reptiles of the Carolinas and Virginia. 2nd ed. Univ. of North Carolina Press, Chapel Hill.

Buhlmann, K., T. Tuberville, and J. W. Gibbons. 2008. Turtles of the southeast. Univ. of Georgia Press, Athens.

Burger, J., and R. T. Zappalorti, editors. 2011. The Northern Pine Snake (*Pituophis melanoleucus*): Its life history, behavior and conservation. Nova Science Publishers, Hauppage, NY.

Conant, R., and J. T. Collins. 1998. A field guide to reptiles and amphibians of eastern and central North America. 4th ed. Houghton Mifflin, New York.

Dorcas, M. E. 2005. A guide to the snakes of North Carolina. Davidson College, Davidson, NC.

Dundee, H. A., D. A. Rossman, and E. C. Beckham. 1996. The amphibians and reptiles of Louisiana. Louisiana State Univ. Press, Baton Rouge.

Ernst, C. H., and J. E. Lovich. 2011. Turtles of the United States and Canada. Johns Hopkins Univ. Press, Baltimore, MD.

Gibbons, J. W., and M. Dorcas. 2005. Snakes of the Southeast. Univ. of Georgia Press, Athens.

Gibbons, W., J. Greene, and T. Mills. 2009. Lizards and crocodiles of the Southeast. Univ. of Georgia Press, Athens.

Huheey, J. E., and A. Stupka. 1967. Amphibians and reptiles of Great Smoky Mountains National Park. Univ. of Tennessee Press, Knoxville.

Jensen, J. B., C. D. Camp, J. W. Gibbons, and M. J. Elliot. 2008. Amphibians and reptiles of Georgia. Univ. of Georgia Press, Athens.

Johnson, T. R. 2000. Amphibians and reptiles of Missouri. Rev. ed. Missouri Department of Conservation, Jefferson City, MO.

Martof, B. S., W. M. Palmer, J. R. Bailey, J. R. Harrison III, and J. Dermid. 1989. Amphibians and reptiles of the Carolinas and Virginia. Univ. of North Carolina Press, Chapel Hill.

Mitchell, J. C., and K. K. Reay. 1999. Atlas of amphibians and reptiles in Virginia. Virginia Department of Game and Inland Fisheries, Richmond, VA.

Mount, R. H. 1975. The reptiles and amphibians of Alabama. Alabama Agricultural Experiment Station, Auburn Univ., Auburn.

Palmer, W. M., and A. L. Braswell. 1995. Reptiles of North Carolina. Univ. of North Carolina Press, Chapel Hill.

Redmond, W. H., A. C. Echternacht, and A. F. Scott. 1990. Annotated checklist and bibliography of amphibians and reptiles of Tennessee (1835–1989). Center for Field Biology, Austin Peay State Univ., Clarksville, TN.

Redmond, W. H., and A. F. Scott. 2011. Atlas of reptiles in Tennessee. Center for Field Biology, Austin Peay State Univ., Clarksville, TN, and references therein.

Scott, A. F., and W. H. Redmond. 2002. Updated checklist of Tennessee's amphibians and reptiles with an annotated bibliography covering primarily years 1990 through 2001. Center for Field Biology, Austin Peay State Univ., Clarksville, TN.

Tilley, S. G., and J. E. Huheey. 2001. Reptiles and amphibians of the Smokies. Great Smoky Mountains Natural History Association, Gatlinburg, TN.

Trauth, S. E., H. W. Robinson, and M. V. Plummer. 2004. The amphibians and reptiles of Arkansas. Univ. of Arkansas Press, Fayetteville.

Reptile Taxonomy

Boelens, B., M. Watkins, and M. Grayson. 2011. The eponym dictionary of reptiles. Johns Hopkins Univ. Press, Baltimore, MD.

Collins, J. T., and T. W. Taggart. 2009. Standard common and current scientific names for North American amphibians, turtles, reptiles, and crocodilians. 6th ed. Publication of the Center for North American Herpetology, Lawrence, KS.

Crother, B. I., editor. 2011. Scientific and standard English names of amphibians and reptiles of North America north of Mexico. Vol. 6.1. Herpetological Circular 37. Society for the Study of Amphibians and Reptiles. http://www.ssarherps.org/pages/comm_names/Index.php/.

Powell, R., J. T. Collins, and E. D. Hooper. 1998. A key to amphibians and reptiles of the continental United States and Canada. Univ. Press of Kansas, Lawrence.

Reptile Biology and Natural History

Greene, H. W. 2000. Snakes: The evolution of mystery in nature. Univ. of California Press, Berkeley and Los Angeles.

Jackson, D. C. 2011. Life in a shell: A physiologist's view of a turtle. Harvard Univ. Press, Cambridge, MA.

Mitchell, J. C., R. E. Jung Brown, and B. Bartholomew. 2008. Urban herpetology. Herpetological Conservation 3. Society for the Study of Amphibians and Reptiles.

Pianka, E. R. 2003. Lizards: Windows to the evolution of diversity. Univ. of California Press, Berkeley and Los Angeles.

Pough, F. H., R. M. Andrews, J. E. Cadle, M. L. Crump, A. H. Savitsky, and K. D. Wells. 2004. Herpetology. 3rd ed. Pearson Education, Upper Saddle River, NJ.

Vitt, L. J., and J. P. Caldwell. 2008. Herpetology: An introductory biology of amphibians and reptiles. 3rd ed. Academic Press, San Diego, CA.

Videos

BBC Warner. 2007. Planet Earth: The Complete BBC Series.

BBC Warner. 2008. Life in Cold Blood.

BBC Warner. 2009. Life: The Complete BBC Series.

Alligator Calls

Britton, A. 2011. Crocodilian communication. Crocodilians: Natural History and Conservation. Available at http://www.flmnh.ufl.edu/cnhc/croccomm.html/.

Academic Journals

Catalogue of American Amphibians and Reptiles (Society for the Study of Amphibians and Reptiles)

Herpetologica (Herpetologist's League)

Herpetological Conservation and Biology (http://www.herpconbio.org/)

Herpetological Monographs (Herpetologist's League)

Herpetological Review (Society for the Study of Amphibians and Reptiles)

Journal of Herpetology (Society for the Study of Amphibians and Reptiles)

Southeastern Naturalist

Regional and Professional Societies

American Society of Ichthyologists and Herpetologists (http://www.asih.org/)

Georgia Herpetological Society

Herpetologists' League (http://herpetologistsleague.com/)

Kentuckiana Herpetological Society

Kentucky Herpetological Society

Missouri Herpetological Association (http://www.moherp.org/)

North Carolina Herpetological Society (http://www.ncherps.org/)

Partners in Amphibian and Reptile Conservation (http://www.parcplace.org/)

Society for the Study of Amphibians and Reptiles (http://www.ssarherps.org/)

Southeast Partners in Amphibian and Reptile Conservation (http://www.separc.org/)

Tennessee Herpetological Society (http://www.tnherpsociety.org/)

Virginia Herpetological Society (http://fwie.fw.vt.edu/VHS/)

Other Relevant Websites

Brakes for Snakes: Promoting snake conservation (http://brakesforsnakes.org)

Center for North American Herpetology (http://www.cnah.org/)

HerpLit (http://www.herplit.com/)

HerpNET (http://www.herpnet.org/)

The Reptile Database (http://www.reptile-database.org/)

Tennessee Amphibians and Reptiles (http://www.herpetology.us/tnherps/)

Tennessee Atlas of Reptiles (http://apbrwww5.apsu.edu/reptatlas/)

Tennessee Wildlife Resources Agency (http://www.state.tn.us/twra/)

Selected References

Beltz, E. 2006. Scientific and common names of the reptiles and amphibians of North America—explained. Available from: http://ebeltz.net/herps/etymain.html/. (Accessed: 2008–2012)

Brimley, C. S. 1926. Revised key and list of the amphibians and reptiles of North Carolina. J. Elisha Mitchell Sci. Soc. 42:75–94.

Collins, J. T. 1991a. Amphibians and reptiles in the upper Mississippi River Valley: Systematic and distributional problems. J. Tenn. Acad. Sci. 66:149–152.

Collins, J. T. 1991b. Viewpoint: A new taxonomic arrangement for some North American amphibians and reptiles. Herpetol. Rev. 22:42–43.

Conant, R., and J. T. Collins. 1998. A field guide to reptiles and amphibians of eastern and central North America. Houghton Mifflin, New York.

Davenport, J. M., and A. F. Scott. 2009. Amphibians and reptiles of Fort Donelson National Battlefield, Stewart County, Tennessee. J. Tenn. Acad. Sci. 84:83–89.

Dodd, C. K. 2004. Amphibians of Great Smoky Mountains National Park. Univ. of Tennessee Press, Knoxville.

Endsley, J. R. 1954. An annotated listing of a herpetological collection mainly from Tennessee. J. Tenn. Acad. Sci. 29:36–41.

Etnier, D. A., and W. C. Starnes. 1993. The fishes of Tennessee. Univ. of Tennessee Press, Knoxville.

Garton, J. S., and R. W. Dimmick. 1969. Food habits of the copperhead in Middle Tennessee. J. Tenn. Acad. Sci. 44:113–117.

Gentry, G., R. M. Sinclair, W. Hon, and B. Ferguson. 1965. Amphibians and reptiles of Tennessee. Tenn. Game and Fish Comm., Nashville. 28 pp.

Gerald, G. W., M. A. Bailey, and J. N. Holmes. 2006a. Habitat utilization of *Pituophis melanoleucus melanoleucus* (Northern Pinesnake) on Arnold Air Force Base in Middle Tennessee. Southeastern Nat. 5:253–264.

Gerald, G. W., M. A. Bailey, and J. N. Holmes. 2006b. Movements and activity range sizes of Northern Pinesnakes (*Pituophis melanoleucus melanoleucus*) in Middle Tennessee. J. of Herpetol. 40:503–510.

Gibbons, J. W., D. E. Scott, T. J. Ryan, K. A. Buhlmann, T. D. Tuberville, B. S. Metts, J. L. Greene, T. Mills, Y. Leiden, S. Poppy, and C. T. Winne. 2000. The global decline of reptiles, déjà vu amphibians. BioScience 50:653–666.

Griffith, G. E., J. M. Omernik, and S. H. Azevedo. 1998. Ecoregions of Tennessee. Map poster. U.S. Geological Survey, Reston, VA.

Horn, D., and T. Cathcart, editors. 2005. Wildflowers of Tennessee, the Ohio Valley, and the southern Appalachians. Lone Pine Publishing, Edmonton, Alberta.

Huheey, J. E., and A. Stupka. 1967. Amphibians and reptiles of Great Smoky Mountains National Park. Univ. of Tennessee Press, Knoxville.

Johnson, R. M. 1958. A biogeographic study of the herpetofauna of eastern Tennessee. Ph.D. diss., Univ. of Florida, Gainesville.

Johnson, T. R. 1987. The amphibians and reptiles of Missouri. Conservation Commission of the State of Missouri, Missouri Department of Conservation, Jefferson City.

King, W. 1939. A survey of the herpetology of Great Smoky Mountains National Park (Tennessee). Amer. Midl. Nat. 21:531–582.

Lawson, R., J. B. Slowinski, B. I. Crother, and F. T. Burbrink. 2005. Phylogeny of the Colubroidea (Serpentes): New evidence from mitochondrial and nuclear genes. Mol. Phylog. Evol. 37:581–601.

Mathews, R. C., and A. C. Echternacht. 1984. Herpetofauna of the spruce-fir ecosystem in the southern Appalachian Mountain regions, with emphasis on the Great Smoky Mountains National Park. In P. S. White, editor, pp. 155–167. The southern Appalachian spruce-fir ecosystem. National Park Service. Research/Resources Mgmt. Rep. SER-71.

Means, D. B. 2009. Effects of rattlesnake roundups on the Eastern Diamondback Rattlesnake (*Crotalus adamanteus*). Herpetol. Conserv. Biol. 4:132–141.

Miller, B. T., J. W. Lamb, and J. L. Miller. 2005. The herpetofauna of Arnold Air Force Base in the Barrens of Tennessee. Southeastern Nat. 4:51–62.

Mitchell, J. C., T. K. Pauley, D. I. Withers, P. V. Cupp Jr., A. L. Braswell, B. T. Miller, S. M. Roble, and C. S. Hobson. 1999. Conservation status of the southern Appalachian herpetofauna. Virginia J. Sci. 50:1–36.

Moriarty, J. J., and A. M. Bauer. 2000. State and provincial amphibian and reptile publications for the United States and Canada. Herpetological Circular 28. Society for the Study of Amphibians and Reptiles.

Nicholson, C. P. 1997. Atlas of the breeding birds of Tennessee. Univ. of Tennessee Press, Knoxville.

Niemiller, M. L. 2005. The herpetofauna of the upper Duck River watershed in Coffee County, Tennessee. J. Tenn. Acad. Sci. 80:6–12.

Niemiller, M. L., and B. M. Glorioso. 2010. The Indo-Pacific Gecko (*Hemidactylus garnotii*) in Rutherford County, Tennessee. Tennessee Herpetological Society Newsletter.

Niemiller, M. L., and R. G. Reynolds. 2011. The amphibians of Tennessee. Univ. of Tennessee Press, Knoxville.

Niemiller, M. L., R. G. Reynolds, B. M. Glorioso, J. Spiess, and B. T. Miller. 2011. Herpetofauna of the cedar glades and adjacent habitats of the inner central basin of Middle Tennessee. Herpetol. Conserv. Biol. 3:203–210.

Norton, V. M., and M. J. Harvey. 1975. Herpetofauna of Hardeman County, Tennessee. J. Tenn. Acad. Sci. 50:131–136.

Palmer, W. M., and A. L. Braswell. 1995. Reptiles of North Carolina. Univ. of North Carolina Press, Chapel Hill.

Parker, M. V. 1937. Some amphibians and reptiles from Reelfoot Lake. J. Tenn. Acad. Sci. 12:60–86.

Parker, M. V. 1939. The amphibians and reptiles of Reelfoot Lake and vicinity, with a key for the separation of species and subspecies. J. Tenn. Acad. Sci. 14:72–101.

Parker, M. V. 1948. A contribution to the herpetology of western Tennessee. J. Tenn. Acad. Sci. 23:20–30.

Parmalee, P. W., and A. E. Bogan. 1998. The freshwater mussels of Tennessee. Univ. of Tennessee Press, Knoxville.

Pyron, R. A. and F. T. Burbrink. 2009a. Neogene diversification and taxonomic stability in the snake tribe Lampropeltini (Serpentes: Colubridae). Mol. Phylog. Evol. 52:524–529.

Pyron, R. A., and F. T. Burbrink. 2009b. Systematics of the Common Kingsnake (*Lampropeltis getula*; Serpentes: Colubridae) and the burden of heritage in taxonomy. Zootaxa 2241:22–32.

Ramsey, L. W. 1953. The Lined Snake, *Tropidoclonion lineatum* (Hallowell). Herpetologica 9:7–25.

Redmond, W. H., A. C. Echternacht, and A. F. Scott. 1990. Annotated checklist and bibliography of amphibians and reptiles of Tennessee (1835–1989). Misc. Pub. No. 4. Center for Field Biology, Austin Peay State Univ., Clarksville, TN. 173 pp.

Redmond, W. H., and A. F. Scott. 2011. Atlas of reptiles in Tennessee. Center for Field Biology, Austin Peay State Univ., Clarksville, TN, and references therein.

Rhoads, S. N. 1895. Contributions to the zoology of Tennessee. No. 1, reptiles and amphibians. Proc. Acad. Nat. Sci. Philadelphia 47:376–407.

Row, J. R., G. Blouin-Demers, and P. J. Weatherhead. 2007. Demographic effects of road mortality in Black Ratsnakes (*Elaphe obsoleta*). Biol. Conserv. 137:117–124.

Schlaepfer, M. A., C. Hoover, and C. Kenneth Dodd Jr. 2005. Challenges in evaluating the impact of the trade in amphibians and reptiles on wild populations. BioScience 55:256–264.

Scott, A. F. 1991. The history and literature of Tennessee herpetology. J. Tenn. Acad. Sci. 66: 145–148.

Scott, A. F., and W. H. Redmond. 2002. Updated checklist of Tennessee's amphibians and reptiles with an annotated bibliography covering primarily years 1990–2001. Center for Field Biology, Austin Peay State Univ., Clarksville, TN.

Scott, A. F., and D. H. Snyder. 1968. The amphibians and reptiles of Montgomery County, Tennessee. J. Tenn. Acad. Sci. 43:79–84.

Snyder, D. H. 1972. Amphibians and reptiles of Land Between the Lakes. Tennessee Valley Authority, Golden Pond, KY. 90 pp.

Stevenson, H. M. 1959. Some altitude records of reptiles and amphibians. Herpetologica 15:118.

Todd, B. D., and K. M. Andrews. 2008. Response of a reptile guild to forest harvesting. Conserv. Biol. 22:753–761.

Townsend, T. M., D. G. Mulcahy, B. P. Noonan, J. W. Sites Jr., C. A. Kuczynski, J. J. Wiens, and T. W. Reeder. 2011. Phylogeny of iguanian lizards inferred from 29 nuclear loci, and a comparison of concatenated and species-tree approaches for an ancient rapid radiation. Mol. Phylog. Evol. 61:363–380.

Withers, D. I. 2009. A guide to rare animals of Tennessee. Division of Natural Heritage, Tennessee Department of Environment and Conservation, Nashville.

Wofford, B. E., and E. W. Chester. 2002. Guide to the trees, shrubs, and woody vines of Tennessee. Univ. of Tennessee Press, Knoxville.

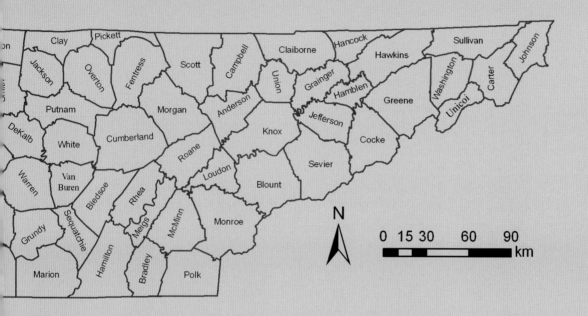

Index of Common and Scientific Names

Page numbers in **boldface** refer to species accounts and illustrations.

The Reptiles of Tennessee was designed and typeset on a Macintosh computer system using InDesign software. The body copy is set in Kepler Light 10/15 and display type is set in Century Gothic and News Gothic. This book was typeset and designed by Chad M. Pelton, and manufactured by Everbest Printing Company Limited.